THE INFORMATION TRADE

THE INFORMATION TRADE

How Big Tech Conquers Countries, Challenges Our Rights, and Transforms Our World

ALEXIS WICHOWSKI

HarperOne
An Imprint of HarperCollins*Publishers*

HarperOne

HarperCollins books may be purchased for educational, business, or sales promotional use. For information, please email the Special Markets Department at SPsales@harpercollins.com.

FIRST EDITION

Designed by Joy O'Meara and Lucy Albanese

Library of Congress Cataloging-in-Publication Data has been applied for.
ISBN 978-0-06-288898-3

20 21 22 23 24 LSC 10 9 8 7 6 5 4 3 2 1

To Jonathan, for being mine.
To Gerome, Novi, & Leo, for being all your own.

Where, after all, do universal human rights begin?
In small places, close to home—so close and so small
that they cannot be seen on any maps of the world . . .
Unless these rights have meaning there,
they have little meaning anywhere.
Without concerned citizen action to uphold them close to home,
we shall look in vain for progress in the larger world.

—ELEANOR ROOSEVELT,
former first lady of the
United States

We don't completely blame Facebook.
The germs are ours, but Facebook is the wind,
you know?

—HARINDRA DISSANAYAKE,
Sri Lankan presidential adviser

CONTENTS

INTRODUCTION

On January 9, 2007, 45,000 software developers, computer engineers, and everyday tech enthusiasts gathered in Silicon Valley's go-to conference spot, San Francisco's Moscone Center, a three-story, glass-enclosed conference space that shared a block with the Yerba Buena Ice Skating and Bowling Center and the Dosa Brothers Indian restaurant. The occasion: the 22nd annual Macworld Expo. The highlight: bearing witness to their patron saint, Apple visionary Steve Jobs.[1]

Wearing his signature uniform—black turtleneck, wire-frame glasses, white sneakers, and blue jeans—Jobs took to the stage. A giant backlit Apple logo loomed on a wall-size screen behind him.

Twenty-two minutes into a speech sprinkled with updates about various Apple products, Jobs stopped. A moment of silence passed. "This is a day I've been looking forward to for the past two and a half years," he announced.[2] Scattered applause peppered the room, but Jobs waved it away.

With something like defiance, he declared, "The most advanced phones are called 'smartphones,' so they say." The audience burst into laughter. In 2007, when most people still carried flip phones and PDAs, the very notion of such a thing seemed absurd. Jobs went on to blast then-current "smartphones"—BlackBerries and Nokias, namely—as being difficult to navigate, even for basic functions. "What we want to do is make a product that's way smarter than any cell phone *and* that's easy to use. This is what iPhone is."

The iPhone launch is worth cherishing. It may very well have been our last mass-magical tech moment, a time when the entire world got truly excited over a technological breakthrough.

This was a time before tech got scary.

It was almost four years before WikiLeaks released 251,287 diplomatic cables to the press, which contributed to the bloody and largely unsuccessful Arab Spring and drove home the terrible power and scale of leaks now possible in the digital age.[3]

It was six years before Edward Snowden's revelations shattered public trust in the US government by unveiling the National Security Agency (NSA) mass covert data collection program that sought info on American citizens.

It was almost a decade before the Russian military's Information Research Agency infiltrated the 2016 US presidential election through misinformation warfare, peeling away the belief that our social networks consisted of our friends, or at the very least, our compatriots.

And it was eleven years before Facebook was outed for giving political consulting firm Cambridge Analytica access to 87 million users' data, finally tipping the world's wide-scale disillusionment with the tech industry into outright anger.[4]

In 2007, we still loved our tech and its keepers. The proof is in the purchases. Half a million people bought iPhones the first weekend they were available.[5] Buyers lined up around the US—for days, in some places.

"I feel wonderful. It's exhilarating," reported 51-year-old engineer David Jackson as he finally held an iPhone in his hands, having waited in line more than 24 hours for the moment. "Man, that was cool. I was shaking at the counter. I couldn't even sign my name."[6]

With the iPhone, Apple gave us what seemed like one of the greatest godsends of the digital era: a keyboardless, full-color, internet-enabled, do-everything device—one that was pretty and sleek and fit in your pocket, to boot.

We may not have recognized it at the time, but Apple did more with the iPhone than create a next-generation personal computer. They created the first *wearable* computer: a device that you could keep on your body, in your pocket, at all times. In 2019, this was a reality for roughly 2 billion smartphone users, whether they carried an iPhone or its chief competitor, an Android (Google) phone.[7] The smartphone didn't just make life easier; it didn't just make us, as Apple's '90s-era slogan urged, "think different." It made *life* different.

ALMOST EXACTLY 10 YEARS AFTER JOBS INTRODUCED THE IPHONE TO the world, another tech luminary addressed a similarly massive audience at the Moscone Center—for quite different reasons.

On Valentine's Day 2017, Brad Smith—the affable, sandy-haired president of Microsoft—took the stage at the annual RSA Conference, the tech industry's premier security conference. "Cyberspace," he declared, "is the new battlefield."[8]

"The world of potential war," he warned, "has migrated from land to sea to air and now cyberspace. As a global technology sector, we need to pledge that we will protect customers." He paused. "We will focus on defense."

Let's take a moment to digest this. The president of Microsoft—*Microsoft*, the company whose products are virtually synonymous with corporate cubicle culture—announced to 40,000 of the tech industry's frontline programmers that they were, for all intents and purposes, at war.

"Because when it comes to attacks in cyberspace, we not only are the plane of battle, we are the world's first responders." He continued, "Instead of nation-state attacks being met by responses from other nation-states, they are being met by us."

Let's see that again: "They are being met by *us*."

Who is "us"?

Smith was talking about something new—some higher-order embodiment of digital power. These new entities are tech companies' next stage of evolution, a giant technological leap from Jobs's iPhone.

These tech entities are no longer simply making spreadsheet software and calendar apps and gadgets. They are battlefields. They are weapons. And, most important, in this speech Smith declared that these new entities should be—must be—a force for good.

The problem here is that no one knows what to call these new things. As I first introduced in a 2017 *WIRED* article, I propose that we call them "net states."[9]

Why not just keep calling them "the tech industry"? The short answer is that the tech industry is no monolith, with all its companies pursuing the same goals with the same business practices.

As hard as it may be to think of the world's newest industry as traditional in any way, a handful of "traditional" companies have undergone a metamorphosis. And, in the same way we don't keep calling butterflies "caterpillars" once they've transformed, these particular companies—Amazon, Apple, Facebook, Google, Microsoft, and Tesla, specifically—have morphed into something altogether different from "the tech industry."

They no longer only make products and offer services. They're reaching beyond their core technologies to assert themselves in our physical world. They're inserting digital services into our lived environments in ways both unseen and, at times, unknown to us. And, most important, they're exerting formidable influence over the way our world works on individual, societal, and geopolitical levels. *These* tech companies are unlike anything we've encountered before.

Net states vary in size and structure but generally exhibit four key qualities: They enjoy an international reach. Their core work is based in technology. Their pursuits are influenced, to a meaningful degree, by beliefs, not just a bottom line. And, perhaps most significant, they're actively working to expand into areas formerly the domain of governments, areas that fall outside their primary products and services—areas they pursue at times separate from and even above the law.

Simply put, net states are not just out to make widgets or get people hooked on a single product. (This is why Tesla and its world-building businesses are included in the book and Twitter, with its single, standalone platform, is not.) Net states are out to change the world—not just in theory, but in defense, diplomacy, public infrastructure, and citizen services.

Net states are tech entities that act like countries. By acting like countries, net states alter our experiences as citizens. And they alter countries' experiences as geopolitical powers.

Two examples—Silk Road and Project Maven—show this in action.

"IT IS WITH A HEAVY HEART THAT I COME BEFORE YOU TODAY," WROTE a user, code-named Libertas, in his farewell letter.[10] "A heart filled with sadness for the infringements of our freedoms by government oppressors." He wrote, "Silk Road has fallen."

Libertas was the Roman embodiment of freedom; the inspiration for the Statue of Liberty. It was also the pseudonym Gary Davis used on Silk Road—not the historic trading route in Central Asia; the illegal marketplace on the dark web that freely sold everything from drugs to hacking-for-hire services to humans from its launch in January 2011 to October 2013, when the FBI shut it down.[11]

Davis worked as a low-level site administrator for Silk Road—the Mafia equivalent of a bookie. Admin though he was, Davis hardly looked like a stereotypical hacker, showing up at his trial sporting a trim suit, well-groomed chinstrap beard, and seemingly well-rehearsed thousand-yard stare.[12] While Davis was a minor figure in the Silk Road case, Silk Road itself was a major problem that had dogged the FBI for years. It thrived, selling illegal wares in plain view of the authorities to its consumers, who spent more than $1.2 billion in its two-plus years of operation.[13] Yet with all transactions encrypted via Bitcoin, the authorities couldn't figure out who was running it. When the FBI finally cracked the case, they scooped up everyone they could find associated with the site, including lowly admins like Davis.

On December 19, 2013, at 8 p.m., at the behest of the FBI, Irish authorities swooped into Davis's hometown of Wicklow, a sleepy seaside town about an hour southwest of Dublin.[14] Finally, after two long years of failed attempts to shut down the site, it looked as if the Silk Road case was under control: the FBI had found their suspects, and it was only a matter of time before they gathered the material evidence needed to put them away.

Then the investigation hit a brick wall. While Davis used an encrypted browser called TOR for his Silk Road–related work, he preferred Microsoft for his personal emails, which meant that Microsoft, whether they were aware of it or not, had been safeguarding content for an international drug trafficker.

As a matter of routine, the FBI got its subpoena for the emails and handed it over to Microsoft. But then, something unusual happened: Microsoft stalled. Because while Microsoft now knew they harbored content belonging to a probable felon, technically they were *allowed* to: since Congress had passed the Communications Decency Act in 1996, tech companies couldn't be held legally responsible for the content on their platforms.[15]

The issue for Microsoft wasn't the particular user the FBI was after, or even the potentially incriminating content of his emails. The problem was that the emails weren't stored in United States territory. They were on a server in Dublin, Ireland. And whether an American subpoena had jurisdiction over data physically housed on machines in another country simply wasn't clear.

So while Microsoft handed over the Davis emails physically stored in the United States, they declined to turn over the ones housed in Ireland.

In sum, Microsoft, an American-based tech firm owned and operated by American citizens, refused to comply with the American government's subpoena. And amazingly enough, they weren't breaking any laws, because none existed at the time that made clear what the appropriate course of action should be.

In 2013, the US Department of Justice sued Microsoft to retrieve the Dublin-stashed emails.[16] That case turned into a fiasco. What might have been a simple paper chase became a many-year legal crusade. Because for Microsoft, it was never about Davis, or even the content of his emails. This was the case that would set precedent for the US government's jurisdiction over global digital communications for years to come. As of this writing in 2019, the case is still making its way through appeals courts.

On one level, the Silk Road story is about citizens' rights online: who gets to decide what happens to digital information, the tech companies who manage user data or the countries in which the users reside. But on another, it's about citizens' rights in real life: which entity gets to decide the fate of that user, a fate that might come with stakes as high as physical imprisonment.

Surprisingly, of the three players involved in this fight—two countries (the United States and Ireland) and one tech company (Microsoft)—the tech company, not the countries, took the lead on safeguarding citizens' rights.

To be fair, the American government is legally beholden by the Constitution to "pursue justice" on behalf of *its* citizens. Which they did, attempting to convict perpetrators whose illegal marketplace harmed untold numbers of victims. But on the other hand, the American government is also constitutionally prevented from conducting "unreasonable searches and seizures." Microsoft could conceivably argue that this is what the US government was doing, as the desired objects of the searches were physically outside US territory.

The elephant in the room is, of course, that while this may *technically* be the case, this is not in any way *practically* the case. It's not like Microsoft would have had to send a team of experts across the Atlantic to excavate documents with a trowel. It could have conjured up the

Dublin emails with the mere click of a button, never having to leave its headquarters in Redmond, Washington.

The question is, then, why Microsoft went through the bother. Legal cases are extremely expensive, even for tech empires. And they're time-consuming; this case has dragged on for over six years already. But most important, unlike governments that are constitutionally bound to look after their people, Microsoft—or any other tech company, for that matter—has no obligation to put up any sort of fight for citizens' rights.

Then a rationale begins to emerge: Microsoft's not actually protecting *citizens*; they're protecting *users*. They're not securing citizens' physical belongings from "unreasonable search and seizure." They're protecting their users' data.

In this way, "citizen" and "user" merge in some information-age mashup, becoming something new: the "citizen-user." Because whether Microsoft is really protecting "citizens" and their rights or protecting "users" and their data is almost irrelevant. The end result is the same: the tech company is standing up for the individual. And in this particular case, the United States, the most powerful country on Earth, can't do a thing about it.

This shows how major tech companies can outmaneuver countries—how they operate above the laws of individual countries. This is how tech companies become net states.

OUTMANEUVERING COUNTRIES ISN'T ALWAYS FUELED BY LEGAL MURKIness or precedent-setting. Sometimes, it's a matter of plain old principle: net states refusing to work with governments because it goes against their beliefs, even when it means losing money.

"It's so exciting that we're close to getting MAVEN!" wrote Google Cloud's chief scientist for artificial intelligence (AI), Fei-Fei Li, in an email obtained by the Gizmodo Media Group.[17] The "MAVEN" Li refers to would be "Project Maven," the plain English moniker for the Department of Defense's (DOD) exploratory artificial intelligence program.[18] Maven's mandate was essentially to put AI capabilities on drones.

"I think we should do a good PR on the story of DoD collaborating with GCP from a vanilla cloud technology angle (storage, network, security, etc.), but avoid at ALL COSTS any mention or implication of AI," Li's email continued, urging her colleagues to steer clear of what could be a public relations nightmare. She instead pitched Project Maven as a "vanilla"—in other words, harmless—cloud storage partnership. This shows how even in those early planning days, Li was conscious, and nervous, about what would happen if the public thought Google was collaborating with the Department of Defense on anything to do with artificial intelligence. As her email suggested, it wouldn't be difficult for this to quickly spiral into "killer robots" headlines splattered across the news.

Turns out the media wasn't Google's main problem. Rank-and-file Google employees would prove to be the project's undoing.

Before getting to Google employee protests, it's important to dig beyond the perception of Project Maven and look at what Project Maven actually aimed to do. Maven's AI was supposed to aid human operators at DOD sift through massive troves of information, "to help a workforce increasingly overwhelmed by incoming data, including millions of hours of video." And no one within DOD had developed AI to the point where it could be deployed in this way. As Colonel Drew Cukor, chief of the Algorithmic Warfare Cross-Functional Team responsible for Project Maven, announced at the Defense One Tech Summit in 2017, "You don't buy AI like you buy ammunition. This effort is an announcement . . . that we're going to invest for real here. The only way to do that is with commercial partners alongside us."[19]

It started as a tiny project, in Google terms. The $9 million contract, which launched in 2017, involved just 10 employees, a minuscule allocation of resources from Google's 88,000-person workforce.[20] But as word about Project Maven spread throughout the company over the next several months, outrage ensued. About a dozen AI researchers at Google resigned in protest, the first mass resignation over a matter of principle in Google's history.[21] This was shortly followed by a petition signed by 4,000 staffers demanding that Google cease its AI contract with the military immediately.

"We believe that Google should not be in the business of war," began the one-page letter to Google's CEO, Sundar Pichai.[22] "Building this technology to assist the US Government in military surveillance—and potentially lethal outcomes—is not acceptable."

Most notable, however, was not the fact of the petition or even that it demanded an end to Google's partnership with the Pentagon; it was the workers' rationale for the protest. "We cannot outsource the moral responsibility of our technologies to third parties," the letter stated. "Google's stated values make this clear: *Every one of our users is trusting us. Never jeopardize that. Ever.*"

The resignations and petition worked. Despite the potential financial windfall future military contracts could bring the company, on June 1, 2018, Google announced that it would not be continuing the Project Maven contract once it expired at the end of the year.

WE'RE IN A WORLD STILL DOMINATED BY NATION-STATES, BUT INCREASingly influenced by the actions of net states. Nation-states continue to own the physical territories within their borders, but net states wield significant power both within and across country space, guiding events that affect us both on an individual and on a global level. Therefore, we need to get smart about what net state power really looks like, and quick.

One country that's excelling in its efforts to do so is Denmark. In 2017, it opened a door that has the potential to radically alter our existing geopolitical order: it appointed a new ambassador to capital-T Tech itself. Ambassador Casper Klynge is the world's first-ever tech ambassador. His mandate: to establish diplomatic relations between Copenhagen and Tech. And what exactly *that* looks like is all fresh territory, yet to be discovered. Fittingly, his office operates as a virtual embassy, with three physical manifestations: one in his home base of Copenhagen and two in the most powerful tech hubs on Earth—Silicon Valley, California, and Beijing, China.

I arranged to interview Ambassador Klynge from his Silicon Valley office in late April 2019. I'd given his communications director my cell number and sat on the sofa in my basement home office in Brooklyn, waiting for the call. The lights were off, the only source of illumination being the gray-white glow from my computer screen. (This was deliberate; frankly, I didn't want to be distracted looking at laundry piles in the corner during the interview.)

Suddenly, my phone rang. But in addition to ringing, the screen lit up, serving up an image of my own face as well.

Oh. We're having a *FaceTime* interview, I realized. I should have expected it—tech ambassador, after all. Because I didn't want to miss the call, I answered before I had time to turn on the light. Two tanned, cheerful people greeted me on what appeared to be a sunny day in a naturally lit office somewhere in Silicon Valley, their friendly faces framed by a whiteboard with vague scrawl in the background.

"Are you . . . um, is this still a good time?" Ambassador Klynge asked encouragingly. I could see in the tiny window FaceTime provides of one's own reflection the eerie computer light cast on my face, giving me a ghostlike appearance. Despite this less than ideal setup, it had taken quite a bit of work to get on the ambassador's calendar, and I didn't want to miss my chance. So, there we were—I in my dark Brooklyn basement and he in his sunny California diplomatic outpost—talking, smiling through our cell phones.

I went straight for my most pressing question first: He's the world's first ambassador to the tech sector—how had that sector received him?

Ambassador Klynge's expression made it seem as if this question brought a story to mind. Having worked with diplomats in my past, I doubted I would hear it in an on-the-record conversation like this, though, and didn't press. After a moment, he said that "some companies" had been "very forward-leaning." He paused. "Then you have the other side of the spectrum," he said, "where some companies have been enormously difficult to deal with. . . . We deliberately say we want more or less to come in at the top level. That means C-suite level, and they sort of offer the oldest intern. . . ."

The ambassador trailed off for a second. On the screen, I could see Klynge lean slightly forward at his conference room table, pinning his index finger to some invisible spot on it. "I think *reluctant* is a very diplomatic term."

I wanted to ask which companies sent an intern to greet him—an ambassador!—but didn't think he would be at liberty to disclose. I figured I could probably guess for myself anyway: for this book, Microsoft and Google happily accepted my interview requests; Facebook, on the other hand, was a vault.

What about governments? I asked. Were they reluctant as well?

Total opposite, he said without hesitation.

"There has been enormous—I would almost say *unprecedented*—interest from other capitals in basically learning from us, getting our experiences from dealing with the tech industry." He said, "They tell us, 'We would love to do something similar [to your tech embassy], but our bureaucracies are so large and so difficult that we would never be able to do it; . . . the distance from *flash* to *bang* is simply too big.'"

Then Klynge's whole face lit up; he was clearly pleased to reflect on what his country had been able to accomplish. "That's one of the areas where being small is a little bit of an advantage."

This comment revealed a distinct advantage Denmark and the two other countries who've since appointed their own tech ambassadors—Estonia and Australia—have over their larger nation-state counterparts. When it comes to tech, the smaller nations have proven to be like speedboats amid a sea of ocean liners. Nation-state behemoths may still have more firepower and financial might, but, essentially moored in place by their unwieldy mass and unforgiving bureaucracies, they can't seem to keep up with net states nearly as well as countries like Denmark and Estonia.

I had time for one last question. How should we—governments, societies, people like you and me—be thinking about technology? How do *you*, Mr. Tech Ambassador, see it?

He took just a moment, barely a beat—it could have been a hiccup in our connection, really. But then he said, "The freight train is coming."

Klynge continued, "It might not be everybody who's seeing the massive impact of technology also on international relations, but one of the reasons why we gather . . . countries [is] to try and help shape thinking in capitals all over the world."

Governments need to understand, Klynge explained, that technology is much more than "an add-on." "It's not the IT office that needs to deal with technology; it's mainstream foreign and security policy."

I invited him to explain why he thought so, and this time he responded immediately. "Technology will have a massive impact on international relations. It will have a massive impact on the convening power of the West. It will have a massive impact on the balance of power in the future," he said.

Then he added one last thought. "For *that* lesson, it's high noon for many, many countries all over the world."

Casper Klynge and his fellow tech ambassadors exist because at least three of the world's most forward-leaning countries have come to recognize, in the most formal and official way a country can, that net states occupy a substantial role in our geopolitical, social, and personal worlds. This book describes the various ways in which net states exert influence on those worlds and each of us in them, as well as what we can—and must—do to ensure that they do so responsibly.

This book shows us our tech in a new light: not just as services we access or devices we use but as forces of personal, social, and geopolitical power. From this new vantage point, we gain additional ground for exploration: the capacity to ask questions about tech's impact that we've yet to even consider.

This book is not an exposé of any individual net state, though the major tech companies serve as our main characters. It's the story of what net states are up to, both as they engage with us—their citizen-users—and as they expand out of the digital and into the physical world.

THIS BOOK STARTED WITH AN ARTICLE I WROTE IN 2015 AND LEFT IN A file for two years. I wrote it to try to make sense of what had happened following the November 2015 terrorist attacks in Paris. The day after those attacks, the hacker collective Anonymous launched a campaign, Operation ISIS, in which they claimed to have taken down upwards of 20,000 ISIS-related social media accounts in a single day.[23] By comparison, the social media companies themselves had taken down only around 800 ISIS accounts over the prior 18 months.[24]

It occurred to me that the social media giants and Anonymous both had a bigger role to play in fighting terrorism than I'd seen discussed. But I ran into a problem: *how* to discuss it. What *was* Facebook? And Google? And Anonymous? And the other major tech companies and movements? They clearly weren't nation-states, like the US and France. But they weren't nonstate actors, like ISIS or al-Qaeda, either. We simply didn't have the language at the time to categorize them. Despite that, it was becoming increasingly evident that these . . . somethings . . . were forces to be reckoned with—not just as commercial entities but as significant players in defense, diplomacy, and other geopolitical arenas.

When I shared the draft article with my most trusted readers, it elicited raised eyebrows—one Anonymous campaign did not seem quite sufficient to support a new theory. So I shelved the piece but kept collecting evidence: examples of incidents in which tech companies had reached beyond their core services and into governmental areas. By 2017, I felt I had gathered enough data to warrant dusting off the article and making the public case for net states. *WIRED* magazine agreed: in November 2017, it published "Net States Rule the World: Ignore Them at Your Peril," introducing the term "net states" to the lexicon.

Since 2017, evidence that tech companies are acting like countries has only continued to amass. In June 2019, Facebook announced the launch of its own monetary project: a cryptocurrency, Libra. As technologist Micah Sifry observed in his newsletter, *Civicist*, "If you're going to be a country, you might as well have a currency, right?"[25]

The Information Trade is both a near history and a profile of what it means to live a tech-enabled life. It celebrates how technology enables us to share the stuff of life—information, data, stories, knowledge, sorrows, silliness, and ephemera. It informs us about what happens to that data when we do share, both with and without our knowledge. And it cautions us against giving up our ability to influence the balance of power between ourselves, our governments, and our net states.

"NET STATES" IS KIND OF LIKE "SEA CREATURES" OR "THE EUROPEAN Union": it's a label that represents a larger group. Its members share features, but there's a lot of variation among them. In the same way you wouldn't want to read a book about sea creatures without learning about sharks and whales, or a book on the EU without touching on Germany or France, this book is structured around five major net states—Amazon, Apple, Facebook, Google, and Microsoft—as well as the net state activity exhibited by Elon Musk's Tesla and its sister projects, and the political movement represented by independent Pirate Parties in various countries.

Chapter 1 looks at how we transformed from audience members to computer users to citizen-users, starting with the launch of Microsoft's Windows 95. While widely associated with office drudgery now, Microsoft broke onto the scene more than twenty years ago with a revolutionary suite of tools that radically transformed what was possible for the average computer user with no technical knowledge: the ability to navigate a computer (via Windows) and get work done (via Office), thus contributing to the information-sharing norms in which we operate today. But information-sharing, for some people, is not simply a feature; it's a *right*, something to believe in—and a cause to fight for. And, over the past twenty years, we have become more than simple recipients of content. We've morphed into something altogether new: citizen-users. Chapter 1 shows how this came to be, how we became citizen-users for whom technology is not just a tool we use but an ideology that dictates how we engage with—and take part in—our governments.

While citizen-users engage with digital content, they're still grounded in a physical landscape. Chapter 2 situates citizen-users in their physical landscape, examining what net states are doing "IRL"—in real life—starting with Tesla's and Google's interventions in Puerto Rico after Hurricane Maria in 2017. It grounds the ethereal internet, "the cloud," in the physical world, tracing how our data is bound to Earth through undersea cables and data centers. The chapter then moves to net state activity in key areas of the physical world. By tracking net state activity IRL, this chapter lays the foundation for a new way of looking at power: distributed not according to borders on a map, but through information flows, investments, and physical assets.

Chapter 3 looks at the battle over our privacy. Privacy is no longer a given. We're engaged in a global battle over who gets to determine the degree of privacy we retain over our content and activities via tech. This chapter explores how our understanding of privacy has evolved and the possibility that its current iteration may be an "anomaly." It also considers how net state partnerships with data brokers create profiles that "know" us, and how some countries are fighting back against these practices. It traces how Europe is leading the way for wielding government regulations over tech companies in defense of citizen-user rights and considers what options Americans have in our currently unregulated landscape.

Chapter 4 considers our physical security, showing how net states like Google became integral to the fight against modern-day enemies. Exploring the differences between the tech ethos and military ethos, this chapter explores how the expertise so prized by the security agencies has become a kind of disadvantage in the fight against terrorism. It shows how net states are uniquely capable of engaging in security issues, through counterterrorism activities via Google's think tank, Jigsaw. The chapter then shows the possibilities for net state/nation-state cooperation through acts of diplomacy and looks at how net states, led by Microsoft, have begun to forge ahead in this domain.

Chapter 5 examines how we use net state tools to curate idealized versions of ourselves—our profiles and activities online—and how nation-states may enact real-world consequences on what we are or are not permitted to access based on them. Using China's Social Credit Score system as an example, the chapter considers what the networks of connections we create, as well as the carefully managed personas that we upload, say about our needs as individuals, citizens, and citizen-users.

Chapter 6 delves further into our daily life by examining the tech we use in our homes and our public spaces with the Internet of Things (IOT). This chapter explores the ways that the IOT currently influences and is likely to affect our daily existence. It then examines developments in user profiling, starting with Amazon's recommendation systems and "smart" technologies that gather data on our health, our environment, the information we seek, the music we play, and even our sleep habits.

Chapter 7 moves the focus from our actions to our cognition, examining what happens to our minds as we interact with net state technologies. It explores the impact of the uniquely immersive qualities of this tech and how that impacts our thinking, our behaviors, and ultimately, our awareness of ourselves and the world around us. Starting with the most ubiquitous net state tech—the smartphone—this chapter looks at how increasingly immersive properties of our technology affect how we learn, what we remember, and how we perceive the world. From the current tools at our disposal to emerging tech like augmented reality, the chapter considers what an increasingly personalized view of the world might mean for people and societies.

Finally, chapter 8 takes a hard look at where we've been in recent history and where we are now, with staggering rates of depression, addiction, and—unique to America—acts of gun violence. It gives us options for how to reconcile our feelings of empowerment via tech with our sense of powerlessness in the face of life-altering challenges. This chapter explores what we as citizen-users must do to ensure that we remain actively engaged in the development of net states, their relationship with our nation-states, and how they relate to our own lives, offering a citizen-user pact with net states.

The book concludes with an assessment of how net states have begun to engage with one another and recommendations for governments to join with them or face irrelevance. It argues that citizenship, whether in a nation-state or a net state, requires engagement, and the consequences for failing to engage are dire. In a democracy, failing to vote means losing out on the chance to be represented by those who protect our interests. In net states, failing to engage means losing out on personal privacy, the implications of which are only starting to be understood.

PUBLIC OPINION ABOUT THE ROLE OF TECHNOLOGY IN OUR LIVES SWAYS. In the first 15 years of the new millennium, tech was going to save us all, make the world more democratic, level the playing field, and provide a platform for the disenfranchised to make their voices heard. Then the elections of 2016 hit; Americans were manipulated en masse by Russian misinformation campaigns. Facebook gave 87 million users' data over to the political consulting firm Cambridge Analytica. "Technology addiction" and "Facebook depression" became well-known conditions. Tech suddenly seemed dangerous.

But public opinion on tech will likely swing again. Tech just does too much good for people not to notice eventually. For example, in 2018, in New Delhi, India, a police department instituted a pilot project using experimental facial recognition software. Within four days, more than 3,000 missing children were located.[26] That same year in Australia, two teenage boys caught in rough waters 2,500 feet offshore were rescued when a remote-controlled drone delivered an inflatable rescue pod to the swimmers within 70 seconds of launch. Traversing the same distance would have taken a lifeguard up to six minutes, during which time the boys could have drowned.[27] And back in the United States, a young woman was warned by her Apple Watch that her heart rate had skyrocketed to 190 beats per minute. This prompted her to get to an emergency room, where she was immediately diagnosed as undergoing what would have been fatal kidney failure.[28]

And so on. Tech can literally save lives. Even beyond its lifesaving capacity, its presence in our daily realm can facilitate better living, with faster answers to our queries, better suggestions for what will be our most beloved book or film, and easier access to aids of all kinds, from maps to encyclopedias, from cookbooks to cameras.

We are not victims here. We've invited tech into our lives for a reason. It makes life easier. It makes life more convenient. Sometimes, it makes life safer. Sometimes, it makes life better. And since tech isn't going anywhere, we owe it to ourselves to know what it *means* that it's here: what our data is worth and how it's used. Just as being a responsible citizen of a nation-state requires paying attention to who's in charge and what they're up to, we need to become responsible citizen-users of our net states, paying attention to who's in charge and what *they're* up to.

Eventually, public opinion will settle somewhere in the middle regarding tech. We will no longer be starry-eyed about its promises or frightened by its possibilities. But until then, we should harness our outrage and our passions to demand that net states take great care with us, their citizen-users. Because while there may be only one Facebook and one Google and one Apple now, those will not always be our only options. Remember, not long ago, there was only one Myspace and one Napster.

Henry David Thoreau once noted, "Is a democracy, such as we know it, the last improvement possible in government? There will never be a really free and enlightened State until the State comes to recognize the individual as a higher and independent power."[29] Net states create tools that elevate the individual, and—just as in our political system—it's up to the individual then to leverage or leave idle that power.

We users have more power over net states than we've yet to claim. *The Information Trade* shows how to be present in the midst of technology, aware of the new ways it controls our world, and able to manage its impact on our lives. We do not need to stop technology from evolving to ensure that it does so responsibly. *The Information Trade* explores what it means to be a responsible citizen-user, engaged with and unafraid of the world that we're building with our tech—and that tech is building for us.

ONE

RISE OF THE CITIZEN-USER

In the 1983 film *WarGames*, a baby-faced Matthew Broderick plays an underachieving teen who develops his hacker chops by altering grades in his high school's mainframe. Trying to impress the doe-eyed Ally Sheedy, he accidentally hacks into a live military operation at NORAD, suddenly finding himself engaged in a computer-simulated war exercise to prevent World War III.

The movie was a huge success. It was the fifth-highest-grossing film of the year and garnered three Academy Award nominations. But its biggest impact was felt by the computer industry, which desperately needed the boost. In the early 1980s, tech still seemed mystifying and cultish to mainstream America; people didn't really know what to make of computers or the rare few who tinkered with them. In 1983, only 8 percent of Americans owned a computer. Apple's first personal computer, which didn't go on sale until 1984, cost an eye-popping $2,500—a third of the price of a brand-new car at the time.[1] Cell phones were clunky, ugly affairs, also prohibitively expensive at about $4,000 a pop, or about $9,520 in 2019 terms.[2] For the average American in the 1980s, "technology" consisted of TVs, cassette tapes, and Ataris. Until *WarGames* came along, personal computers were, by and large, a curious luxury.

But the American imagination had now gotten a taste of computers as tools worth their attention, and pop culture responded accordingly. The ultimate manifestation of this moment of tech awakening was when Apple barreled into mainstream American consciousness with their now famous 1984 Super Bowl commercial, directed by *Blade Runner*'s Ridley Scott. In the commercial, an athletic heroine races past the dull-eyed masses as she wields a sledgehammer. She launches the hammer at a massive television screen that had enraptured its audience, symbolically destroying the means of control over passive television consumers and introducing them to a tool designed to reinvigorate and empower the individual: the Apple home computer.

In addition to boosting sales for home computers, another gift *WarGames* gave the '80s was the stereotype of the hacker: the image of the obsessive, scrawny teen squirreled away in his parents' basement conducting virtual break-ins for personal gain or juvenile kicks.[3]

This depiction was almost an affront to actual hackers—originally a term reserved for self-motivated technology tinkerers. In reality, most hackers were serious computer scientists, gainfully employed by prestigious research universities like MIT and Stanford. Hackers had been around since the 1950s in a loosely connected community of like-minded programmers. And they changed history: hackers built the US Department of Defense's ARPANET, the predecessor of the World Wide Web, and mainframes at IBM. Hardly goofballs digitally breaking and entering classified data warehouses, hackers were among the early architects of today's technological infrastructure.

The group of original hackers included Sir Tim Berners-Lee, aka TimBL, the British computer scientist who invented the World Wide Web, its first web browser, *and* HTML, the dominant programming language for websites. While studying physics at Oxford in 1976, Berners-Lee cobbled together a computer using an old TV and a soldering iron—the very portrait of a hacker in action.[4] His contemporary Richard

Stallman (aka rms), a software engineer and digital activist, launched the Free Software Movement and the GNU operating system, which would later become a part of Linux, the most widely used operating system on the planet. Android phones all run on a version of Linux: that's 88 percent of internet-enabled mobile devices—approximately 4.4 billion worldwide.[5] Stallman, who on his personal website lists among his hobbies "affection," "international folk dance," and "puns," is known as much for his philosophical intensity as his programming chops. His Free Software Movement gave rise to open source software—that is, software whose inner workings aren't proprietarily protected, like the web browser Firefox and the website builder WordPress—and he's arguably one of the forefathers of the very concept of information-sharing as an ideal, not just a practice.

Over the years, dozens of hackers—almost all of them university professors or professionally employed engineers, with the exception of Bill Gates in the early years of Microsoft—contributed to the rise of the web as we know it today. It wasn't until the 2000s that the teen hacker college-dropout trope would become a reality, with Facebook founder Mark Zuckerberg as poster child. And these programmers were serious about their work and serious about their culture: the "hacker ethos" was a code to live by, a topic of debate and deliberation, and, most of all, a point of pride. To be a hacker was to uphold a set of values and a way of life.

The hacker ethos, which was both pragmatic and idealistic, consisted of six basic tenets.[6] First, hackers believed that access to computers should be universal, regardless of skill level or intent for use. Because they viewed computers as tools for empowering the individual, they believed that every individual should have access to one. Second, they fervently adhered to the notion that information should be free. This is reflected in the early days of the internet, when, indeed, all information online *was* free. It was only after the World Wide Web was commercialized in the mid-1990s that websites began charging for content—a move that was anathema to hacker ideals.

Third, hackers held a deep mistrust of centralized authority of any kind. This is also reflected in the way the internet works: it's a decentralized system, running on millions of computers across the globe. There's no one person or organization who can "turn off" the internet—redundancy is a safeguard built into the very foundation of the web. Fourth, hackers believed that they should be judged by skills and abilities, not official credentials. Being a college dropout is worn almost as a badge of pride in the software industry, and over 50 percent of employed programmers in 2015 didn't have a computer science degree.[7]

The final two tenets are the most idealistic: that one can create art and beauty with code, and that computers should be used to change life for the better. While there is no shortage of malevolent hackers now, nor was there in the early days of the web, the hacker ethos took seriously the idea that tech is a tool that can be used for good or ill, and it is up to the coder him- or herself to make the moral choice to apply programming skills for creative, artful, and positive ends.

As universities across the country began to gain access to ARPANET, hackers started collaborating with one another virtually, sharing code and problem-solving tactics. And so by the '80s, serious hackers had started to band together, resulting in a frenzy of invention and innovation. This energy and the hacker ethic were captured by journalist Steven Levy when he published his 1984 book, *Hackers: Heroes of the Computer Revolution*. Thirty-five years later, this book is still lauded as the manifesto of its era. But when it was published, critics viewed the "hacker ethic" as a historical anomaly, an oddball set of ideals that died before they even got a chance to get going. The *New York Times* review recoiled at the book's account of programmers plying their skills on games like Frogger. Christopher Lehmann-Haupt concluded that "if the point of the entire computer revolution was to try to get a frog across a road . . . then it's not only unsurprising that the hacker ethic died; it isn't even sad."[8]

Hackers themselves disagreed. Far from seeing the hacker ethic as dead, they took the book *Hackers* as a catalyst that inspired them to, for the first time ever, physically come together, bringing the hacker ethic to the table for discussion and celebration. This took the form of the first-ever "Hackers Conference," organized by publisher-activist Stewart Brand, Apple cofounder Steve Wozniak, and others. On November 1, 1984, 150 of the most talented programmers, engineers, and designers gathered at the Headlands campus of the Yosemite National Institutes in Sausalito, California, to meet face-to-face and discuss their craft.[9]

Most people have at least heard of Apple and its cofounder Steve Wozniak; Stewart Brand is less well known, but worth knowing about. Brand wasn't a hacker. He didn't even know how to code back then. But he'd launched something called the *Whole Earth Catalog* in 1968, and in its way it epitomized the hacker ethic.

It was, on one level, a traditional catalog; you could mail-order things from it just as you could from the Sears or JC Penney (or any other) catalog. But it stood out from others of its kind in key ways—first, for what it sold. For a world Brand described as needing to go "back to basics," his catalog offered, fittingly, a range of back-to-the-land type stuff. Wares had to fit at least one of four criteria: they had to be useful as tools, relevant to independent education, high quality or low cost, and easily available by mail. Under this umbrella, *Whole Earth* sold materials for and published articles on everything from "earthworm technology" (for aerating farm soil) to "cooking with fire" (for outdoor and off-grid living); it also offered—under the "useful as a tool" and "relevant to independent education" categories—ads for the first Apple computer.

It was really the articles accompanying the goods it sold that defined *Whole Earth* as the start of a movement that empowered individuals—not just as part of collectives or communes, but as people capable of existing wholly and fully on their own two feet. *Whole Earth* promoted the individual on every level: logistically, with teachings on how to build fires and yurts; physically, with articles on DIY agriculture and hydration devices; and intellectually, with essays from the most forward-leaning and controversial thinkers of the day, from Buckminster Fuller and Carl

Sagan to the Dalai Lama and members of the Black Panther Party. *Whole Earth* was recognized as revolutionary in its time: for example, it's the only catalog to ever win the National Book Award.[10] Brand's publication elevated the citizen not as a consumer, but as a vessel of power, a being capable of shedding the trappings of "modern" (1960s) life and finding fulfillment by going back to basics. A reflection on our place in the universe, *Whole Earth* acknowledged how even 1960s technology was emerging as the next force of nature we would be forced to reckon with.

So it's not surprising that Brand, of all people, came up with the mantra for a generation of citizen-users. At that 1984 Hackers Conference, Brand, clad in a tan leather vest over a black-and-white gingham button-down shirt, made an offhand comment in a panel discussion with Wozniak that perfectly put into words an idea whose time had come.

"On the one hand," Brand said, "information wants to be expensive, because it's so valuable. The right information in the right place just changes your life." And then he added, "On the other hand, information wants to be free, because the cost of getting it out is getting lower and lower all the time."[11]

The comment was casual, a nod to an audience member who had just voiced frustration over the rise of proprietary software shutting down collaboration opportunities. But the words themselves—"information wants to be free"—struck a collective nerve. That statement would go on to become the rallying cry for a generation.[12]

At the time, Brand was talking about how it would be increasingly difficult to charge money for information once it was digitized and thus easily copied. But as global networked computing became a reality, tech activists adopted the idea as a literal one.

The reasoning behind it goes something like this: Information, once digitized, is easy to share. And digitized information is also easy to manipulate and search, from basic everyday Google queries to sophisticated data mining. This digital information searching reveals all kinds of valuable things, and shockingly fast—from patterns and research material to regular old know-how on how to do things. Since digitized information can be shared with many people simultaneously, and since it can reveal useful and beneficial things, many people should be able to benefit from it as a kind of public good. As such, information should be free, and freely shared. Thus, information *wants* to be free.[13]

Not everyone agrees with this, least of all net states whose business models today rely on monetizing user content. Ironically, though, it was the forefathers of those same net states who first promoted the "information wants to be free" ethos and hacker code. You can see that ethos in Steve Jobs's 1980 Apple mission statement: "To make a contribution to the world by making tools for the mind that advance humankind."[14] It's reflected in Google founders Sergey Brin and Larry Page's mission statement in 1998: "Don't be evil."[15] It's in Mark Zuckerberg's "Move fast and break things" motto, which he adopted for Facebook in 2004.[16] Notably, all three companies have since moved on to more conservative versions of their mission: Google's is now "Do the right thing"; Facebook's is, only partly jokingly, "Move fast with stable infrastructure"; and Apple's has plummeted from inspiring to anodyne, now reading, "Apple designs Macs, the best personal computers in the world, along with OS X, iLife, iWork and professional software."

Even Stewart Brand himself has tempered his early antiestablishmentarianism. Reflecting on the *Whole Earth Catalog* and its associated movement from a 2018 vantage, he was quick to qualify that it was very much a reflection of its time. "'Whole Earth Catalog' was very libertarian, but that's because it was about people in their twenties," he said in a *New Yorker* interview.[17] "Everybody then was reading Robert Heinlein and asserting themselves and all that stuff. We didn't know what government did. The whole government apparatus is quite wonderful, and quite crucial. [It] makes me frantic, that it's being taken away."

The hacker ethos matters because it inspired the generation of computer programmers and technologists who would go on to found the net states we interact with today. It is the reason that their companies—

Google, Facebook, Amazon, Apple, Microsoft, and Tesla, to name the biggest among them—are driven not solely by their bottom lines, but in addition by beliefs that their products and services create some form of good in the world. While these firms may not adhere to the hacker ethos in all of their business decisions, it is still an influential force that drives many of those who work at net states and, perhaps at times, still even occupies the minds of their founders.

BEFORE INFORMATION COULD BE FREE, HOWEVER, PEOPLE NEEDED DEvices to process it.

The problem was, even by the late 1980s, computers had yet to become commonplace home products. Gradually, they were becoming more affordable and more interesting. But still, only a small percentage of the population picked them up. By 1989, 15 percent of American households owned a computer.[18]

Perhaps that's because, without the internet, computers didn't do all that much. You could type, edit, and store documents—a huge improvement over the typewriter—but that was only marginally exciting. You could play games, of course; but those weren't terribly sophisticated just yet. Computer use at home didn't really take off until the World Wide Web landed in the early 1990s. But even then, uptake started slowly.

"If people are to be expected to put up with turning on a computer to read a screen," mused Microsoft founder Bill Gates in a 1996 essay, "they must be rewarded with deep and extremely up-to-date information that they can explore at will."[19] But even that wouldn't be enough to keep users happy, he wrote. Imagining what a future internet might look like, Gates went on to suggest, "They need to have audio, and possibly video."

Keep in mind that in 1996, "turning on a computer to read a screen" was pretty much the most you could look forward to. Even then, with the World Wide Web just a few years old, it wasn't something everyone was eager to experience.

Even for those who had web access then—about 18 percent of US households—going online was a huge pain. Dial-up internet connections seemed to take forever—a web page took roughly 30 seconds to load, even with a 56K modem, which was state-of-the-art for the time.[20]

However, the biggest problem wasn't getting online. The problem was that there wasn't much to do once you got there. In 1996, there were only about 100,000 websites, most of which featured text and text only. Some offered a few low-resolution graphics, but not too many, as that would have caused the pages to take even longer to load. Without much of interest to keep people online, it's not surprising that the average American in 1996 didn't bother with the internet much, spending about 30 minutes online a *month*—an average of a minute a day.

Given this state of affairs, it makes sense that tech pioneers like Gates spent a lot of time worrying about how to get more people to "put up with" turning on their computers. That phrase summed up most people's relationship with technology back then. It wasn't yet the touch-of-a-button pocket device we enjoy today. With the exception of enthusiasts—20 years ago, the most likely demographic online was white men over the age of 50; only 14 percent of women under 30 used the internet on a regular basis—tech simply wasn't a big part of people's lives. Circa 1996, tech was the Motorola StarTAC flip phone, with its green pixelated text and black screen. Tech was Tamagotchis: virtual "pets" attached to keychains that activated themselves to demand "feeding"—which involved pushing one of three identically mundane-looking buttons—so that they wouldn't die on you. Tech was Hollywood fiction and teenage toys. In sum, tech didn't matter yet; it hadn't yet graduated from mild distraction to grown-up necessity.

Before the turn of the century, what self-respecting grown-ups really focused on was TV. This was the height of the *Friends* era, the years of *The X-Files* and *ER* and *Law & Order*. TV was king, and audiences ate it up: the average household tuned in for more than seven hours a day.[21] Oprah reigned supreme, launching her now-legendary book club in 1996. And people still read actual books. *TIME* magazine praised Amazon, which launched in 1994 selling *only* books, as one of the top websites of 1996, primarily because you could search by nifty features such as "author" or even "subject or title"—or, best yet, you could "read reviews written by other Amazon readers and *even write your own*" (italics added).[22]

And being able to do this—"write your own" review—signifies why this is where the web starts getting interesting: interactivity arrives. As noted, most 1996-era websites were little more than digital brochures. Interactive features that allowed users to shape their experience didn't become common until "Web 2.0" emerged almost a decade later.[23] So the option of submitting your own review—contributing your *own* voice to anyone who happened to be on the World Wide Web—was something totally novel. All of a sudden, a person didn't have to be famous or a news producer to get their opinions in front of the masses; they just had to go online.

While several variables influenced how tech changed for the average user, one of the biggest contributing factors came down to a single product: Microsoft's Windows 95 operating system, released in 1995. Pre–Windows 95, your computer probably had a black or dark-green background with yellow or bright-green text or, worse, an oversaturation of hyper-rich colors ("pretty" not being the forte of '90s-era computer engineers—see figure 1.1 below). Windows 95 radically transformed this by bringing a sane-looking design to computing (figure 1.2). It also introduced key features that made the user feel in control, like the task

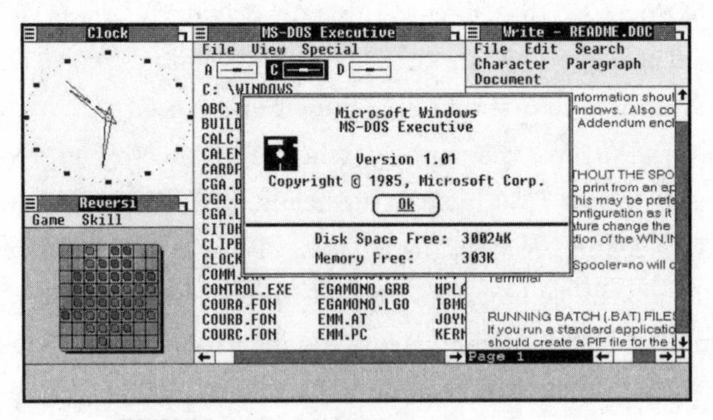

FIGURE 1.1. Microsoft MS-DOS Interface, 1985

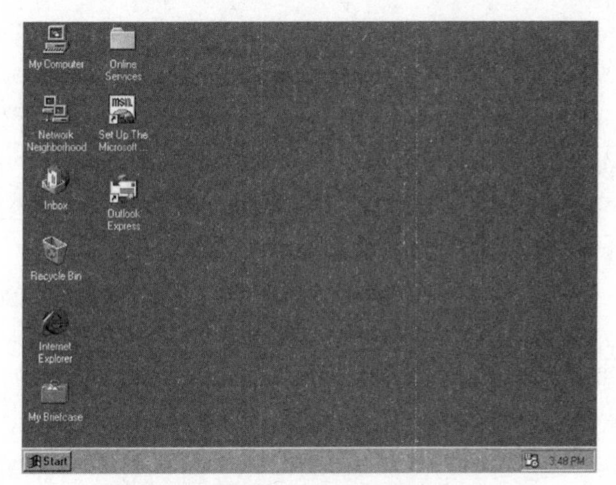

FIGURE 1.2. Microsoft Windows Interface, 1995

bar along the bottom of the screen and the now-well-known "Start" menu button on the lower left-hand side.

Windows 95 and its accompanying internet browser, Internet Explorer, catapulted technology to the next level. It made computing much easier for the average user. With the launch of Windows 95, the cultural attitude toward technology in the United States transformed. All of a sudden, instead of only weird or nerdy types using computers, *everyone* could be a computer user. Not only was it no big deal; it started to become the norm.[24]

By simplifying the browsing experience on your computer and the World Wide Web for the masses, Windows 95 democratized computing. As one reporter reflected, even the introduction of what seems like a simple feature, the "Start" button, brought about transformative change: "In 1995, computers were still mostly for the office and productivity. Windows 95 brought with it a word that consumers understood: 'Start.' Start what? Start *anything*."[25]

One of the things this new feature started was the idea of the computer user as a person with power. Compare computer use to television-watching, for instance. In contrast to how solitary TV-viewing may be today, it used to be a communal experience, a reason for families to gather together. Broadcast networks scheduled set times for shows, and families sat around the TV, together, at exactly the same time, watching the nightly news or prime-time programs. And with just three major channels to choose from, it was likely that your neighbors were watching the same shows you were—extending the community experience of television-watching from your own home to your broader network. More important, television-watching in the 1990s required consensus: you and your brother and sister and parents all had to agree on what you'd watch. TVs had *audiences*, groups—we as individuals were just some subset of a larger body.

Computers, on the other hand, had *users*. The internet offered us all the gift of personalized choice. We didn't have to confer with our siblings over what website to go to; we just went, by ourselves. It was like hogging the remote control, every time we logged on. In the early days, with one computer in the house, people still had to take turns going online, which necessitated some level of interpersonal interaction. But once online, our experiences were our own; we were the master of our browsing, the driver of our curiosity fulfillment.

If after Windows 95 people became computer users, then with the explosion of websites people became computer *citizens*, empowered entities interacting with other empowered entities. We weren't just website audience members; we were "visitors," each one singular and unique. It's in the language itself: on websites, each set of watching eyeballs is measured as a "unique visitor." We may not have known it yet, but we began to matter to content producers not just as part of a larger audience, but as independent units whose actions could be tracked and monitored and learned from. In less than a decade, we went from television audiences to computer users to website visitor, singular.

Through our interactions with early operating systems like Windows 95, during the AOL / GeoCities / Myspace days of the web, we were not only deepening into our identities as computer users. We were testing out the waters of being citizen-users.

And so Microsoft itself, for a moment, was king, not only in how widely its products were used, but in popular culture. People, briefly, loved Microsoft.

Things went south fast.

By any measure, Windows 95 was a smashing success;[26] it sold 7 million copies in its first five weeks.[27] By 1998, industry experts estimated 90 percent of computers ran on some version of Windows; by 2018 this number included over 1 billion devices.[28] And Microsoft's internet browser, Internet Explorer (IE), which users installed, perhaps unwittingly, when installing Windows, was so successful that it essentially killed the other browsers. The first full-color web browser, Mosaic, got folded into what would become the other dominant browser, Netscape Navigator.[29] Netscape enjoyed market dominance for a hot minute, but once the masses got Windows 95 with its bundled IE, Netscape started to tank. That browser's user base declined almost in lockstep with IE's rise.[30] AOL, which had acquired Netscape for a massive $4.2 billion in 1998, was forced to shut it down just five years later.[31]

The meteoric rise of Microsoft and its college-dropout boy-genius founder Bill Gates—he was crowned by *Forbes* magazine the richest man in the world by 1995, a title he would hold for 13 of the next 17 years[32]—got the attention of just about everyone, especially government regulators. On March 3, 1998, the Senate Judiciary Committee called several tech industry leaders to a hearing, including Gates.

Senator Orrin Hatch asked Gates the question at the heart of the hearing—mirroring questions still being asked 20 years later to newer quasi-monopolies like Facebook and Google. Hatch asked, "Is there a danger that monopoly power is or could be used to stifle innovation in the software industry today or, perhaps more importantly, looking forward?"[33]

Turns out the hearing just laid the groundwork for what was to come. Two months later, Microsoft got slapped with an antitrust lawsuit by the Department of Justice and 20 state attorneys general for, in effect, holding a monopoly and engaging in anticompetitive practices.

While companies as large as Microsoft get sued with some

regularity—by conservative estimates, Microsoft has been sued over 50 times (for patent infringement, by its competitors, by the US government alone at least five times, and even by companies it's invested money in)[34]—an antitrust suit is a big one. Antitrust cases have a history of taking down giants: they're what forced the breakup of the telecommunications behemoth AT&T ("Ma Bell") in 1982[35] and oil industry titan John D. Rockefeller's Standard Oil in 1911.[36] In short, while some lawsuits are regarded by massive corporations as flies to be swatted, or, more to the point, settlements to be paid out (Microsoft has paid out an estimated $9 billion in settlements over the years), an antitrust case is rightly regarded as a potential bear on your doorstep, with the power to take down even a colossus like Microsoft.

Microsoft fought this case for 19 years and, if you ignore the hundreds of millions of dollars in lawyers' fees, actually ended up getting little more than a legal slap on the wrist.[37] But in the eyes of the public, the Gates/Microsoft trial was damning. While their products were as popular as ever, Microsoft and Gates went from being held aloft as America's ideal for innovation to just another big bad business out to bilk the American consumer.[38] Microsoft simply wasn't cool anymore.

But that didn't actually matter. America was already hooked.

By the early 2000s, the internet, though still accessible to only 43 percent of Americans and just 5.8 percent of the global population, was well established as the place to be. Web 2.0—websites that permitted interactive engagement versus just being digital brochures—had finally arrived. But within this same decade, a new trend emerged that threatened to upend the order we'd just begun to get used to: a collection of content producers such as newspapers, magazines, and music services started experimenting with charging money for their content.

This did not sit well with "information wants to be free" believers. So in June 2003, a group of friends in Sweden became official internet activists when they launched Piratbyrån—"the Bureau of Piracy." Originally, Piratbyrån was little more than a protest in response to Sweden's establishment of Antipiratbyrån, the official copyright enforcement agency.[39] The protest organization's mission was not to organize internet piracy per se, but rather to encourage the spread of information regardless of intellectual property rights.

Its mantra was strikingly similar to Brand's "information wants to be free" movement: "Use what you know for good. Spread it further. Sow what you want. Add, delete, change and improve."[40] Self-described as a "loosely organized think-tank, a website, a philosophical greenhouse or FAQ guide to digitization," Piratbyrån pioneered what would become one of the most popular activities online in the early 2000s: free file-sharing.[41]

Twenty-five years into the internet era, it might be difficult to fathom how much work went into sharing intellectual property such as music, movies, TV shows, and games before the web. You had to physically go to a store, buy the original whatever, then take however many hours needed to make a physical copy onto a video or tape cassette or, eventually, CD or DVD in order to make a single copy to share with one friend. Most Americans who were online at the time will likely remember the 1999 rollout of Napster as world-changing: here was a music-sharing service that allowed, for the first time ever, massive and free file-sharing online with strangers from all over the world, thus eliminating the need to physically store copies of your favorite movies or playlist on anything but your computer.[42] Shortly thereafter, The Pirate Bay, or TPB, was born—the largest BitTorrent site in history (BitTorrent being the name of the protocol that permits the transfer of massive files such as movies and albums). With 300 million users and counting, and despite having been taken down multiple times over the years for copyright infringement, TPB is still operational to this day.[43]

As might be expected for an organization that blatantly encourages what is technically intellectual property theft, TPB has had its ups and downs with the law. More than 60 police officers raided TPB's Stockholm main data center in 2006, promoting hundreds of protesters to take to the streets in Stockholm and Göteborg.[44] The raids successfully took the site offline—but only for three days. Its devoted followers moved it to another reserved domain name to get it up and running again (using one of the about 70 domain names that TPB reported they have reserved for such contingencies).[45] Then came the real crackdown: in 2009, the police came after the TPB for copyright infringement.[46]

Undeterred, the movement created first by Piratbyrån and advanced by The Pirate Bay—the organization of the "information wants to be free" principles set forth by Stewart Brand decades earlier—went on to do something almost unprecedented in modern social movements: it made the leap from merely conducting online activism to inspiring a set of international political parties that have won hundreds of elections worldwide.

As might be expected with a group of antiauthority activists, these parties don't all coordinate with each other. And some actively disavow the others. But they all operate under the same umbrella: the Pirate Parties International.

Since the first Pirate Party officially formed in Sweden in 2009 on a platform of the "protection of human rights and fundamental freedoms in the digital age,"[47] the movement has spread to 68 other countries. And they've actually managed to insert themselves into the traditional political establishment. To date, the Pirate Party has racked up 547 separate electoral victories across the globe at the local, state, national, and even international organizational levels. At the time of this writing, it even has four seats in the European Parliament, the elected legislative body of the European Union.[48]

Thus far, the Pirate Party's biggest victory has been in Iceland. In 2016, the so-called Panama Papers revealed that the family of the prime minister of Iceland had apparently been hiding millions of dollars in offshore accounts, triggering public accusations that they were dodging Iceland's substantial personal income tax rate of up to 46 percent of one's earnings. The prime minister resigned as a result of the public uproar.[49] Shortly thereafter, running on an antiestablishment platform, Iceland's Pirate Party won 15 percent of the vote, which was sufficient for an invitation to form a government. (For context, seven parties ran in that election, making a 15 percent win a substantial victory.)[50]

"Information wants to be free" had clearly transcended the hacker ethos to become an organizing principle for citizens around the globe who wanted to make the leap from protesting government to becoming a part of it.

The hackers were now in charge.

AT THE SECOND ANNUAL WASHINGTON IDEAS FORUM ON OCTOBER 1, 2010, Eric Schmidt, Google's CEO for sixteen years, reminded an audience of journalists, policy-makers, and politicians what was going on. "With your permission," he said, "you give us more information about you, about your friends. And we can improve the quality of our searches."[51]

"We don't need you to type at all," Schmidt continued. "Because—with your permission—we know where you are. With your permission—we know where you've been. And—with your permission—we can more or less know what you're thinking about."

Nervous laughter broke out across the room, prompting Schmidt to quickly interject, "Now, was that over the line? Is that *right* over the line?"

The "line" Schmidt referenced alluded back to something he had said earlier in his remarks. "There's what I call 'the creepy line,'" he had said. "And the Google policy about a lot of these things," referring to far-future technologies, like brain implants, "is to get right *up* to the creepy line, but not cross it."

Back in October 2010, Google had not quite crossed the creepy line. None of the tech giants had: in the fall of 2010, the global love affair with Facebook—still primarily a friends-connection network—was in full effect. In the fall of 2010, half a billion people logged on to Facebook to play FarmVille and Mafia Wars and to make use of the "Like" feature (introduced only the previous year) on each other's posts, photos, and comments.[52] In the fall of 2010, Facebook still felt innocent and hopeful, epitomized by the December 2010 launch of the Arab Spring. That movement's early protests were largely organized via Facebook, which *Atlantic* author Rebecca Rosen referred to as "the GPS for this revolution."[53]

In 2010, technology was still exhilarating. We were enamored of our smartphones, Apple's iPhone being less than four years old and still in the category of craved-for tech, owned by just 33 percent of Americans.[54] The notion that personal technology use might be bad for us—as suggested by early research into "internet addiction" and "Facebook depression"—was, back then, still a novel and academic debate, not yet something taken seriously in popular culture.[55] Social media, especially once accessible through personal devices like smartphones, was going to be a democratizing force, we thought, with the promise that it comprised "long-term tools that can strengthen civil society and the public sphere."[56]

In 2010, our technology—our iPhones and Facebook and Google—was still going to empower us. Amazon, which had morphed from bookseller to everything-seller, was just going to make it easier for us to buy anything we wanted. Microsoft was just going to power our office work. And then there was Tesla, bursting onto the scene with moonshot projects to get us into space and traveling on Earth at hypersonic speed—the Jetsons of the pack, futuristic and sexy and exciting.

In 2010, technology had yet to become creepy. It was glorious. We were blissfully unaware of the complications it would bring.

By 2019, we've become well aware. We're aware that we are more than net states' user bases; we've become their populations as well. Our real lives are becoming more integrated with our digital ones. "With our permission," we've allowed our lives to become reliant on net states in certain areas, trusting them to manage our data rights, defend us from cyberattacks, and sign on to diplomatic treaties for our protection. Our relationship with net states comes with unenumerated benefits and unexpected responsibilities.

As net states "know" us more—as Schmidt said, knowing where we've been, where we are, and what we're thinking about—we are increasingly dependent on them in ways we couldn't have anticipated. Thus, our roles as citizen and user are merging.

To understand how citizen-users engage with their *net* states, it's helpful to first look at how citizens engage with their *nation*-states. In our social media–fueled age, we commonly hear how citizens make up "the public sphere." This phrase in its current usage can be traced back to German sociologist Jürgen Habermas,[57] who coined it in his dissertation, which was published in German in 1962 and translated into English in 1989.[58] In this work, Habermas described the history of how citizens came to emerge as a real check against government.

The story goes that by the mid-1800s, middle-class educated citizens in Europe ("the public," as opposed to the aristocracy) began to engage in discussions about not just their daily lives, but also subjects relating to the broader public good: what Habermas called "rational-critical debate." En masse, these conversations would emerge as what we generally refer to as "public opinion": the broad set of ideas and sentiments that a nation holds about political issues. Public opinion would, in theory, serve as a check on government. With legislators informed and influenced by what the public thought, they would legislate in such a way as to reflect what the public wanted. And so goes the theory of democratic societies in general: the public expresses its opinion and then—critical step here—*votes* for people who will make laws in accordance with those opinions, resulting in a happy, healthy society.

There are some problems with Habermas's version of the public sphere, not least of which is the issue of inclusion. His 1960s manuscript about activities in the 1830s considered "the public" to be, frankly, wealthy white men. But the issue I want to direct your attention to is not *who* is part of the public sphere, but *what we do* as members of the public sphere.

Electorally speaking, Americans are notoriously bad at taking action. Only 61 percent of Americans voted during the 2016 presidential election,[59] a 20-year low for that type of election.[60] Worse, only 36 percent of Americans voted during the 2014 midterm election, an abysmal 72-year low.[61] Put another way, when asked whom we want to represent us nationally, 4 out of 10 opt out. When asked whom we want to represent us locally—these are the representatives who are ostensibly members of our communities, our cities, our states—7 out of 10 of us don't bother to weigh in.

These turnout rates have serious real-world implications, among the largest being that our elected leadership is decided by a handful of people, statistically speaking. For instance, Donald Trump's victory in the 2016 presidential election is credited to approximately 80,000 votes in three states: that's smaller than the population of an average three-square-mile neighborhood in Brooklyn, New York.[62] While the uniquely American quirks of the electoral college influence these outcomes as well, if every eligible citizen voted, the political landscape both locally and nationally would likely look very different.

It may seem an odd comparison to make, but contrast these voter turnout rates to cell phone purchase rates. As of 2019, 96 percent of Americans over the age of 18 own cell phones, and 81 percent of those are smartphones.[63] We go through the bother of upgrading our smartphones every 21 months, or about once every two years.[64] Compare this to the act of voting in presidential elections, which takes place once every four years.

The difference between going through the trouble of getting a better phone and going through the trouble of getting a better elected representative is pretty basic: one is tangible, the other abstract. Our phones fulfill many in-the-moment purposes: they're navigation devices, music players, cameras, internet access points, and, of course, actual telephones. On the other hand, voting doesn't feel connected to most of our lives—on any basis, let alone a daily one. As a culture, our country is increasingly less connected to other people in general: a third of Americans haven't even met the neighbors who live directly next door to them.[65] If we don't even share words with the people who live right next to us, how many of us, then, interact with our congressional representatives, who represent roughly 700,000 people in a district? In short, we are very much in touch with our technology. We are far less so with our democracy.

One obvious follow-up question about citizen engagement is whether merging elections *with* technology might help. There's no obvious answer, however, and understanding why requires considering what it takes to be a citizen and what it takes to be a tech user.

WE USE OUR PHONES FOR MANY REASONS, BUT THOSE CONSIDERATIONS generally boil down to trying to improve something about our lives: to check the weather, to find information we need, to reach out to someone we care about, to read articles or books or the news, or to discover something we didn't know before. We also use our phones for less lofty reasons, such as to avoid boredom (93 percent of 18- to 29-year-olds) or even specifically to avoid having to interact with people around us (47 percent).[66] Given how much we already use our phones, then, the big question is this: If we could vote in elections on our phones, would we?

It turns out that the question of why people do or don't vote is complex, and introducing technology into the mix only makes it more so. According to a RAND Corporation report in March 2018, plenty of countries around the world already have e-voting, but that doesn't necessarily translate into better voter turnout.[67] Research tells us that people who vote generally do so not because it's convenient, but out of a sense of civic duty.[68] Conversely, people who don't vote aren't generally deterred by having to physically go to a polling site. Rather, they don't vote because they don't feel as if their vote matters.

In other words, whether we vote comes down to power—specifically, whether we feel that we have the power to effect change. Some of us do feel powerful with respect to our votes, as though by voting we're acting as civically engaged participants in our democracy. Others among us feel the opposite—powerless—as if we as individuals are ultimately irrelevant to the outcome of the vote and so there's no point in even bothering.

Power matters, because if there's anything net states give users, it's a sense of power. Look no further than Apple's branding of its wildly successful series of tech gadgets: the iPhone that kicked off the smartphone revolution in 2007, the iPad, iTunes, and so on. There's a key indicator there: "I." Me. You. Unlike political representation, tech is not an abstract concept. No, it's tech for you and you alone. You are the center of your digital universe, and you've got the products that make it so.

This is a massive shift from how technology was experienced by users in the past. As recently as 2007, when the iPhone hit the market, 90 percent of American homes shared landlines. A caller phoned, and anyone in the household could answer. Nowadays, people sharing a household—the most intimate social unit we have—generally still have to divvy up their physical space and everything in it. But not our tech. We have to share the contents of our refrigerator—literally, our food supply—with other humans; but not our tech. Not anymore.

Next to our clothing and shoes, tech is the sole area of the home that is, in 2019, entirely personalized. And, like our clothing and shoes, we increasingly wear our tech on our person—in our pockets or, more often than not, simply in our hands: 50 percent of millennials report holding their phone in their hands not just when in use, but throughout the entire day.[69]

Which brings us back to voting and citizenship. For most of our recent history, we were *always* citizens, regardless of whether we thought about it and regardless of whether we exercised our rights. Conversely, we were only *occasionally* tech users. To use tech, we had to take some sort of action to engage with a digital device.

That all changed in the past decade. Increasingly, even when we're not taking action we are tech users. Google tracks our location through our cell phone even when it's in our pocket (and, in some cases, even when we've opted out of location-tracking or our phone is turned off).[70] The apps we've downloaded collect our data, even when we're not using the apps. "The more data [tech companies] get, the more useful it is," reported Abhay Edlabadkar, founder of Redmorph, a company that develops apps to block trackers on your phone. "Within the limits that your app has asked for, it can collect and scoop up as much data as it can."[71]

All of this happens in the background, without needing our action to initiate use or even requiring us to pay attention. Information about where we are, who we're with (by our physical proximity to other users), the information we seek (through our searches), and even our mood (by the nature of the content we post) is streamed, tagged, and, more often than not, bought and sold by the tech companies we've invited into our lives: 7 out of 10 smartphone apps share our personal data with third-party services (more on that in chapter 3).[72]

Here is the key: *we* did this. We bought the devices. We signed up. We logged on. We signed terms of service or user agreements with every bit of tech that we own.

Net states have their own version of our Constitution's Bill of Rights: the terms of service. We just don't generally bother to read them—and for good reason. According to a study done by Norway's Consumer Council, it would take, on average, 31 hours to read all the terms of service on an average person's smartphone[73]—more time than it would take to read the New Testament of the Bible.

What's more, terms of service and user agreements change, often and unseen. Even if we paid attention to such things—and we historically have not—we may not have known they were changing.[74] Until the internet is subject to some sort of regulation by the US Congress and international equivalents, this is unlikely to change—unless we, as citizen-users, make a change.

The good news is, we've already overcome one of the biggest hurdles to effecting change with respect to net states—a hurdle we've not yet overcome with our nation-states: engagement. Americans may not be engaged with our political process, but we are very much engaged with our tech.

We are the masters of our universe of tech. We don't have to rely on some proxy to represent our interests—*we* are the keepers of our relationship with our technology. It is specifically ours, after all: like clothes and shoes, our particular profiles are molded to our highly personalized habits and preferences.

What's more, engagement doesn't mean taking any special action. We don't have to vote to effect change with our net states. We—as members of the citizen-user public sphere—create public opinion through every post, every "like," every tweet, every search, every website we visit and shop from and access. Our *habits* are our votes. Individually, changing our habits has an enormous impact on our relationship with net states in daily life. A collective change of our habits can make or break the very existence of a tech company, for they are only as strong as their user base. Their population. Their citizen-users. Us.

Inventor Buckminster Fuller frequently contributed to Stewart Brand's *Whole Earth Catalog*, sharing his musings on everything from life on Earth to an interplanetary future. In one essay, he wrote, "Whether humanity will pass its final exams for . . . a future is dependent on you and me, not on somebody we elect or who elects themselves to represent us. We will have to make each decision both tiny and great with critical self-examination—'Is this truly for the many or just for me?'"[75]

In our net state citizenship, our every decision is simultaneously for ourselves *and* for the many. We simply need to remember that we are more than ourselves. We are part of a public sphere, influencing, in this case, the net states that govern our existence digitally and the ways that the digital world extends into real life.

We're no longer a people "putting up with" computers. We're wearing them and inhabiting them. This gives the keepers of our digital lives great power over us. It is only fitting that we, in turn, demand that this power be used judiciously. Content may still be king for users, but as *citizen-users*, we owe it to ourselves to pay attention to more than content. In some cases, net states like Microsoft are taking action to protect us. If we keep watch, we can also take note of when they fail to.

Like the citizen, the citizen-user has responsibilities as well as rights—to keep informed, to keep engaged, and to vote—in this case, with our actions and with which tech we use. As Adlai Stevenson once said, "As citizens of this democracy, you are the rulers and the ruled, the law-givers and the law-abiding, the beginning and the end." With so much of our lives played out in the digital sphere, we must remember that we are both in charge and overseen: the rulers and the ruled. In our hyper-individualized existences, we'd have no one to blame but ourselves if we didn't keep our rights as well as our responsibilities in mind.

TWO

NET STATES IRL

Citizen-users may engage with digital content, but they're still grounded in a physical landscape. This chapter explores what net states are doing "IRL"—in real life—situating the ethereal internet, "the cloud," in the physical world and tracing how our data is tethered to Earth through undersea cables and data centers. By tracking net state activity IRL, this chapter lays the foundation for a new way of looking at power: distributed not according to borders on a map, but through information flows, investments, and physical assets.

Never has Puerto Rico's ambiguous status in the American experience appeared in sharper relief than in the aftermath of Hurricane Maria. From 6:15 a.m. on Wednesday, September 20, 2017, when the category 4 hurricane made landfall in what would be the worst storm the island has ever seen—and the fifth-most-powerful storm to ever hit the United States—the island suddenly seemed to be on its own.[1] The storm ravaged Puerto Rico, battering its 3.4 million residents with winds of 155 miles per hour and more than 30 inches of rain in a single day. As a point of comparison, Hurricane Katrina's rainfall maxed out at nine inches after making landfall on the Gulf Coast; its major source of damage came from floodwaters.[2]

"It was as if a 50- to 60-mile-wide tornado raged across Puerto Rico like a buzz saw," reported meteorologist Jeff Weber from the National Center for Atmospheric Research. "It's almost as strong as a hurricane can get in a direct hit."[3]

Just three weeks earlier, in another part of America, Houston, Texas, had been hit by Hurricane Harvey, an equally powerful storm in its own way. Though less severe in intensity (it had been downgraded to a tropical storm by the time it hit Houston), its rains were relentless, dumping 40 to 60 inches on the 2.3 million residents of America's fourth-largest city over the course of 117 hours. Harvey flooded 40 percent of the city's buildings and residences and broke the all-time record for hurricane rainfall in the United States.[4] An estimated 82 people perished in the storm.[5]

The federal government's response to the flood damage in Texas was swift and massive. The Federal Emergency Management Agency (FEMA) coordinated and deployed over 31,000 personnel from multiple agencies and organizations to the city even before the storm made landfall.[6] President Trump personally toured Houston four days after the storm hit.

To Puerto Rico, with a population double that of Houston, FEMA sent fewer than 500 staffers.[7] The president didn't appear for almost two weeks. And yet the damage was far more severe than what had befallen Texas. Immediately after landfall, the entire island lost electricity. More than 95 percent of cell service went out. The chief executive of the government-owned Puerto Rico Electric Power Authority, Ricardo Ramos, told CNN, "The island's power infrastructure had essentially been destroyed."[8] Hurricane Maria's death toll from the storm and its aftermath is estimated to be 4,645 people—more than 50 times higher than the loss of life in Texas following Hurricane Harvey.[9]

Time went by. Things got worse. A week after the storm, almost half the population still lacked access to drinking water. Ten days later, that number increased to 55 percent.[10]

Reporters covered the disaster from every angle imaginable—mostly doom and gloom: the loss of life, the dramatic absence of the federal government, the potential looming food shortages, and the consequences of long-term lack of electricity on a population of over 3 million.[11] But one reporter took a different tack, identifying a rare opportunity. At 2:45 p.m. on October 4, 2017, Brian Kahn, a reporter with the environmental news website Earther, filed a story titled "Puerto Rico Has a Once in a Lifetime Opportunity to Rethink How It Gets Electricity."[12] A separate tweeter with about 9,000 followers then posted the story with the comment "Could @elonmusk go in and rebuild #PuertoRico's electricity system with independent solar & battery systems?"

Elon Musk read that tweet and responded. Given that he has 23.7 million followers, it was rather extraordinary that Musk personally replied to the post. He tweeted, "The Tesla team has done this for many smaller islands around the world, but there is no scalability limit, so it can be done for Puerto Rico too. Such a decision would be in the hands of the PR govt, PUC, any commercial stakeholders and, most importantly, the people of PR."

Approximately eight hours later, word of Musk's tweet had reached Puerto Rico's governor, Ricardo Rosselló. Rosselló tweeted back, "@elonmusk, let's talk."

Tech entrepreneur Elon Musk had long made headlines for his almost preternatural ability to plant a flag in future-leaning endeavors. Before most people were even using the internet, Musk cofounded PayPal, an online payment system, in 1998.[13] He designed a proposed "Hyperloop" for high-speed mass transit between San Francisco and Los Angeles, with plans resembling schematics from *Star Trek*.[14] He created SpaceX, which has successfully completed multiple restocking trips to the International Space Station, cornering the rocket launch market.[15] And he formed Tesla, whose electric cars can travel upwards of 400 miles on a single charge.[16] Musk is one of those rare people who can claim to "make life multiplanetary" within his lifetime and be taken seriously.[17]

In 2015, Tesla broke yet more new ground when it plowed into the energy business. It launched Powerwall, a company that makes a rechargeable lithium-ion battery-pack kit—Powerpack—which stores solar power for homeowners. The kit can be bought outright (or with loan financing) or leased.[18] "We have this handy fusion reactor in the sky called the sun," Musk noted at Powerwall's inaugural press conference.[19] But existing batteries, he noted, "suck." Powerwall, Musk promised, was going to make traditional batteries obsolete.

In the two years following the Powerwall launch, Musk lobbied hard to get consumers, businesses, and governments alike to adopt solar energy storage via his Powerwall system. He achieved only moderate initial success. In 2015—launch year—Queensland, Australia, which already had one of the world's highest rates of household solar panel systems (with more than 88,000 such systems), entered into a year-long trial with Powerwall to test how it could integrate Powerwall with the state's energy infrastructure.[20] Gradually, Powerwall installations began to make gains beyond Queensland and into other territories in Australia, albeit on the consumer rather than the governmental level. In short, despite receiving a positive reception conceptually, Powerwall had yet to really gain a foothold in a large-scale energy grid. Its first big foray into a national public grid was in 2016, when Powerwall provided energy to the entire island of Ta'u in American Samoa. But benefiting a population of fewer than 600 residents, this hardly gave Musk the large-scale proof of concept he craved.[21]

And for Musk, Powerwall wasn't just some side-job, do-gooder passion project to save the environment. His work on solar energy was part of a larger plan, one that Musk began when he bought the solar energy company SolarCity. During a joint SolarCity-Tesla product launch in 2017, Musk spoke to the crowd of approximately 200 people. "This," he said, gesturing to solar panels on the roofs of nearby houses, "is the integrated future. You've got an electric car, a Powerwall, and a Solar Roof." As if to shoo away any naysayers in the crowd, he concluded, "It's pretty straightforward, really."[22]

With Powerwall, Musk wasn't just launching another business. He was adding to his vision of a future in which energy and transportation would be fundamentally altered from the systems our governments have traditionally relied upon. As noted above, Musk is already transforming transportation: with Tesla (electric cars), the Hyperloop (high-speed urban transit), and SpaceX (space transit). He's transforming energy: with SolarCity (solar panels) and Powerwall (solar energy storage). Combined, these endeavors form puzzle pieces that create an interconnected infrastructure, one that controls how future humans physically will move from place to place and gain access to energy. In so doing, Musk is building a pseudo-public utility. It's not "public," in that government won't own it; Musk will. But these projects in many ways *act* like public utilities, in that they theoretically supply "the public" with basic needs: energy and transportation.

With this backdrop in mind, it makes sense that when Hurricane Maria came along, Musk saw an opportunity, the chance he'd been waiting for to bring Powerwall to scale. Weeks after FEMA's much-criticized response to Puerto Rico's island-wide power outage, Musk stepped up with a tantalizing offer: not only would he donate Powerpack systems to Puerto Rico free of charge, but they could be used as a first step in rebuilding the entire energy infrastructure of the island.

The US federal government—historically responsible, at the bare minimum, for providing basic infrastructure such as power, roads, and water to its people—barely showed up to aid 3 million of its citizens. A tech company—historically responsible for nothing but its bottom line—stepped in with an offer not only to donate equipment but to assume management of the island's energy, an essential piece of critical infrastructure.

Just five days after the Twitter exchange between Elon Musk and Governor Rosselló, Tesla shipped hundreds (no exact number could be confirmed by reporters) of Powerpacks to the island, each of which stores up to 210 kilowatt-hours (kWh).[23] For context, according to the US Energy Information Administration, the average American home uses about 900 kWh of energy per month.[24] In other words, one Powerpack could provide electricity to one home for about a week—assuming it was never recharged, which, in a sunny environment such as Puerto Rico, was unlikely to be the case.

And Musk's Powerpacks continued to deliver benefits to Puerto Ricans even beyond the immediate aftermath of Hurricane Maria. During an island-wide blackout seven months later, in April 2018, Musk's Powerpack batteries generated electricity at 662 sites across the island.[25] The Powerpacks may not yet have turned into the island-wide energy infrastructure transformation Musk had hoped for. But at the very least, they put themselves on governments' radar.

VERMONT IS PERHAPS THE MOST STRIKING EXAMPLE OF HOW "PUBLIC" infrastructure is being transformed by net states. With just over 600,000 people, Vermont is a small state, population-wise. But it's got a national reputation for punching above its weight when it comes to enacting progressive change. Back in 2000, Vermont was the first state to legalize same-sex marriage—a full eight years before a second state followed suit.[26] As of 2015, Burlington, Vermont, became the first city in the nation to run completely on renewable energy.[27] The state has one of the most technologically advanced energy grids in the country. In 2016, nearly all of its in-state electricity was generated by renewable energy, including hydroelectric, biomass, wind, and solar sources.[28] In fact, the state's campaign to get residents to install renewable energy systems has been so successful that the utility commission recently cut back on awarding financial compensation for doing so.[29] "Renewable energy is flourishing in Vermont," said the commission, "and has reached a level of maturity where it can continue to be deployed with lower incentives."

Perhaps one of the reasons Vermonters take their energy supply so seriously is that they'll quite literally freeze if they don't: with about 81 inches per year, Vermont's snowfall is among the highest of any state in the US.[30] While this may be great for ski slopes, it also means that most Vermonters have to repeatedly weather winter storms accompanied by extreme cold and winds that down trees and power lines—in fact, storm-related power outages occur an average of once every 13.5 days from October through March. According to the US Department of Energy, power outages affect an average of 69,732 Vermonters each year, roughly 10 percent of the state's population.[31] Some winters are harder than others, of course. In December 2013, for instance, storms dumped 30.7 inches of snow on Burlington, leading to eight power outages that month alone—a blackout once every 3.75 days.[32]

Given these climate conditions, it's probably no surprise that when the energy arm of Elon Musk's Tesla—Powerwall—proposed a pilot home-generator program in Vermont in 2015, utility companies jumped at the chance to participate.[33] By 2017, Vermont's Green Mountain Power offered residents Powerwall backup batteries for $15 a month for a 10-year contract.

This makes immediate sense for the utility companies: once Powerwall batteries hit 2,000 users, energy use goes down so much that it's like taking 7,500 homes off the electrical grid. It makes sense for homeowners as well. Unlike generators, Powerpacks are integrated into the home energy system.

The advantage for individual users whose homes have solar panels is also pretty clear. You pay $15 a month and get a state-of-the-art backup battery that—unlike a more conventional generator—doesn't require that you remember to gas it up or go out in the snow to get it going. Your solar panels juice up the battery without requiring any activity from you, the user.

And Vermonters aren't the only ones benefiting from this sort of deal: Tesla and its offshoot, SolarCity, have deployed solar arrays and Powerpack systems to the city of Manchester in the UK[34] and, as mentioned, to tens of thousands of homes in Australia,[35] as well as utility companies within several US states. In short, Tesla is edging into what used to be government turf: the electrical grid. In addition to Vermont, Powerpacks are now integrated into the electrical grids in parts of Hawaii,[36] Colorado,[37] New York,[38] California, and Connecticut[39]—Tesla is now responsible for providing the infrastructure to light and heat the homes of over 2 million American citizens. The question is whether those citizens are aware of this.

Again, as with the response in Puerto Rico, chances are that so long as the lights stay on and the house is a comfortable temperature, people won't give much thought to who's providing the energy to their homes. Yet, as with Puerto Ricans, it would benefit Vermonters and residents of other states to remain vigilant about the fact that their "public" infrastructure—their energy grid—may not be in fact be publicly owned and operated, but rather outsourced to a tech company.

Tesla wasn't the only tech company exploring public utilities. A short time after Tesla showed up in the wake of Hurricane Maria in Puerto Rico, Google came, too. Google (or rather, its parent company, Alphabet) has also been developing numerous infrastructure-related projects in recent years—notably, Project Loon. Launched in 2013, Project Loon emerged with a veritable army of what look like hot-air balloons intended to be deployed in the stratosphere around the globe. The goal: provide free high-speed internet access to rural and remote areas—or, as Project Loon's tagline says, "Balloon-powered internet for everyone."[40] Since Project Loon launched its first balloons, it has partnered with governments in Brazil, France, Sri Lanka, and Indonesia. Successful as the partnerships have been, though, Google continues to label them "tests."

Hurricane Maria offered Google a new opportunity: the chance to supply telecommunications and internet—not as a pilot program or an add-on service, but as part of the critical infrastructure to an entire region. Less than a month after the hurricane hit, Project Loon filed an application with the Federal Communications Commission to partner with the local telecommunications provider, AT&T, to deploy its balloons.[41] The application was cleared the same day. Cynically, one might see this as the federal government's eagerness to get rid of the responsibility for providing relief to Puerto Rico. But Federal Communications

Commission chair Ajit Pai insisted the decision was simply pragmatic. "More than two weeks after Hurricane Maria struck, millions of Puerto Ricans are still without access to much-needed communications services," he said. "That's why we need to take innovative approaches to help restore connectivity on the island."[42]

Just eight months after the hurricane, Project Loon and AT&T had indeed restored connectivity, with cell service and internet connections to the majority of the island.[43]

Again, the residents of Puerto Rico probably didn't care who fixed their electrical grid or got their cell phones working. Chances are, they were just glad they worked. With the infrastructure problems sorted, at least for now, it would be tempting to move on without considering the intentions or implications of the whos or hows or whys of it all.

But if people had the luxury of stopping to think about what transpired here, they might note that the infrastructure fixers were not governments, but rather tech companies. This matters. While net states acted in this instance the way we've come to expect governments to act, they differ from governments in two key ways. First, unlike governments, net states are under no legal obligation to maintain in perpetuity the infrastructure they implement. The Powerpacks and Project Loon in Puerto Rico may have been just another pilot project for Tesla and Google: one that could morph into a permanent program, or one that they draw to a close as soon as they've amassed the data they're looking to acquire. Second, while net states do engage in contracts for services, they are not mandated by law to assist any other states or territories that should find their energy and telecom infrastructures wiped out by storms in the future. In other words, net state activities in Puerto Rico don't set a precedent for net state responsibilities in the future.

The result of these two key differences is that net state–sponsored aid may not last, or even be offered in the first place, regardless of the needs of the people. Project Loon's CEO, Alastair Westgarth, wrote in a blog post that Google would keep the Loon balloons flying over Puerto Rico "as long as it is useful and we're able to do so."[44] It's hard to imagine a contract between citizens and a public utility company like Con Edison or AT&T either (1) providing free internet or cell phone services at all, or (2) pledging to continue to provide such services for a period of time as vague as "as long as we're able to do so."

This statement is a reminder that just because net states offer infrastructure services doesn't mean they'll *always* offer them. And because they're free, we, as consumers, don't get a say in how long net states will continue to provide these services or under what conditions. In Puerto Rico, Tesla's and Google's offerings have not been infrastructure as a public good, but infrastructure as charity: an endowment by net states to consumer, granted or removed as and when net states see fit.

PROJECT LOON'S 50-FOOT-WIDE BALLOONS AND TESLA'S SHED-SIZE Powerpacks serve as useful reminders that even invisible resources—in this case, internet access and energy—originate from something physical. This holds true with other invisible resources as well—most notably, our data.

Ninety-nine percent of our data is transmitted online through 428 undersea cables that collectively stretch more than 700,000 miles—enough to circle Earth 28 times. (There are also underground cables, but they only transmit about 4 percent of the total global data traffic, with the remaining 95 percent managed by the undersea cables.)[45] The fiber-optic cables are thin, roughly the diameter of a garden hose.[46] As technology reporter Jameson Zimmer aptly put it, "The Internet is commonly described as a cloud. In reality, it's a series of wet, fragile tubes."

It's not surprising, then, that the cables are vulnerable to physical damage from any number of things. In 2008, off the coast of Dubai, an errant ship anchor cut off internet service for 75 million people in Europe and the Middle East.[47] An earthquake off the coast of Taiwan in 2006 knocked out connectivity for most of Southeast Asia.[48] Amazingly, sharks are a problem as well—they apparently have a real taste for fiber optics. By 2014, so many sharks were biting through Google's underwater network that the company started wrapping its trans-Pacific cables in Kevlar.[49]

Internet cables are sort of the plumbing of technology. When they work, we rarely think about them, no more than we think about how water runs through our pipes when we turn on our kitchen faucet. When we're watching YouTube, if a video loads it's easy to ignore *how* data is transmitted through the many data centers, servers, and undersea cables that bring internet-enabled content to our device.

As it turns out, the plumbing matters quite a bit when we're talking about data. Entertainment may not be the noblest reason to care about it, but for many people, it's certainly one of the most relevant. Americans work pretty hard: full-time-employed adults put in an average of 47 hours a week, almost six full workdays.[50] Factor in an average of 14 hours a week actively parenting[51] and 12.67 hours of cooking, cleaning, laundry, and other household chores, and this adds up to more than 73 hours of work, paid and unpaid, per week. There's little time left for leisure, but by cutting out sleep—as a nation, we're down to around seven hours a night[52]—we manage to find two to four hours per day to escape reality: notably, by watching TV.[53]

From personal entertainment to professional communications, net states are the mediators of our attention, at work and at home. They produce the devices that we use to go online—our iPhones and our laptops and our internet-enabled televisions—as well as the platforms on which we spend our time online: watching videos on Google-owned YouTube, reading the news on social media sites like Facebook, or getting work done via productivity packages such as Microsoft Office.

Given that net states supply our online destinations and the devices to get us there, it's not surprising that they are increasingly interested in controlling our *means* of access as well: the physical cables that provide internet access and the data centers that store our data. The cables matter because they're how we stream data. And we use a lot of it: 190 gigabytes per month for the average American household.[54] Twenty hours a month on Facebook alone consumes approximately 2.5 gigabytes of data.[55] Net states want more control over ensuring that your data actually gets to you. Until recently, Google, Facebook, Amazon, and other net states leased bandwidth from telecommunications companies that owned undersea cables.[56] But starting in 2016, net states started buying the cables themselves, either in whole or in part, to ensure that they retain greater control over the cables' maintenance, usage, and upkeep: by the end of 2017, Amazon owned or co-owned 2 cables; Microsoft, 4; Facebook, 8; and Google, 14.[57]

That may not sound like much considering that, as noted earlier, there are more than 420 undersea cables all told. However, the significance lies not in the net total; it's in the pattern. Three years ago, net states owned zero cables. Now, in 2019, Google alone owns or co-owns 63,605 miles of cabling—about 8.5 percent of the total length of all undersea cables—enough to wrap around the earth's equator two and a half times and still have cable to spare.[58] To go from owning zero miles of undersea cabling to just under 64,000 miles in just three years is an astonishing growth rate, and one that begs the question: Why?

The answer is simple: the cloud. With net states more involved in being cloud service providers in some shape or form, controlling the *means* of data transmission moves from being a simple technical convenience to an absolute business imperative. Because the more that net states and their services push data to and from the cloud, the more we as users gobble up bandwidth. From 2016 to 2017, data flowing through the cable network increased 52 percent. Globally, we transmitted 689 terabytes of data every second. So the more cables that net states can own themselves, the more opportunities they acquire to expand from simply providing digital services to controlling how fast we can get the *content* moving in and out of those services.

The cloud, where this content is increasingly stored, is, like the internet, also physically bound to Earth. While the very name "the cloud" suggests that data is something intangible, a swarm of electrostatic bits drifting in the air somewhere, our cloud-based data is still located in a storage center in a physical location somewhere, in some country.

Google's cloud platform, for example, operates caches at 90 data centers in North America, South America, Europe, Asia, and Australia.[59]

The cloud's physical footprint serves as a reminder that while net states may be headquartered in one country, they operate internationally—in terms not only of who uses their products and services but also of who houses their infrastructure. We may not know whether our data is stored in Europe, which now has the most stringent user data protection laws in the world, or in Taiwan, whose laws are far less strict. But it's handy to keep in mind that when we send an email or store a document in the cloud, this data is, in fact, stored somewhere, physically, in the world. What's more, the laws of the country or countries in which the data is stored, sent, or received may factor into who ultimately has access to it. And as of this writing, net states are by far the most influential parties in making that determination.

All of these various activities combine to reveal a pattern: net states are migrating from the digital to the physical. As we'll soon see, they're easing their way into our everyday experiences through the energy grid that powers our cities, the communications satellites that facilitate net use, the food services we rely on, and the healthcare industry that manages our physical well-being. While governments may still serve as the primary holders of public infrastructure, they're facing increasing competition from net states that want to be part of the unremarkable yet powerful scaffolding that orders our lives—and collect valuable data about us while they're at it. (More on why that data matters in chapter 3.)

Consider just one particular sector: health. In November 2018, Google hired the CEO of a healthcare system, Dr. David Feinberg, to manage its move into the health business.[60] His mandate is to consolidate strategy and operations across various Google health-related initiatives, including its search engine, its cloud storage business, the Google Brain artificial intelligence program (just a slice of folks working on AI at Google), Nest (owned by Google), and Google Fit wearables. Of particular interest to citizens are the possibilities that emerge when combining the personal data collected across these groups.

For instance, with Google Fit, Google captures your physical health data, such as heart rate, blood pressure, activity frequency, and physical routes traversed.[61] With Nest's suite of products, they track your movements outside and inside your house, not only via Nest security cameras that you've installed but also via motion-detecting sensors in your smart appliances—anything connected to the internet—such as, potentially, your thermostat, fridge, coffee maker, television, and anything else you've connected to the web.[62]

Google's search engine already customizes your search results based on prior searches, the content of your emails (if you're on Gmail), and your web-based activities. With the health data it can now acquire through Fit and Nest, Google can customize your search results even further, adding in not only the questions you ask *about* your health—like "What are common flu symptoms?" and "Which are the best running routes near me?"—but your *actual* health. Because it now has data beyond what you report. It has the data that it observes: information about your bodily functions, like your heartbeats and breaths per minute; your activities, like how often you open the fridge; and your movements, like when you leave the house, where you go, and when you come home again. Compiled, Google's health programs act like a personal private investigator on your tail, tracking your actions both indoors and out. Google and net states like it are no longer limited to a person's online activities. With smart home products and wearables, they're asserting a presence into people's physical lives, a new market they're only beginning to explore.

DESPITE GOOGLE'S AND TESLA'S ADVANCES INTO PUBLIC INFRASTRUCTURE projects—which used to be the domain of only governments—they're clearly not governments, let alone countries or, to use the wonky term for such entities, "nation-states." Outside policy circles, "nation-state" is not a term you run into often. In some ways, it's just a fancy word for "country." But there's a bit more to it. Nation-states are defined by three key components: First, they have a physical territory with distinct borders. Second, they have a "people"—individuals and communities bound together by an overarching culture, which may include several ethnicities and subcultures. Third, they have a government—a body that provides communal utilities (things such as roads, bridges, and power systems), protection from harm (whether from other citizens or from foreign actors), public services such as access to education, and some sort of financial system.[63]

No matter what country—what nation-state—we live in, it's easy to take for granted that governments actually provide these things for us. For instance, I can't remember the last time I reflected in the moment on how lucky I was to be driving on a paved road rather than a dirt one. Even in developed countries, there remain millions of miles of unpaved roads. (Fun fact: the United States still has 1,417,903 miles of dirt roads that compose 34 percent of all American roadways.)[64] As for these kinds of background services—roads, bridges, military protection, a functional banking system—it's probably a good sign if you *don't* notice them. As *New York Times* columnist David Brooks once wrote, "Government is essential, but . . . it's the stem of the flower, not the bloom. The best government is boring, gradual and orderly."[65]

Net states operate similarly: in the background. Digital platforms are stems. Both in digital and in physical life, the bloom is people, living their lives. The bloom is the actions, interactions, thoughts, hopes, worries, and silliness of us all. The stem, to use Brooks's metaphor, *enables* life, but it's the people who flower.

Net states get this. They are always and forever trying to get people to use their services to create and recreate life in digital form: to post updates about their day, share stories they found interesting or funny or weird or annoying, or upload visceral images that capture the feel of a moment, a creature, or a tragedy. Net states, like nation-states, are background: they provide the foundation for our online communities, the space for a people to come together. They also provide the services that make our interactions—our online lives—possible. In these ways, what net states give us—services that enable some portion of the goings-on of life—is not so very different from what nation-states provide for us.

However, there is a point where the similarities between net states and nation-states end and the two sorts of entities become very different in important ways. While we may be *of* the digital world, we still reside *in* the physical one. In other words, so long as we have bodies, we need to park them somewhere. And, while net states permit us to connect with like-minded people—*our* people, a community that shares our culture and values and beliefs—across the globe, in places like Niger and Peru and Afghanistan and Brazil, each of us is subject to a very different physical existence precisely because of where our body is parked. No amount of sharing of beliefs or values or culture can change the differences in our lived, physical experience. For instance, net states can't change that a like-minded friend in Niger will experience a power outage an average of 22 times a month (if they even have electricity at home in the first place). Living in Brooklyn, I almost never worry about this happening.[66] Net states also can't do anything about the fact that a like-minded friend living in Afghanistan is 87 times more likely to be harmed in a terrorist attack than another friend who lives in Brazil.[67]

Or can they? What happens when net states enter terrain formerly the primary domain of nation-states?

IN 1943, ABRAHAM MASLOW FAMOUSLY POSITED THAT HUMANS UNIversally long to fulfill a hierarchy of needs. Using a pyramid to illustrate his point, Maslow showed how we humans must first meet our physical needs—food, water, warmth, and shelter—before we can move up the hierarchy to our emotional-social needs, such as interpersonal connection. Finally, at the top of the pyramid come self-fulfillment and creative needs: what he termed "self-actualization."

Skip ahead to the information age. Tech companies and collectives started out solely focused on the tip of the pyramid, enabling users to find, manipulate, store, share, and otherwise wield information that would help us reach self-fulfillment. Base-order needs such as food, water, warmth, and the rest were left to governments to address.

Until recently, that is. Net states have quietly been moving down the pyramid. Entities like Facebook and Twitter, for instance, have long dominated the middle, allowing people to connect and fulfill their need for belonging and even, perhaps, their need for emotional connection. Increasingly, they're moving farther down the pyramid toward our most basic needs: Amazon recently entered the food business with its purchase of Whole Foods. Elon Musk's Powerwall marks his flag-staking in the energy storage business. Net states of all kinds are funding projects that will directly harvest natural resources. And so, while tech dominated the top of our hierarchy of needs when we entered the information age, they're increasingly investing in and seeking to control aspects involving our base needs as well: things like food, shelter, and warmth—areas that historically have sat squarely in the wheelhouse of the nation-state.

Take infrastructure, such as Google's and Tesla's forays into hurricane-ravaged Puerto Rico, two of the most high-profile examples of tech's increasing investment in IRL domains. To confirm that these ventures weren't just anomalies, I decided that I needed to understand more about what net states are up to. That required looking beyond the stories here and there and relying on a more systematic assessment. As the saying goes, I followed the money.

In May 2018, I examined every single acquisition and investment made by the most influential net states—Apple, Google, Microsoft, Tesla, Facebook, and Amazon—over the past 20 years, from the earliest available data in 1998 to May 16, 2018.[68] Using publicly available transaction data from Crunchbase, the premier tech industry database, I came up with a total of 1,227 investments and acquisitions.[69] Next, I identified which of these investments and acquisitions seemed like distinct departures from online-focused activities—that is, that dealt specifically with goings-on in the physical world. This included investing in and buying companies working on the energy grid, on public transportation systems, on agriculture and food markets, on medicine and healthcare, on the telecommunications infrastructure, and so on. (See this book's appendix for a complete list of sectors and companies.)

SECTOR	# OF INVESTMENTS / ACQUISITIONS	% OF NONCORE TECH ACTIVITY (N=232)
Health & wellness	51	22%
Biotechnology	29	12%
Energy supply & grid	21	9%
Telecommunications infrastructure	21	9%
Manufacturing	16	7%
Agriculture & food markets	16	7%

TABLE 2.1. Net State Investments & Acquisitions in Physical Sectors, 1998–2018

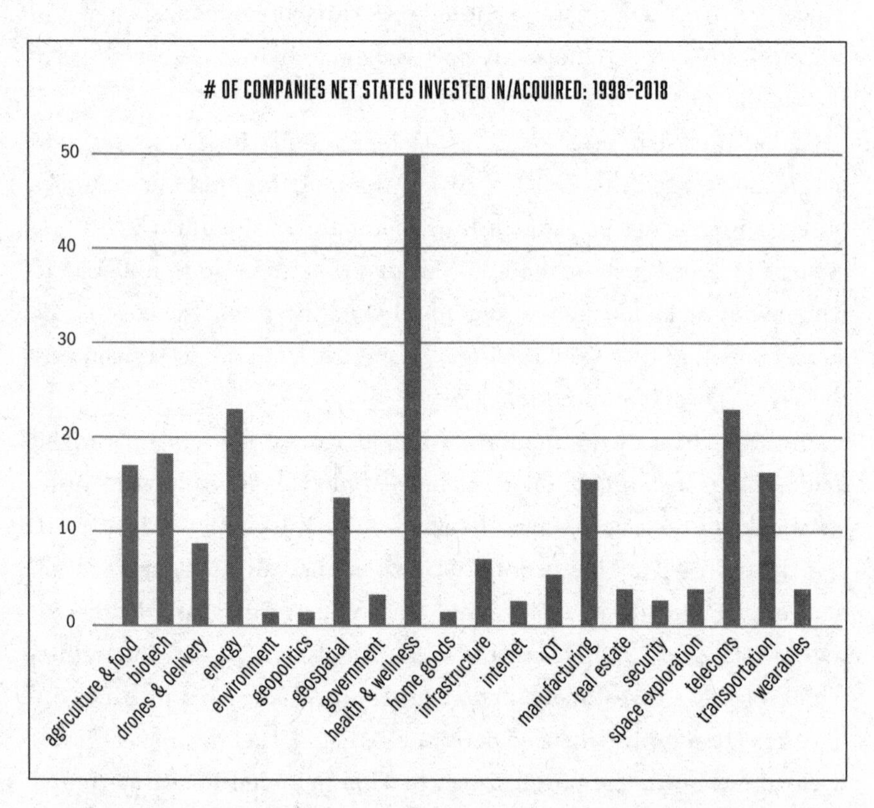

FIGURE 2.1. Net States Investments & Acquisitions, 1998–2018, by Category

Of the 1,227 companies the six net states invested in or acquired over the 20-year period, 232 met the criteria for physical-world activities. Put another way, of all the money Big Tech spent from the late 1990s to 2018, almost 20 percent went to products, services, and activities focused on the physical world.

The primary areas where net states put their money in physical sectors are shown in table 2.1, with those sectors broken into more specific categories in figure 2.1.

The six most influential tech companies in history—Microsoft, Apple, Google, Tesla, Amazon, and Facebook—spent nearly 20 percent of their money buying or investing in products, services, and activities that exist offline, firmly rooted in the physical world. Google invested in entities such as Impossible Foods, a company dedicated to developing organic "meats and cheeses" derived entirely from plants. Facebook bought Endaga, a company that creates telecommunications-in-a-box devices, while Google cofounder Larry Page bought Kitty Hawk, a flying car company. While this was Page's personal purchase, not a Google investment, the activities of the net state founders—Page and Brin for Google, Musk for Tesla/SolarCity, Bezos for Amazon, and so on—often illuminate the kind of future tech their companies then adopt. For instance, in 2008 Jeff Bezos purchased Rethink Robotics, described as "a leader in collaborative robotics, specializing in . . . robots that can work alongside humans."[70] Four years later, Bezos's company Amazon bought Kiva Systems (renamed Amazon Robotics), whose specialization is robotic fulfillment systems in Amazon warehouses, working alongside human operators. For this reason, I included in my analysis the investment and acquisitions activities of the net states themselves as well as those of their research arms (for example, Google X), subsidiaries (for example, Elon Musk's SolarCity), venture capital funds (for example, GV, an entity of Alphabet, Google's parent company), and founders (for example, Mark Zuckerberg).

These purchases reveal a pattern: net states are quickly expanding from the online to the world in real life, and not just conceptually:

they're doing so through substantial financial investments. In addition to shaping our digital experiences, net states are building their ability to shape our real-life experiences as well—and they're willing to bet nearly 20 percent of their acquisition and investment dollars that they can.

With 232 investments and acquisitions composing the data pool in question, there are simply too many variables to permit sweeping claims about why. It's not immediately clear, for instance, why Amazon bought the organic food chain Whole Foods in 2017, or why Facebook put money into Ocean's Halo, a seaweed-processing food outfit, or what Google was really going for when it bought the wind-turbine company Makani Power in 2013. There are likely excellent business reasons for all 232 physical world industries that meet some primary need, including infrastructure, health, and food. For example, Facebook most likely invested in telecommunications infrastructure to ensure that more people can go online to use Facebook. Amazon probably fed money into drone delivery to ensure that people continue to buy their wares from Amazon. There are surely sound bottom-line reasons for their investments.

Yet I suspect there's more to the story, for two reasons. First, think of the scale. It makes sense that tech companies would spend the majority of their acquisition and investment dollars—80 percent, as it turns out—on their core technologies: better geolocation services for Google Maps, for instance, or software that facilitates faster processing for Microsoft. It would also make sense that tech companies in particular—organizations that found success precisely because of a future-leaning innovative bent—would spend at least a little money on research and development of new technologies to nudge them along.

But close to 20 percent of their total investment and acquisition funds is more than a *little* money. It represents *billions* of dollars. The single transaction to purchase Whole Foods cost Amazon $13 billion, a heck of a lot of money to spend on something just because it's shiny and new, even if it probably benefits your bottom line.

The second reason has to do with the changing nature of digital technology. With the advent of the smartphone in 2007, we went from going online on a desktop or laptop to carrying the internet on our person. "Online" is no longer wholly distinct from "in real life"; it's integrated *into* real life. It makes sense, then, for the keepers of our digital lives to branch into our physical and personal spaces. If nothing else, doing so means they have more control over when and how their products and services will be put to use.

Thus, while net states are still largely digital beasts, through expansion into the daily activities of our physical environments, they are becoming, slowly but surely, more closely tethered to the physical world.

Before they can expand much further into our physical environments, however, they need to shore up the supply of material resources required to produce, operate, and maintain the core products and services they already manage within the digital realm. To do so, they've had to be creative. They're also looking to outer space.

In 2012, Google's former CEO Eric Schmidt and cofounder Larry Page became principal investors in an asteroid mining company called Planetary Resources. Asteroid mining may seem like far-flung future tech—and, at present, it may still be—but it's also one of the fastest-growing space-related investment areas out there. Peter Diamandis, cofounder of the X Prize, likened asteroid mining to the next "gold rush," with space rocks supplying a potentially limitless supply of rare earth metals essential to modern consumer electronics, including everyday items like your laptop and cell phone.[71] The 17 elements classified as "rare earth metals"—from the lithium in your iPhone battery, to scandium, essential for televisions and digital screens, and neodymium, an integral part of your headphones—are currently harvested from Earth's crust.[72] However, like most precious natural resources, they're expensive to mine and process, and at some point risk scarcity in supply.[73]

Elon Musk is also in on the space game, but for other reasons: interspace travel. In another testament to his prescience, Musk founded the spacecraft manufacturing and ferrying service SpaceX in 2002, more than a decade before other private-sector entities began investing in space. Despite multiple failed launch attempts in its early years, SpaceX has finally hit its stride: in 2017, it successfully launched its first fully recycled mission to the International Space Station, using previously flown rockets and spacecraft. Since then, it has advanced so rapidly that analysts estimate SpaceX now controls 50 percent of the rocket launch market globally, a share so substantial that Russia's chief space-flight official, Deputy Prime Minister Dmitry Rogozin, conceded in a recent interview with Russian television that it "isn't worth the effort [for Russia] to try to elbow Musk . . . aside."[74]

Space exploration has historically been so expensive, so specialized, that no commercial entities had either the capital or the technical expertise to try it. Historically, only countries could harness the resources to pull it off. This looked like it might change in 1982, when Space Services' *Conestoga 1* made history as the first privately funded spacecraft to perform a test launch that reached space.[75] Despite this effort, for the next 30 years spacefaring remained primarily the domain of nation-states.

Even for entities as rich in resources as nations, outer space is a reach. Since the early days of the space race, only nine countries—the United States, Russia, China, France, the United Kingdom, India, Israel, Iran, and North Korea—have developed the capacity to launch anything into orbit. About 60 countries maintain satellites in space,[76] but they have to hitch them to a spacecraft from one of the nine spacefaring nations to get them there. In short, until relatively recently, the only robust investments into rocket launchers, spacecraft, and other orbitals were restricted to deep-pocketed superpower nation-states: select countries in Europe, Russia, the United States, and China.

In recent years, SpaceX, with its innovative rocketry, has dominated the commercial market. No other private spacefaring company has successfully recycled rockets, a feat that revolutionizes the possibility of reducing the cost of space travel. And yet, being the sole entity currently capable of executing this feat, SpaceX recently raised the ticket price to NASA for restocking the International Space Station—a sobering example of what might happen if governments continue outsourcing space exploration to the private sector.[77]

And Musk likely won't be alone in dominating space for long. In 2018, Amazon founder and CEO Jeff Bezos announced that he's investing $1 billion a year of his own Amazon stock in his company Blue Origin, with claims that they would be transporting tourists to the edge of space by 2019.[78]

Asteroid mining and the new race into space are reminders that the material resources that create our tech world are not a given. The electronics that we rely on and take for granted—our cell phones, laptops, tablets, watches, televisions, and the like—are not magically produced out of thin air. We've become used to their being built at the rates they are today, and to their getting faster and generally cheaper with each passing year,[79] but these historic norms are not assured in the future, nor is the very existence of the devices themselves—not by any stretch of the imagination. These gadgets require physical resources, on Earth or from space, that are far from easy to harvest.

Similarly, the investment by net state leadership from Microsoft, Google, Amazon, and SpaceX suggests that outer space is not merely an interesting new frontier to explore. Rather, it is likely essential for the continued dominance of net states. That investment may indicate that conditions on Earth aren't quite as hospitable for expanded business opportunities as we have thought. It also may explain why net states are expanding their businesses from the purely digital into the physical realm—into areas where they might not be quite so reliant on precious minerals from a single country, and an authoritarian one at that.

Because, as of today, between 90 and 95 percent of processed rare earth metals come from land owned by China.[80] The US, for instance, gets the vast majority of these materials from the Chinese, a fact that likely makes the Department of Defense, which itself consumes 800 tons of rare earth metals each year, rather nervous.[81] While the US has its own rare earth minerals reserves, they're barely mined at all: as of

May 2019, there was only one rare earth minerals mine in operation, in Mountain Pass, California.[82] China's overwhelming monopoly on terrestrial-sourced rare earth metals is so valuable that they've created a shipping port in landlocked Kazakhstan—a shipping port so far from the sea that it's practically sitting on the Eurasian "Pole of Inaccessibility," the official term for the most remote regions on Earth.[83] It's part of China's planned global "New Silk Road" trading route, hailed by Chinese president Xi Jinping as "the Project of the Century." Also known as the "Belt and Road Initiative" (formerly "One Belt, One Road"), BRI involves ports, train routes, pipelines, and other infrastructure throughout 70 countries.[84] Through these routes they're selling all manner of goods, key among them being their processed rare earth metals.[85] China has already invested over a trillion dollars in its infrastructure.[86]

With China processing upwards of 90 percent of all rare earth metals and constructing the largest shipping route in history, the US tech industry's ability to produce digital devices would be absolutely decimated should China decide not to sell these resources to Western net states. With so much of the necessary raw material needed for tech in the control of a single country, the idea of net state entrepreneurs investing in asteroid mining suddenly doesn't seem so crazy after all.

Here's the final reason why asteroids and the new space race matter: net state investments and activities may feel academic and distant, but behind the scenes they're exploratory moonshots, literally—projects so grand that they would take us into space, onto asteroids, and who knows where else. We may think of net states as benign and cool—making search engines, social networking sites, and smartphones. But they're doing so much more than this. They are seeking to transform our experience on Earth and, in a handful of years, above it.

"THE WORLD IS MUCH BETTER. THE WORLD IS AWFUL. THE WORLD CAN be much better."[87] Max Roser, an economist at Oxford University, wrote this headline to a blog post explaining how the content of his online publication, Our World in Data, can be construed. "All three of these statements are true," he wrote. But he countered that while "it is important that we know what is wrong with the world[,] . . . given the scale of what we have achieved already and what is possible for the future, I think it's irresponsible to only report on how dreadful our situation is."

Roser's publication tracks long arcs—historical trends over centuries—about the most important questions in life: How do our lives today compare to those of people who lived four hundred years ago? One hundred years ago? Fifteen years ago? He uses statistical data to tackle dozens of variations on this question. Are we living longer? Are we fighting less? Are our children stronger? Are we suffering less?

And the answer to all of these—*all* of these—is that yes, over long periods of time, life has improved. We are living longer. We are fighting less. Our children are stronger. We are suffering less.

But we still have so much work to do. Our progress hasn't traced a straight line from bad to better. Nor has good fortune been evenly distributed across the globe.

It's sometimes tempting to see today's technology as somehow qualitatively different from previous epochs of tech; its transformation of our experience on Earth as a wholly new, and often bad, thing. Yet "modern" technology has been in the process of transforming our experience on Earth for over a hundred years, since long before net states came to be. With the widespread adoption of radio in the 1920s, for example, we could hear human voices from across the world, and with that new experience, expand our ability to humanize those with different lives than our own. With the mass distribution of television in the 1940s and '50s, we could see human faces across the world, and with that, grow in our capacity to empathize with hardships lived by others.

In some ways, net states have made our world better. In others, they have made our world awful. And in yet others, they have the capacity to make our world *much* better in the years to come. The first Powerpacks Musk took to Puerto Rico less than two weeks after Hurricane Maria had wiped out power to over 3 million residents were delivered to a San Juan children's hospital, Hospital del Niño,[88] a central medical facility that served over 3,000 children and housed more than 30 seriously ill children who required round-the-clock care. One could cynically argue that Tesla helped out just for good publicity, which they certainly got in spades. But regardless of their motivation, they made things happen at a time when the people's own government was failing them. "I've never seen a team arrive so fast and work so fast. They built this in a week," Rafael Pagán, the hospital's chairman of the board, said to Telemundo.[89] Compare Tesla's activities those weeks to the American government's; not many days before Tesla's donation, the American president threw paper towels to a crowd assembled at a press conference. Paper towels. Between net states and the nation-state, it's not hard to see who really made a difference to the people of Puerto Rico in this instance.

Tech moves fast; it evolves at dizzying speeds. But it's still trackable and traceable over time. In another few decades, we will have the luxury of looking back at net states' longer arc of progress or decay, their contributions to or degradation of humanity. But even with the data we have thus far, I would argue that the trend points toward progress. The question then becomes, What is *our* role, as citizen-users? With government, our role as citizens is clear: we vote; we support or oppose laws; we ensure that our peers receive a fair trial when called upon to do so.

But with net states, our role is not only less clear; in some ways, it doesn't yet exist. Thus far, we've yet to vote for their leadership. We've yet to assert our personal values on their data and product policies. We've not substantially swayed their investment decisions. Throughout this book I argue that we have a responsibility to change all that; that unless we demand a more active, contributory role in net states' use of our data and their investments in what has been, up to now, *public* infrastructure, we will miss out on the opportunity to shape how they influence our lives: socially, politically, and even physically.

Net states' expansion into real life is not on the horizon; it's underway. This could make life better. It could make life awful. Both of those statements may end up being true to some degree, depending on how you look at things. But what is absolutely true, regardless of the vantage point, is that tech in our lives *could* be much better if it included our active participation, both in guiding how it integrates into our real lives and in having a say about what happens with the data this integration generates.

As Roser concludes in his post, "We know that it is possible to make the world a better place because we already did it. It is because the world is terrible still that it's so important to write about how the world became a better place." For us, it's possible to see how tech has made our real lives better. And it's possible to imagine how it could make our real lives unbearable. Our job is to pay attention: not just to the content our net states produce—the TV shows and the memes and the headlines—but to the net states themselves. We're their citizen-users and their product consumers and their watchdogs, all at once. Playing all three roles is a lot of work. But the consequences of ignoring any one of those roles would be worse. We need to evolve, ourselves, from passive consumers of net state resources to active participants—to grow out of our passive role as users and step into an active role as citizen-users.

THREE

PRIVACY ALLIES AND ADVERSARIES

Privacy is no longer a given. We personally, and regulators on our behalf, are engaged in a global battle over who gets to determine the degree of privacy we retain over our content and activities via tech. This chapter explores how our understanding of privacy has evolved and the possibility that its current iteration may be an "anomaly." It considers how net states' partnerships with data brokers create profiles that "know" us, how Europe is fighting back against these practices, and how these interventions can and can't help the citizen-user.

On August 17, 2018, for the cost of a $400 court filing fee and undisclosed legal expenses, Napoleon Patacsil, a San Diego–based security guard at CSE Security Services, filed in California's Northern District Court a class-action lawsuit on behalf of all US users of Android phones and iPhones against Google.[1] *Patacsil v. Google, Inc.* argued that Google illegally tracked users' locations, regardless of user privacy settings, and without adequate disclosure.[2]

Patacsil leaped to action after reading an Associated Press report just days earlier unmasking Google's location-tracking practices. The AP report, which was corroborated by researchers at Princeton University, ran with a detailed map of all the places postdoctoral researcher Gunes Acar had visited over three days in July while his phone's location-tracking capability was turned off. Jarring in detail, the map is accurate down to Acar's presence at specific buildings on a city block.[3]

Patacsil himself, despite mounting a court case to take down one of the most powerful web companies on Earth, remains an almost invisible entity online; he granted no interviews regarding his motivations to sue Google. The only details that can be gleaned about his intentions come from the official August 17 court filing itself. It notes that Patacsil owned an iPhone from 2016 through 2018 and had attempted to turn location-tracking services by Google off on this device. "Nevertheless," reads the filing, "Google continued to track his location information." Despite Google's public assurances that turning off location-tracking means that "the places you go are no longer stored"—that "when you turn off Location History for your Google Account, it's off for all devices associated with that Google Account"—the filing characterized Google's representations as "false." Citing the AP report, the filing reads, "Turning off 'Location History' only stopped Google from creating a location timeline that the *user* could view. Google, however, continues to track the phone owners and keep a record of their locations."

Immediately after the Associated Press exposé, Google changed the language in its terms of service to read, "Some location data may be saved as part of your activity on other services, like Search and Maps." Privacy advocates weren't impressed with Google's update. The damage to users—unknowingly having their location data tracked and stored for, potentially, years—had been done. As Princeton computer scientist and former chief technologist for the enforcement bureau of the Federal Communications Commission, Jonathan Mayer, was quoted as saying in Patacsil's court filing: "If you're going to allow users to turn off something called 'Location History,' then all the places where you maintain location history should be turned off. That seems like a pretty straightforward position to have."[4]

Perhaps most significantly, Patacsil is one of only 39 million people—

the residents of California—out of the 325 million people who live in the United States who is even capable of filing a lawsuit against Google, or any other tech company, and having any hope of winning. That's because his home state of California had adopted, a month before Patacsil's filing, the California Consumer Privacy Act of 2018, or AB 375,[5] the most stringent data protection law in the country. Though it wouldn't take effect until January 2020, the spirit of the law and the increasing demand for the regulation of net states emboldened Patacsil to file his class-action suit. Outside the state of California, however, there are currently no other laws on the state or federal level in the US substantially regulating what net states like Google or Facebook can and cannot do with a person's digital information. Nor are there any laws preventing third-party data brokers from collecting, cataloging, and selling our personal, medical, and financial information. Americans—outside the state of California—simply don't have significant legal protections with respect to the privacy of their digital data.

UP TO THE 1850S, MOST AMERICAN HOMES DIDN'T HAVE INTERNAL walls. Families lived, slept, washed, ate meals, and conducted every other activity in a single room;[6] even the livestock were housed indoors.[7] There were advantages to this arrangement: it meant only one room to heat, so everyone could keep warm from a single heat source. With animals to tend and fields to plow, most of the family would spend the day outdoors anyway. Families generally huddled together just to eat and sleep—again, a warmer arrangement than being on one's own—so mornings and evenings were the only times when lack of privacy might have even been an issue.[8]

Physical privacy emerged as years went on for several reasons, health being key among them. With contagious diseases among the prime factors killing off 30 percent of all children in the year 1900, public health experts at the time urged parents to separate their children into their own beds if possible.[9] Beds, being expensive, then also became status symbols: only the wealthy could afford to have separate beds for various individuals in their homes. However, privacy in the home is also a culturally specific phenomenon. In a study of 186 cultures, anthropologist John Whiting determined that 67 percent of all children were sleeping in beds with at least one other individual, child or adult, as recently as the 1950s.[10]

Physical privacy matters to this discussion because it's very difficult to have *informational* privacy without it. Imagine your spouse and siblings and children all sitting on the sofa with you. It'd be difficult to have a private phone conversation in that sort of setting.

While informational privacy is not a uniquely American phenomenon, it has its roots in the very founding of the country. The practice of forcing Americans to house and feed British troops, codified in the Quartering Act of 1774, emerged as part of a series of measures instituted by the British following the Boston Tea Party as a means to get the colonists back in line. Better known by locals in America as "the Coercive Acts" or "the Intolerable Acts," the colonists deeply resented these measures.[11] The very personal nature of housing troops from what was seen as an occupying force especially irked them, leading them to explicitly abolish the practice in the Third Amendment to the US Constitution—lesser in importance only than what's stipulated in the First and Second Amendments: the right to bear arms and freedom of speech. The Fourth Amendment also speaks to early America's concern with personal privacy: ensuring protection of one's property from illegal search and seizure.

Despite the interest in personal privacy in early America, until very recently the realities of life meant that privacy was still hard to come by, as noted above. Just as physical privacy was available only to the very wealthy in past centuries—those who could afford homes with walls separating all their rooms—informational privacy was available only to those who could afford to hire message carriers to deliver notes with relative certainty that their contents would remain at least somewhat secure.

It wasn't until well after the advent of modern technologies such as the telephone that Americans en masse could really experience privacy in the sense that we experience it today. Until the 1960s, most telephones were connected to what was called a "party line": a single number that you shared with your neighbors.[12] If you wanted to use the phone, you had to pick it up, determine if someone else was already using the line—maybe eavesdrop a little—then wait your turn. It wasn't until the popularization of the cell phone barely two decades ago that each individual could have a dedicated phone number that wasn't also shared with family members or coworkers. It wasn't a given, though: only 50 percent of Americans had cell phones as recently as 2002.[13]

One of the most startling observations about privacy in the information age comes from Vint Cerf. Cerf is widely considered one of "the fathers of the internet," having developed the file transfer protocol (the technique that lets you upload and download web pages, files, photos, and so on). He led DARPA (the Defense Advanced Research Projects Agency; the people who invented the early internet, called ARPANET) and founded ICANN (the Internet Corporation for Assigned Names and Numbers, the group that makes it possible for us to keep track of names like www.google.com instead of IP addresses like http://74.125.224.72/). He's now Google's chief internet evangelist—that's his actual job title—working on the company's artificial intelligence team. In short, Cerf's no slouch. When he speaks, people tend to pay attention.

At an event at the Federal Trade Commission on November 19, 2013, Cerf, sporting his trademark three-piece suit—a habit that makes him one of the very few sartorially sophisticated tech gurus in the world—made a prediction that rattled the industry. "Privacy," he announced, "may be an anomaly."[14] Technology is what made privacy in the modern sense of the word possible in the first place, he argued, with "the industrial revolution and the growth of urban concentrations that led to a sense of anonymity." And privacy today is under threat, mostly because of our own actions. "The technology that we use today has far outraced our social intuition. . . . [There's a] need to develop social conventions that are more respectful of people's privacy." But we're not anywhere close to that yet, he said. "This is something we're gonna have to live through."

Privacy matters, and not just as something we view as an inherent right. We behave differently when we know we're being watched than when we believe we're in private. It's called, fittingly, the "observer effect" or, for reasons explained below, the "Hawthorne effect."

Back in the 1920s, many Americans made their living in factories—34 percent of us, compared with just 5 percent today.[15] Taylorism, or the practice known as "scientific management"—designed to derive maximum efficiency from the workforce—had swept across industries. As part of this trend, from 1924 to 1932 a massive telephone equipment manufacturing facility in Illinois called Hawthorne Works underwent a series of studies, ostensibly to identify ways to increase labor productivity.[16] (The study was commissioned in order to "prove" that more lighting would lead to greater output, thus helping to sell more electricity.) Scientists noted that under certain conditions, worker productivity increased, while it decreased under other conditions—not solely to do with lighting, but also with other factors, including cleanliness of the work space, timing of breaks, and so on. For decades, these studies influenced practices on factory floors, serving as a benchmark for the ideal quantity and quality of lighting, hygiene, and breaks necessary to eke out the utmost from factory laborers.

About twenty years later, a social scientist named Henry A. Landsberger took a closer look at the Hawthorne studies. He determined that it wasn't the lighting that had made the workers more productive. Rather, it was most likely the fact that workers were being paid attention to, even if only as subjects of a study.

Since Landsberger's observation, the Hawthorne effect has been widely studied, debated, debunked, revived, and otherwise dissected—mostly in an attempt to determine once and for all whether the Hawthorne effect was actually exhibited in the Hawthorne studies, as opposed to whether the effect actually exists at all.[17]

What the observer effect means is pretty intuitive: when we know we're being watched, we act differently than when we believe we're alone. In many cases, this means trying harder, also called the "social desirability bias": in short, people want to look good to others. So if we know other people are looking, we work a little harder to make ourselves appear appealing.

The question becomes, then, What happens when we feel as if there might *always* be someone watching? If all our digital activity is monitored, and—with an increasing deployment of public sensors and surveillance cameras in smart cities—our physical activity is monitored, does this mean we *always* try harder to look good? To contextualize why this matters, it's worth noting that the social desirability bias wields great power over us. It's so powerful that even depictions of eyes, as on a poster with people staring at you, are enough to make you modify your behavior for the better (called, appropriately, the "watching eye effect").[18] In one pilot test conducted in a police precinct in Nottinghamshire, England, hanging posters of watching eyes coupled with police warnings was credited with a 40 percent drop in theft.[19]

Although our behavior may improve due to the observer effect, we don't tend to enjoy the feeling of being studiously observed. Ever been in a crowded room and look up to catch someone who's clearly been watching you? People tend to describe the sensation as uncanny—creepy, even. And you're not psychic, by the way, when you catch an observer—you can thank your visual cortex in your brain, for being able to notice someone's stare.[20] Here's how it works: In any given environment, there's too much stimulation to process all at once. Your eyes take in a vast amount of stimuli, but then your visual cortex helpfully divvies it up and sends it out to around 10 distinct areas of your brain. The region to which these information packets are sent depends on the nature of each stimulus: whether you perceive a threat, a familiar face, or something confusing, they all get processed differently. In one study, a subject who was cortically blind—meaning, unable to consciously see anything—was given a series of faces to look at. Although he couldn't actually *see* anything, the area in his brain that processes emotions around faces, the amygdala, still activated. This shows that our brains are working hard even when we don't know we're being observed, just in case it turns out that we *might* be observed.

Which all sounds really tiring. Think about it: Behave when you know when you're being watched. Straighten up when you know you're being watched. Be smarter, kinder, better, add-desirable-quality-here when you know you're being watched. We're so preoccupied with the possibility of being watched that our brain is alert for signs of observation even when we're not consciously aware of being watched. Add tech to the mix, and we create new opportunities to be watched. Whenever we post anything online—a photo on Instagram, a tweet, a Facebook status, an email—we're essentially throwing up a flag in the ether, saying, "Hey! Over here! Check me out!" And now we can add in the Internet of Things—sensors and surveillance cameras in everyday devices (such as refrigerators) that we don't traditionally think of as being online—as cities become "smarter" (more on that in chapter 6). Opportunities to be watched—both created by us and placed in our lived environment—multiply exponentially. If we actually made ourselves aware of all this watching, where would the cognitive drain leave us?

To better understand cognitive drain and why privacy is so important to us, we need to first understand what our brains are capable of. Scientists estimate our brain processing speed at approximately 120 bits per second.[21] As a frame of reference, to pay attention to one person talking eats up approximately 60 bits per second. To be in the presence of multiple people talking doesn't just *feel* overwhelming; it *is* cognitively, neurologically overwhelming. It exceeds the available bandwidth of our attention.

In this way, the trend toward net state investments in the physical world via the Internet of Things may be unwittingly helping us out. As it is, we don't pay much attention to net states. When we use our iPhone, we're likely not thinking of Apple, the company, or even of the phone as a device itself; we just *use* it. Similarly, we don't call to mind Google, the company, when we send an email; we open up Gmail and just write the message. As net state products and services become more and more integrated into our lives, they become less noticeable. They become, as in Brooks's stem and flower metaphor, background—the infrastructure for our interactions: effective, familiar, and blissfully unremarkable.

This is important because demands on our attention have never been higher, and the trends suggest that these demands will only increase. Yet our attention has yet to evolve to the point where we can effectively process more stimuli, making us feel perpetually overwhelmed or exhausted or stressed out or some combination thereof.

Consider how we pay attention.[22] We do so in different ways depending on the circumstances: We pay attention to things around us—that's spatial attention. And we pay attention to things as they happen, which is temporal attention. Then within that temporal category are lots of subgroups. There's selective attention—as when you have to focus in on a friend talking to you amid a crowded party; you have to tune out other conversations and pick out just what your friend is saying. There's sustained attention—as when you need total silence in order to focus on writing for long periods of time. Then there's alternating or divided attention—as when you cook; you have to go back and forth between checking a recipe and doing the chopping and measuring and what have you, alternating your attention between reading and doing.

The demands that today's technology places on our attention are largely temporal. If you look at your smartphone for a moment, you'll note that it takes up very little of the visual landscape. It's just not terribly taxing of spatial attention. Contrast this with, say, navigating a traffic jam on a busy city street. Strewn with billboards, storefronts, pedestrians, streetlights, traffic lights, cars, trucks, motorcycles, cyclists—that's a packed tableau that'll suck spatial attention dry in a heartbeat.

Back to the phone: despite its relative smallness, it has an outsize ability to absorb us. Media stimuli—social media posts, games, news feeds, emails, texts—all soak up our temporal attention. And we can process only so much. In fact, if we get too many new bits of information in rapid succession, our brains helpfully tune out some of them, in what's called the attentional blink.

The attentional blink is brief—about half a second—meaning that it's not going to help us tune out distracting notifications that pop up on our screen (usually twice, for upwards of 10 seconds). But knowing that our brains engage in attentional blink is an important reminder that we have limits. Our brains, which are astronomically more powerful than the world's fastest supercomputer, hiccup when they're overloaded.

This warrants a quick breakdown of the brain. That organ is more than a machine, of course, but if we were to simply regard the machine-like aspect of it—its processing power—we'd have to admit it's an absurdly powerful machine. In 2013, researchers set out to use the world's fastest computer—then Japan's Fujitsu-built K computer—to simulate human brain activity. They calculated that one second of brain activity involves 1.73 billion nerve cells firing on 10.4 trillion synapses.[23] Simulating one single second of human brain activity took this computer 40 minutes—a computer made up of 82,944 processors. (By comparison, the MacBook I'm writing on has one processor.[24] One.) The K computer used the equivalent of 250,000 PCs' worth of memory. To provide a point of comparison on memory, 250,000 PCs' worth is still small potatoes; scientists believe that the human brain holds about a petabyte of data, roughly the same amount of data floating around in the billion websites of the entirety of the internet.[25] This data is far more than details such as your kindergarten teacher's hair color or the formula for finding the area of a triangle. It encompasses such things as instructions that keep your nervous system functioning, your heart beating.

So that's your brain. Impressive, no? But even your brain—a machine so mighty that a supercomputer roughly 83,000 times more powerful than your laptop can simulate only a tiny fraction of its power—has a measurable limit to its ability to pay attention. And that limit is half a second. If you were driving 60 miles per hour, you'd cover a distance of 44 feet in half a second. That's about three midsize car lengths.[26]

Or, if you experienced an attentional blink, you might miss those three car lengths. This precious resource, our attention, is somewhat fragile. When we've got our attentional wits about us, we can catch sudden changes on the road—a quickly braking car ahead of us, a deer, or a child chasing a ball. But despite how awesomely powerful it is, the brain rations attention, parceling it out to given stimuli depending on how important each one seems. And digital media—colorful, flashing, moving images especially—sure do seem important to the brain.

Net states recognize that we have only so much attention to give, so they are starting to take steps to help us help ourselves preserve it. Not for any altruistic reasons, though; if we burn out on digital media, they lose us as users. To combat attention fatigue, Apple recently rolled out Screen Time, an app that tracks how much time you spend per app and provides a weekly report that tallies it up for you.[27] Google plans a similar feature for its Android phones—this one will go so far as to gray out your screen when you've exceeded the limit you set for yourself.[28]

The digital realm needs to stay "healthy" in order to keep users there. It's in net states' best interests to invest in the health of that ecosystem—for example, by combating terrorists on their platforms (discussed in chapter 4). And an ecosystem is only as healthy as the creatures who inhabit it. We users of the internet need to remain attentionally healthy, so to speak, in order to be fit enough to stay online, let alone thrive online. And only if we thrive online do we generate data—the accumulation of which is the ultimate goal for net states.

"Privacy as we normally think of it doesn't matter," said Aza Raskin, cofounder of the Center for Humane Technology—a technology watchdog group founded by Silicon Valley engineers—in a 2019 interview.[29] It's data the big tech companies are really after, he said. "Imagine it's a stick figure at first, and as you use the system, it's collecting fingernail scraps and bits of hair. What do you care that you lost a fingernail scrap? But they're putting it together into a model of you . . . little models, little avatars, little voodoo dolls of you." The real problem with this lies not in your invasion of privacy, per se, but rather, in the fact that net states use your data to create models of you—not someone *like* you; *you.*

The questions become, then, What happens with these data models? And do we get any say over how they're used?

WHAT WE'VE LEARNED ABOUT THE COLLECTION OF OUR DATA THUS FAR has not made us happy.

Twenty years, a month, and a week after a then-youthful Bill Gates made his infamous appearance before Congress in the late '90s, on April 11, 2018, Facebook founder and CEO Mark Zuckerberg found himself in the very same hot seat. Facebook had been pilloried in the news for almost a solid year. First came the reports about the Russian misinformation warfare carried out on Facebook's platform during the 2016 presidential election. This was followed by the previously mentioned revelation that a political consulting firm, Cambridge Analytica, had gained access to 87 million users' private data without their explicit knowledge or consent.[30]

What was lost on the public is that this was in no way a data breach or hack or leak: Cambridge Analytica's access to user data is just one example of the kind of access scores of companies doing business with Facebook had, according to Facebook's 2015 business model (which they've since modified). Yet the public outrage either missed or ignored this fact. As technology writer Will Oremus pointed out, this may have been in part due to the fact that our collective rage at Facebook was not just because of what Facebook had done with our data; it was also because of our own gullibility for having given up so much of our information in the first place. "It was Facebook," he wrote, "that taught people around the world to freely give themselves online and to accept the use of their personal data in targeted advertisements as the price of admission to the modern internet."[31] And thus it was Facebook that received the brunt of the global ire. We had awakened to the fact that our data wasn't just being "liked" by friends and family members; our data served as the behind-the-scenes currency that businesses traded with

Facebook in order to let Facebook proudly claim in public, "It's free and always will be."

Combined, the outcry over the two scandals torpedoed the public love affair with Facebook in a matter of months. Congress, which had largely ignored Silicon Valley in the two decades since the Microsoft hearings, finally had no choice but to confront its leadership head-on. For the 2016 presidential election scandal, in November 2017 Congress settled with dragging just the lawyers from Google, Twitter, and Facebook before them in what ended up being a blustering non-event with no consequences.[32] But by the spring of 2018, the public backlash following the Cambridge Analytica story had grown so intense, Congress had no choice but to pull in Zuckerberg himself.

Mark Zuckerberg had apparently taken note of Gates's disastrous encounter with the Senate Judiciary Committee in 1998. In his own testimony, which spanned a total of 10 hours and addressed over 600 questions, Zuckerberg attempted to appear sober, respectful—chastened, even.[33] It didn't always work. Over the two-day grilling by both the House of Representatives and the Senate, Zuckerberg sat on the receiving end of much finger-pointing and huffy indignation. But ultimately, the hearings revealed more about Congress's lack of understanding about tech than anything Facebook had or hadn't done with user data. In one epitomizing moment, 84-year-old Senator Orrin Hatch—the same senator who two decades earlier had effectively taken Gates to task—demanded of Zuckerberg: "So, how do you sustain a business model in which users don't pay for your service?" With what has been characterized by numerous news outlets as a "smirk," Zuckerberg replied, "Senator, we run ads."[34]

While much news analysis has been written about the hearings themselves, as of this writing they yielded no tangible legislative change nor any punishment for Facebook or any of the other net states that might be mishandling user data: Amazon, Apple, Google, Microsoft, and Tesla artfully blended into the proverbial wallpaper during the Facebook hearings and their aftermath.

The disappearing act worked for only so long, though. Google's day in court came next—not with the Patacsil case, which is still in preliminary stages, but with Congress. On December 11, 2018, Google CEO Sundar Pichai stood before Congress to defend the company's data-handling practices. But again, Congress missed its chance to reveal anything other than its own misunderstanding of the tech industry. In this hearing, it was Rep. Steve King (R-IA) who put his foot in his mouth, demanding to know why his granddaughter's iPhone displayed certain information. Pichai politely noted, "Congressman, the iPhone is made by a different company." King attempted to backpedal, but in that moment, the optics of who had power irreversibly shifted. What could have been a firm congressional investigation into the tech industry came off instead as a game of political point-scoring, patiently tolerated by the person who appeared to be the real grown-up in the room: Google's Pichai.

News reports following Google's hearing thus drew similar conclusions to those following Facebook's hearings: Congress had had yet another shot at bringing down our net state kings a peg, and they blew it.[35] Thus despite increased public awareness in the United States about our lack of protections of our digital information, there remains no legislation to address it.

Legislation, though, may not be the perfect solution for protecting users' rights. "Tech will always change so fast that hard and fast legislation rules freeze you in the wrong place," Tom Wheeler, President Obama's chair of the Federal Trade Commission from 2013 to 2017, told me in an interview in January 2019. "What's ironic is that the very same tech companies that fought tooth and nail against federal regulations three years ago—here's a story for you; when they fought rules in Congress, they passed around a document detailing how, and I quote, 'privacy regulations would significantly *harm* consumers'—have now come around to *ask* for federal privacy rules."

When asked why he thought this was the case, Wheeler replied, "Now that they're seeing California with its own privacy regulations, they're seeing that other states will likely follow suit. There will be a plethora of rules. So now the tech companies themselves are advocating for one set of federal rules that they can comply with, because if they have to deal with 50 sets of state rules, all of which are different, they won't be able to operate."

If the tech industry itself is advocating for federal regulations, why haven't they materialized? There are several superficial reasons. The first is a perennially gridlocked US Congress that has been so mired in partisan rancor for the past several years that it has been unable to pass sweeping legislation of any kind, let alone a package that might be considered controversial, such as a federal data protection law. Also, the US lacks a central agency that would oversee enactment of such laws even if they did exist.[36]

But the root cause of our lack of data protection laws in the US requires us to look inward, and it's a tad more depressing than partisan bickering: perhaps people just don't care enough about the issue to push for change.

In what *Atlantic* writer Ian Bogost has termed "privacy nihilism," one of the reasons people fail to take the requisite steps to safeguard their digital data is simply preemptive exhaustion, a communal malaise. We sense that our data's already gone, that we lost the privacy battle before we even knew it was ours to fight. As such, there's no point in getting all up in arms about it now.[37] In a 2017 Pew Research Center study on trust in online platforms, experts identified this trend, noting that whether we trust our tech (or not) may not translate into action regarding our *use* of that tech. Trust, or the absence of it, will simply become "irrelevant," they predicted. "Hacking, identity theft, trolling, doxxing"—when someone reveals another person's personally identifiable information publicly online—"will become increasingly commonplace and a daily cost of doing business on the internet."[38]

Research supports that this trend will likely continue. According to a 2018 study by the University of Michigan, people who purchase "smart" home devices like Google Home or Amazon Echo (detailed more in chapter 6) are fully aware that they're risking their private information being collected; they've simply given up on the idea of privacy.[39] "What was really concerning to me," said Florian Schaub, assistant professor at the University of Michigan and one of the study's coauthors, "was this idea that 'it's just a little bit more info you give Google or Amazon, and they already know a lot about you, so how is that bad?'"[40] He concluded, "It's representative of this constant erosion of what privacy means and what our privacy expectations are."

We've become so inured to news stories about data breaches, data leaks, and untrustworthy personal privacy policies that we don't even bother to read them anymore (if we ever did). Which is ironic. More and more privacy notices are being written in plain English as tech companies attempt to comply with the European Union's recent data regulations. So it's now more *possible* to read them than ever. The question is, Will it make any difference if we do?

For instance, if you live in Vermont and you're having a Powerwall installed in your home, as described in chapter 2, you may have the opportunity to read and sign Tesla's privacy notice—which is totally doable; it's not even two pages long and is written in accessible English. But what about Tesla's Powerwall partnerships directly with cities such as Norwich, Connecticut, and San Jose and Santa Ana, California? Will their residents be notified that Tesla may collect their personal information? Will they even know that Tesla is a partner in running their electrical grid?

Here's why this matters. A sentence in the "How We May Use and Share Your Personal Information" section of Tesla's privacy notice reads: "We may share Personal Information we collect with *third parties* when necessary to perform services on our, or on your, behalf" (italics added).

Who are these "third parties"? The notice doesn't say. But if Tesla is anything like the majority of businesses that collect consumer data in the United States, "third parties" most likely include data brokers.

Data brokers are essentially what they sound like they would be: companies that buy and sell your data—for gobs of money. They're a $50-billion-a-year industry in the United States. Yet despite their massive financial assets, they operate largely in the dark; most Americans have never heard of even the biggest among them, which include Acxiom, Nielsen, Experian, Equifax, and CoreLogic.[41] "Most [people] have no idea who these companies are and how they got their data on them," commented Amul Kalia, who works for the Electronic Frontier Foundation, a consumer privacy advocacy organization. "And they would be very surprised to know the intimate details that these companies have collected on people."[42]

While Tesla and Google and Facebook gather personal and behavioral information on their singular platforms, data brokers—so powerful that one reporter referred to them as "the whales of the data ecosystem"—compile it all to create comprehensive and detailed profiles.[43] A 2014 report by the Federal Trade Commission showed that one data broker, Acxiom, had databases storing 3,000 "data segments" on just about every individual consumer in the United States.[44] In some cases, these segments reflect our religious preferences, with labels like "Bible Lifestyle"; our economic circumstances, like those of us who fit into the categories "Financially Challenged" or "Modest Wages"; our marital status, like "Working-Class Mom"; and our entertainment preferences, like "Media Channel Usage—Daytime TV," "Exercise—Sporty Living," and "Outdoor/Hunting & Shooting."

This data segmentation doesn't stop at the behavior associated with the products we buy. We're also sorted into categories based on our overall lifestyle, such as "Rural Everlasting"—that would be single men and women over the age of 66 with "low educational attainment and low net worth"—or "Downtown Dwellers," those of us who are "upper-middle-aged" and with a "high-school" or "vocational/technical" degree working to "make ends meet with low-wage clerical or service jobs."

While this may feel high on the creepy spectrum, the data brokers themselves attempt to paint such specific market segmentation as actually good for us. For instance, on Acxiom's website section "How Personalized Marketing Works," they display two kinds of ads: what you'd see if you were a, say, 46-year-old single male—that would be an ad for a flashy red sports car; and what you'd see without personalized marketing—an ad for, heaven forbid, makeup. Another example: if you've recently moved, Acxiom shows that you'd see ads sharing the opportunity for 20 percent off "housewarming" furniture; without personalized marketing, they display some dreary impersonal credit card pitch that clearly doesn't know how to congratulate you properly on your big move.

With these examples of the ways that "personalized marketing" is "helpful," we can begin to see how our personal data gets monetized. Because if I'm in the business of selling flashy red sports cars, I probably really do want to target single men in the midlife crisis range. Plain old common sense suggests that they're a more likely buyer group than, for instance, women with young children. With the amount of money American businesses spend on advertising—$218 billion in 2018—it's not hard to imagine that they'd want to ensure that those dollars are spent targeting the absolutely most likely buyers, down to zip code, lifestyle, age, parental status, and other relevant variables.[45]

The question then becomes, How can data brokers compile enough information to know not only our demographic details, but how far we went in school, the kind of job we have, how much we earn, and whether we have kids, hunt, or go to Bible study?

That would be from the "data elements" they collect from various companies. They may learn about our vacation habits from the companies we buy our plane tickets from, like Expedia[46] or Kayak.[47] Our weight, height, educational attainment, and parental status may come from dating sites like Match.com[48] or OkCupid.[49] One dating site, Grindr, even shared users' HIV status with data brokers—if the users voluntarily shared this information with the site (it's unclear whether this is in violation of HIPAA protections).[50] Our political affiliation, credit card use, occupation, net worth, and other data elements are regularly collected from just about every company and online service we do business with.

The true value for data brokers isn't individual data elements or even the marketing segments they sort us into. Instead, the value resides in the aggregation of data that allows them to create a finely detailed picture of who you are—again, not someone *like* you; specifically *you*—and what you're likely to need. "The extent of consumer profiling today means that data brokers often know as much—or even more—about us than our family and friends," said Edith Ramirez, then chair of the FTC, in a May 2014 press release.[51]

What's even more alarming about our data profiles is that they're not limited to information available digitally. Thanks to a process called "onboarding," data brokers collect information about our offline habits to include in our profiles as well: things like magazine subscriptions, what we purchase with store loyalty cards, and publicly available data from government records.[52] "The marriage of online and offline is the ad targeting of the last 10 years on steroids," said Scott Howe, chief executive of Acxiom, in 2014.[53] Onboarding is how it's possible for you to stop at Target to buy DayQuil and then, when you log on to your computer a few hours later, see ads popping up on Facebook for humidifiers and throat lozenges.

What's the big deal? you ask. What possible data could a company like Tesla sell to a data broker about your energy use that would matter to you? Think about it: When do you turn the lights on? When you wake up. When do you turn them off? When you leave your house. On again? When you get home from work. Off again? When you go to sleep. And so, using just your energy use data, Tesla and whoever they sell your data to have your daily routine mapped out to the minute. They know, in general, when you will be home. More important, they know when your home will be empty—and potentially vulnerable to any criminal who might want access to it.

This is a problem because all digital data—*all* of it—is hackable. And hacks have happened. Equifax, one of the three largest credit score firms in the United States as well as a major data aggregator and broker, suffered a massive cybersecurity breach in 2017 that left 143 million Americans' data exposed.[54] Whoever stole this information could theoretically use your data profile to find out where you live and—if they have access to your energy use profile—when you'll be out during the day.

Consider the Equifax breach and Tesla Powerwall side by side. Much of your personal information is floating around in the ether, thanks to the Equifax hack (and perhaps other hacks that we're not yet aware of; Equifax waited two months before reporting the breach).[55] This data profile includes who you are, where you live, what you like, where you work and shop, what you earn, and—if Powerwall data were also collected—when you're home and when you're out.

Assume also you're one of the 47.5 million Americans—that's one out of every five of us—who also has access to a voice-enabled assistant such as Amazon Echo, Google Home, or even your iPhone's Siri.[56] Smart speakers are also hackable, and hacks are well documented. For instance, researchers at the vulnerability testing firm Checkmarx demonstrated that it was "surprisingly easy" to gain remote access to an Echo and have it record everything it hears—all without notifying the user.[57]

Hacks, however, are the worst-case scenario. The more likely outcome is that net states willingly give—"share" is the more popular term—your data to third parties. Amazon, for instance, is currently considering giving third-party developers access to the transcripts of user conversations recorded via the Echo. Google Home already does.[58] This means that in addition to the content of your conversations—what you say you want, where you're going, what you're arguing with your spouse about—they have a new piece of biometric data: your voice. The iPhone X already has a fair bit of personal biometric data, including your thumbprint (if you enabled that login option) and your facial features (if you enabled Face ID). With voice-enabled assistance, it adds your voice to the data pool as well. Data brokers already have behavioral profiles on the majority of Americans; with the next generation of technology, they'll be able to add physically identifying characteristics to the profiles as well.

It may be that Google and Amazon and other purveyors of voice-enabled assistants collect user conversations simply to improve their products. Yet regardless of how innocent their rationale may be, the question remains: What happens to our data once we share it? Even though the original data collector may have our permission to possess our information—which we gave when we signed their terms of service—if the language contains any reference to third-party sharing, we may as well kiss our privacy goodbye.

OUR RIGHTS AS CITIZENS ARE PROTECTED THROUGH THE BILL OF RIGHTS in the US Constitution. Most of us are generally familiar with the big ones that pertain to us: freedom of speech, assembly, religion, and so on. But how our rights extend to the data *about* us is unclear. For instance, the Fourth Amendment says that we are protected from unlawful search and seizure. This means that no one can take our property—that we should "be secure in [our] persons, houses, papers, and effects," to be precise—without a legal warrant.[59]

Is our personal data our property? What exactly *is* personal data? These types of thorny questions are currently being hashed out in the halls of the European Parliament. The world's strongest data protection legislation to date came out of the European Union (EU)—the 2016 General Data Protection Regulation, better known as the GDPR, which took effect in May 2018.

The GDPR introduced a new term for the residents whom the legislation protects, calling them "data subjects."[60] By contrast, the tech companies that collect information about data subjects are "data controllers." Under the GDPR, data subjects—Europeans from 28 countries—are granted a slew of rights, including the right to consent to how their data is collected, the right to be notified within 72 hours of a data breach, the right to know what their data is being used for, and, perhaps most important, what's known as the "right to be forgotten"—or "data erasure," to use the technical term. Should a data subject withdraw consent at any time, the data collector must stop using his or her data or face heavy fines: 4 percent of the company's annual global turnover, or €20 million (roughly the same as $20 million).

That a bill of rights for data subjects exists somewhere in the world is a huge step in the right direction for users. Citizens of the EU now have explicit rights with respect to their personal data. Put another way, Europeans have rights on multiple levels. They have rights as citizens of their individual nation-states *and* as members of the EU, the latter role now conferring rights as "data subjects."

This is a totally laudable move; it's great that such legislation exists somewhere in the world. But it's problematic in practice.

As with all diplomatic endeavors, there's the issue of enforcement. At present, each individual member state of the EU—again, 28 different countries—will have its own enforcement mechanism. That's 28 different methods for making sure every tech company that interacts with its citizens complies with GDPR requirements.

To consider what this means in practice, think about our phones. The average user accesses 30 apps on their smartphone each month and has about 90 apps total installed.[61] About 502 million people live in the 28 nations that compose the EU. This means—even if we're just talking about the 30 most commonly used apps a month—that the supervisory authorities are responsible for ensuring that tech companies comply with the GDPR in 15,060,000,000 (more than 15 *billion*) citizen-user tech interactions a month. It would be the understatement of the year to note that oversight authorities in the EU are going to have their hands full.

And it's just not clear how oversight is going to work. At present, the GDPR stipulates that there be "one GDPR supervisor per country." According to Article 54, it's up to each of the 28 member states to determine how to set up its Data Protection Authority (DPA) and Supervisory Authority: how many people should staff their offices, how they should operate, and what level of funding they're allotted by their own government.[62]

Put another way, GDPR supervisors in Belgium, Slovakia, France,

Germany, and Estonia may all operate completely differently from one another, function with different budgets, and be given varying levels of priority within their home country, depending on their local political climate. So, while there's an oversight board for all DPAs—that would be the European Data Protection Board—as with all things GDPR, "many details still needed to be hammered out" (the official GDPR characterization of the situation).[63]

Enforcement issues aside, the GDPR is all well and good for the Europeans. But what does it mean for us Americans? For the tech industry as a whole, it means a lot of compliance catch-up. In the days preceding the May 25, 2018, deadline, you may have noted a flurry of emails in your inbox about privacy policies for various tech services you've registered with (and some, perhaps, that you didn't even know you were registered with). Despite having more than a year's notice that the GDPR was coming, a lot of tech companies simply didn't make the deadline. For instance, when the GDPR took effect, several major American news sites, including the *Los Angeles Times*, the *New York Daily News*, the *Orlando Sentinel*, the *Chicago Tribune*, and several others simply went offline to anyone in Europe.[64] Deciding that GDPR compliance simply isn't worth the hassle, a number of tech companies—for example, the massive multiplayer online game Ragnarok, the advertising services Verve and Drawbridge, the Brent Ozar Unlimited consulting company, the Unroll.me email manager, and the security company Steel Root—simply ceased doing business with EU users.[65]

Meanwhile, a month before the GDPR took effect, Facebook moved 1.5 billion users—more than half its global user base—out of reach of European regulators and to its HQ in California.[66] As privacy researcher Lukasz Olejnik noted, this was a massive undertaking: "Moving around one and a half billion users into other jurisdictions is not a simple copy-and-paste exercise," he said. The end result? "Users will clearly lose some existing rights, as US standards are lower than those in Europe," he noted. Given the scale of the user base moved, as well as the timing, it's not difficult to imagine what Facebook had in mind with the move.

Facebook and other net states can do what they like with user data in the United States. So long as we keep using their products and services, many of them will continue to persist as they have.

In the United States, regulations focus on the *product*—the information itself and how it's handled—not on the user having any inherent rights. Take, for instance, HIPAA, the Health Insurance Portability and Accountability Act of 1996. HIPAA protects a person's medical information from being shared without consent. If you've been to a doctor's office in the past 20-plus years, you've probably signed a HIPAA consent form (probably without reading it) that gave permission to release or share your medical information. Note, however, that HIPAA doesn't give *you* any special rights. Rather, it carves out special protection for one subset of data *about* you—your medical information—and what a certain group of people—medical professionals—can do with it.

Another example is what's known as PII: personally identifiable information. This includes name, date of birth, Social Security number, and other data that, in aggregate, would give away key details about a person's legal identity. PII is protected under the Privacy Act of 1974, which establishes a "code of fair information practices," basically spelling out what federal agencies can and cannot do with the records they keep on citizens.[67] As with HIPAA, though, this regulation focuses on what certain groups of people—in this case, federal officials—can *do* with a certain type of information.[68] Once again, a certain type of information *about* us is regulated, but we as citizens aren't afforded any special protection.

Compared with residents of the European Union, American citizens don't have data protection rights. We are not "data subjects" here. We are citizens of our nation-state with respect to our physical selves and our physical environments. And we are users of our net states with respect to our digital selves and our digital environments. At present, net states have free rein, as stated earlier, in terms of what they wish to do with our personal data (with the exception of medical information). But they're not all reacting to that freedom in the same way.

Some net states are taking a more user-centric approach of their own volition. Microsoft, for instance, is actively expanding "data subject" rights: they've launched a "privacy dashboard" that permits any tech user in the world to sign up for GDPR-level privacy protections.[69] And some Americans are taking advantage of that feature.

As Microsoft president Brad Smith noted during our interview in September 2018, "When we made those data subject rights available to customers on a global basis, in the European Union itself, a market with over 500 million people, there are roughly a million plus people who took advantage of that. In the United States, a market of 320 million, there were two million people who took advantage of it." He said, "It really, I think, should cause people to reflect on what is a long-standing view that people often present the United States, including in the tech sector, that Americans don't care about these issues the way Europeans do. Well, there's something there that Americans got pretty interested in."

AS DISCUSSED ABOVE, NET STATES ARE ESSENTIALLY GIVEN FREE REIN in the US but are subject to new, strict regulations in Europe. But Europe's not the only place that net states are facing backlash from nation-state governments. In May 2018, Uganda, in an attempt to wipe out "gossiping," enacted a tax on social media use:[70] 200 shillings a day, or $0.0531—about $19 a year. It may not sound like much, but in a country with an annual GDP of $615 per capita, it's not nothing. In Papua New Guinea, a small island nation in the South Pacific, authorities banned Facebook for a full month.[71] "The time will allow information to be collected to identify users that hide behind fake accounts," commented Communications Minister Sam Basil. Facebook and other social networking sites such as Instagram and WhatsApp have also been the target of scrutiny by governments elsewhere, such as Cambodia, Myanmar, Sri Lanka, and the Philippines.[72] While there are very likely less than noble reasons for enacting these bans—clamping down on free speech from citizens, journalists, and advocates high among them—the upshot is that not every nation-state is giving Facebook or other tech platforms free rein in their sovereign territories.

In short, in 2018, 33 nation-states—the 28 EU members plus the aforementioned—suddenly became much more difficult places for net states to do business. Given this, it's probably not surprising that net states are looking to set up shop *outside* nation-state borders. All the land on Earth is currently governed by nation-states, but two categories of Earth's territory have yet to be clearly claimed and regulated: outer space and the deep sea.

In 2009, Google was awarded a patent for seafaring data barges: offshore data centers that could traverse international waters.[73] While Google has yet to build any such barges, other net states are beginning to explore what the seas have to offer them.

The ocean offers distinct advantages. Data centers—the machine hubs where our online information is processed, transported, and stored—are physical computers housed in physical structures that consume tons of energy and generate significant heat. According to the United States Data Center Energy Usage Report, data centers in the territorial US are projected to consume approximately 73 billion kilowatt-hours by 2020.[74] To translate that into English, charging your smartphone for a full year requires 1 kilowatt-hour. To keep the internet going, however, requires more than 70 *billion* kilowatt-hours—or, as one reporter described it, "equivalent to [the output of] about 8 big nuclear reactors."[75]

For this very reason, chilly Scandinavia has become a popular destination for net state data centers: companies have opened centers in Denmark (Apple), Iceland (Facebook), Sweden (Amazon), Norway (Microsoft), and Finland (Google).[76]

In 2015, Microsoft launched Project Natick, a pilot program to sink self-contained data centers in the ocean. The idea was first floated in 2013 in a white paper—one of the coauthors had worked on a Navy submarine—and by 2015, Microsoft decided it was worth throwing some money at it.

"When I first heard about this I thought, 'Water . . . electricity, why would you do that?'" said Ben Cutler, one of the engineers for Project Natick, in an interview. "But as you think more about it, it actually makes a lot of sense."[77] Computers crash when they overheat. That's expensive. Cooling them with air-conditioning demands lots of electricity. That's also expensive. Storing all this machinery requires valuable physical real estate. *That's* expensive. The ocean undercuts costs in all three areas. The currents keep the machines cool while also powering turbines. Theoretically, these turbines could convert the ocean currents into electricity, and there you have it: functional, powered data centers that are cool enough to keep from melting down.

There are also potential legal advantages to investing in sea-based data centers. There's certainly plenty of physical real estate: more than 70 percent of the planet is covered in ocean.[78] The United Nations Convention on the Law of the Sea (UNCLOS) established that country borders extend exactly 12 miles offshore. Considering that there are about 332,003,271 cubic miles of ocean, clearly the majority of that undersea sector doesn't fall under the purview and protection of any nation-state.

By now Project Natick is more than just an idea. In June 2018, Microsoft dropped a pilot data center off the coast of Scotland in collaboration with the European Marine Energy Center.[79]

"We see this as an opportunity to field long-lived, resilient data centers that operate 'lights out'—nobody on site—with very high reliability for the entire life of the deployment, possibly as long as 10 years," said a Microsoft spokesperson. The company also projects that it can manufacture underwater data capsules in as little as 90 days, significantly faster than it takes to construct traditional land-based centers.

The possibilities here are more than a little intriguing. Data centers can be created and deployed in just a few months. And there's no reason they can't be dropped, say, 13 miles off the coast—just a smidge outside the bounds of nation-state sovereignty. In theory, the data centers would not be subject to the nearest nation-state laws or regulations. Rather, they would fall under the UNCLOS, which, while well intentioned, is effectively toothless in terms of enforcement. In short, net states could operate their data centers, which could theoretically house all our personal data, beyond the reach of any nation-state. If they did, net states—not nation-states—would have sovereign domain over their citizen-users. The sea would set them free.

FOUR

INFORMATION-AGE WARFIGHTERS

As much as we experience life online, we are creatures of the physical world, vulnerable to physical threats. This chapter explores our physical security, showing how net states like Google can aid in the fight against modern-day enemies. Analyzing the differences between the tech ethos and the military ethos, this chapter considers how the expertise so prized by security agencies has become a potential disadvantage in the fight against terrorism. It shows the possibilities for net state/nation-state cooperation through acts of diplomacy and notes how net states, led by Microsoft, have begun to forge ahead in this domain.

"We Love Death More Than You Love Life," reads the bold-print Facebook post in the video. The shot cuts back to a young man seated in what looks to be a dining room, his face masked by a blue-and-black checkered keffiyeh pulled up so high it almost obscures his eyes.[1] The video names him "Abo Abdullah." He stares at his interviewer just off camera. "A year ago," he says before taking a long pause, "I joined ISIS."

The YouTube video, entitled "Captured ISIS Fighter on How He Was Betrayed," is slick, professionally produced—the product of an American digital news outlet called Vocativ, founded in 2013. "What did your leaders tell you about America?" the interviewer asks off camera. Abdullah hunches forward, lifting his right hand to cradle his head. "They used to teach us that Americans were atheists who deserved to be slaughtered." He rubs his forehead, stops talking.

The video, which runs about five minutes, details how Abdullah was recruited into ISIS from his home country of Saudi Arabia, arriving in Syria under cover of night. He tells how he believed that he would become a martyr "for the crimes committed against Islam"; how he went through a closed-camp, 30-day training program in which he wasn't permitted to communicate with anyone on the outside; how, once in the camp, "they start to indoctrinate you, . . . 'Don't trust this,' 'Don't trust that.'"

In the video, Abdullah wipes sweat from his brow with his sleeve when he gets to the part about how he ended up leaving ISIS. "I was captured," he says, "during a battle between ISIS and FSA [the Free Syrian Army, opponents of Syrian president Bashar al-Assad]." He says, "The FSA told me, 'You can be sold for cash—why did you come here to fight?' That's when I realized I had been sent to my death by ISIS."

This video is one of dozens of ISIS defector videos that Google and a counterextremism organization, Moonshot CVE, leveraged in a pilot test to deter ISIS recruits.[2] For eight weeks in 2018, rather than finding ISIS-related propaganda, users who typed "How to join ISIS" into Google's search engine were redirected to 116 different counternarrative videos detailing the dark side of ISIS—ISIS defector videos like the one described above, for instance, or homemade videos showing ISIS fighters languishing in long food lines, undermining the narrative that the caliphate was a well-provisioned paradise.[3] During the pilot, in addition to serving up anti-ISIS videos, Google featured ads for mental health services alongside user's search results.[4]

The campaign worked surprisingly well: it reached 320,906 users during the test period. Users who had started out searching for terrorist recruiting materials instead ended up watching 500,070 minutes of video of counternarratives.[5] The mental health content worked even better: users who entered the most violent search terms ended up clicking ads for psychiatric support at rates 115 percent higher than a control group.[6] The campaign diverted more than a thousand potential terrorists from recruitment materials and sent many vulnerable people to resources for psychological help.

This isn't an unusual approach for a counterterrorism campaign. What makes it unusual is who was running it: not the US Department of Defense or the CIA or any other national security agency. It was Google. And they weren't doing it for profit. Google made a moral judgment here. Indeed, by placing their own ads on the site (in lieu of selling space), they lost out on potential revenue that they could have earned. They also acted against their primary business practice: to facilitate search for their users. In this case, Google worked with the express purpose of luring users *away* from the content they were searching for.

It's worth noting that Google wasn't simply the business broker in this equation. The company was acting as a quasi-counterterrorism agent, having made a policy decision to intervene in normal business processes to achieve a greater good. Antonia Raithatha, then digital marketing specialist at Moonshot CVE, the counterextremism organization Google partnered with on this, emphasized in an interview that it was Google's understanding of users, not just Google's tech, that made the campaign successful. These campaigns need to tap into "nonideological drivers of extremism that recruiters use to sway vulnerable individuals," she explained. You have to understand users' desire for "the sense of purpose and belonging." This desire—to be a part of something bigger than oneself, to have a clearly defined purpose handed to you—that's what recruiters tap into to lure prospects.

In this respect, Google had an advantage over, say, the Department of Defense. Google's counterterrorism strategy was based not on policy directive, but rather on a deep understanding of what their users want; something in which a search engine—privy to our most private questions—is uniquely qualified.

This particular campaign was a success, though most of us have probably never heard about it. This raises some questions: When *else* does Google intervene in searches? Do they redirect people seeking information about nefarious conduct across the board? Or only in cases of terrorist activity? And who at Google makes that decision? Does the company have some sort of counterterrorism wing, or was this just an experiment cooked up by some do-gooder engineers?

One answer, it turns out, is that Google does indeed have a department devoted to counterterrorism and other campaigns for the public good. It's called Jigsaw.

CHELSEA MARKET IS, ON THE FACE OF IT, AN UPSCALE FOOD HALL. Situated just beneath Manhattan's elegant elevated park, the High Line, its ground floor offers a variety of shops ranging from the Heatonist—specializing in artisanal hot sauces—to the quirky Artists & Fleas. The 51 rather pricey shops nestled in its charming, brick-walled halls attract well-heeled tourists and tote bag–wielding locals alike, giving it a bustling, almost European feel.

What most people don't know is that in February of 2018, Chelsea Market was purchased by Google.[7] And just inside its 10th Avenue entrance, tucked away behind what appears to be an exotic pillow and lamp shop, stands a glass door with stairs leading up a flight to Google's geopolitical think tank, Jigsaw. Formerly known as Google Ideas, Jigsaw was founded in 2010 by then CEO Eric Schmidt. Its mission: to use technology to "tackle the toughest geopolitical challenges, from countering violent extremism to thwarting online censorship to mitigating the threats associated with digital attacks."[8]

I met Jigsaw's managing director, Scott Carpenter, in his office on a Tuesday morning in July 2018. Once inside Jigsaw's suite, you'd never know that a sea of shoppers meandered the floor below. In keeping with standard tech company architecture, an open pit hosted rows of tables that served as desks for multiple people, each work space piled with dual monitors, quirky decorations, and what looked to be at least two cups of coffee per staffer. Carpenter ushered me into his office, one of Jigsaw's few closed-door spaces, which he shared with another senior leader.

I dove in with the basics: What's Jigsaw up to? And what's in it for Google to "tackle the toughest geopolitical challenges"?

"If the internet is our ecosystem in which we live, breathe, and operate and make all of our money," Carpenter replied, practiced and patient, "we need to think about what are the ways in which we need to protect that ecosystem."

A former senior State Department official, Carpenter is mild-mannered and deceptively low-key—an intellectual powerhouse in the guise of a smiley-eyed tech evangelist. He and I were halfway through what felt like an easy conversation when I realized he had been unwinding carefully selected anecdotes to contextualize a grand global narrative. Like great politicians, Carpenter has a way of making you feel totally at ease—charmed, even—while wowing you in a humble, understated way. Like the best of nations, Google chooses its leaders wisely.

"In the same way, you protect the Gulf of Mexico if you're BP," he mused. "Understanding that [terrorists] are a threat to your existence if you don't, [we asked ourselves,] What can we do?"

The tricky part is figuring out the how: How precisely do you go about protecting an ecosystem of over 1.5 billion websites?[9] According to Carpenter, Google is open to discussing the question of how with absolutely anyone. Information-sharing is, after all, core to the hacker ethos. "I led a group to northern Iraq to talk to former ISIS fighters in detention facilities," he reflected. "But we'll talk to the CIA, too."

However, "talk to" is where Google draws a self-imposed line. "We'll talk to anyone," he said. "But we're not going to *collaborate* with governments."

The CIA is certainly part of the American government, I almost pointed out. But then I reminded myself that "talk to" may mean sharing information, exchanging ideas even. But that's just conversation—a far cry from engaging in some sort of active collaboration.

"*Any* governments?" I asked.

"Yeah," he nodded, cool but firm. "And that's kind of a principle."

Whether we like it or not, net states don't need governments nearly as much as governments need net states. One example where this is painfully evident is with attacks on official computer systems.

"We are dealing with a hostage situation," Atlanta mayor Keisha Lance Bottoms noted almost casually at the hastily assembled press conference. Standing flanked by a troupe of commissioners and city council members on March 22, 2018, her team took turns detailing the enormity of the problem their city now faced.[10]

After approximately ten minutes of damage reports, the mayor herself returned to the microphone. "It's an attack on government," she noted, poised, unruffled, "and, therefore, an attack on all of us."

The hostages in this case weren't people; they were computers used by the 424 programs—around 30 percent of which were deemed "mission critical"—that served as the foundation for Atlanta's government. Due to a ransomware attack, the 8,000 employees using the city's computer network were locked out, their machines either compromised or completely disabled.[11] An entire city government—one that served over 5.7 million residents—had been effectively shut down.

Scrambling for solutions, the city of Atlanta did what most local government officials would do—they called in the feds. The FBI and the Department of Homeland Security were immediately engaged to investigate the source behind the attack.

But not even the FBI would be able to bring Atlanta's systems back online; FBI agents are investigators, not computer experts. So Atlanta made another call—to the tech industry itself. Microsoft, Dell, and several other tech firms launched into action to free the hostage data and render the systems operational again.

The ransom was, relatively speaking, a paltry sum: $51,000. By contrast, the tech contracts cost the government of Atlanta $2.7 million.[12] But as the FBI noted, there was no way of knowing whether the ransomers would cease the attack if they got their money. Also, Microsoft was a trusted entity. If government was going to pay, clearly they felt it was safer to spend millions of dollars on Microsoft, a trusted ally, than any amount on unknown criminals.

And the tech firms did indeed deliver. They restored access to the computer network in just six days, recovering the vast majority of the compromised data. As Atlanta recovered from its attack in the months to follow, the overwhelming advice they heard was this: "Back up your files to the cloud"—specifically, to net states cloud services: Google Drive, Apple's iCloud, Microsoft Azure, or Amazon Web Services.[13] Notably, the advice was *not* to create a government version of the cloud; rather, government officials and tech experts alike advised the same approach—to use the cloud that exists, despite the fact that it had been created and was managed and safeguarded not by government but by net states.

As the Atlanta case shows, net states are both governments' digital suppliers and cyber defenders. They create the platforms governments run on, the storage citizen data resides in, and the search-and-rescue forces called upon when public data falls under attack. In the digital age, government—whose essential job is to protect citizens—is finding that it can't always protect even itself. That government is calling on net states like Microsoft to bail them out and Google or Amazon or Apple to back them up reveals a shift in power that is the hallmark of the rise of net states.

Another example can be seen in the case of the Russian misinformation campaigns that infiltrated social media during the 2016 presidential election. These practices were not, in fact, a new phenomenon. According to a comprehensive study of Russian misinformation commissioned by the Senate Intelligence Committee, the attacks started about five years prior. Since then, "disinformation has evolved from a nuisance into a high-stakes information war."[14]

Congress finally took some steps to understand the phenomenon, in the form of commissioning the above-mentioned report (in 2018, two years after the election). But despite multiple US official confirmations that Russia was behind these actions, the federal government has failed to offer any tangible help to the tech companies on whose platforms this warfare is being waged. Net states are essentially on their own to deal with the information wars waged using their services.

To win these information-age wars, countries need to recognize the power of net states as not just allies, but coequals. Net states wield the kind of power and influence necessary to go toe-to-toe on digital warfare with other nation-states, like Russia, as well as nonstate actors, like ISIS. As recently as 2017, the American government didn't seem to be getting this.

"I don't think there's ever been a technological advance that governments have been able to keep up with," reflected Carpenter during our 2018 Manhattan interview. "I think governments still don't understand enough about what they're doing—the brain surgery they're doing without understanding anything about the brain."

As I first described in the 2017 *WIRED* article that inspired this book, the world needs net states because they occupy the digital space where warfare is manifesting, with tactics like disinformation warfare by other nation-states as well as recruitment and organizing attacks by nonstate actors.[15] As such, net states like Google understand the norms and tactics of digital warfare—the brain surgery Carpenter referred to—far better than a Cold War–era strategist ever could. Major General Michael K. Nagata, former commander of the American Special Operations forces in the Middle East, circled around this idea back in 2014. In a leaked confidential conversation about ISIS, he said, "We do not understand the movement, and until we do, we are not going to defeat it. We have not defeated the idea. We do not even understand the idea."[16]

The notion that ideas are a national security consideration is not a new one: since 9/11, the United States military has been trying to win "hearts and minds" of combatants in Iraq and Afghanistan.[17] Despite this, the military establishment has been slow to recognize the potential of net states to assist in the fight against radicalization. It appears that military leaders have instead continued to view net states simply as keepers of the platforms where recruiting and organizing occur, rather than as real assets in the fight—savvy insiders with unique insights into how terrorists move and operate.

Our enemies learned long ago how to exploit the digital landscape for their purposes. Terrorists have proven themselves to be extremely adept at using net state platforms—social networking tools like Twitter and Facebook; encryption tools like Snapchat, WhatsApp, and Telegram; and video-sharing platforms like YouTube—to organize attacks, recruit members, and spread disinformation.[18] In one study, researchers who looked at terrorist activities across 96 countries found that recruiters routinely found new would-be terrorists through Facebook's "Suggested Friends" feature.[19] And Facebook itself admitted that terrorists and other criminals likely used its encrypted messaging platform WhatsApp as a secure communications tool.[20]

But knowing how misinformation spreads or how terrorists recruit online isn't enough. National security agencies and the military need to put themselves into the mindset of tech-savvy digital warfighters. But they can't do this without first becoming tech-savvy themselves. Here I'm not talking about knowing how to *use* technology; I'm talking about having a handle on the tech *ethos*, grasping it deeply and intuitively.

For such a tech-heavy institution, the military just doesn't seem to "get" tech. This is not to say that they don't understand technology. Far from it. They are steeped in extremely advanced technology in most aspects of their work. But using tech is different from having a tech ethos. And the military ethos and the tech ethos couldn't be more different.

In many ways, the tech ethos is an outgrowth of the hacker ethos: but instead of describing how to *make* tech, it describes how to *use* tech. The best expression of the tech ethos I've ever come across is from internet pioneer and MIT professor David Clark. He said, "We reject kings, presidents and voting. We believe in rough consensus and running code."[21] Putting primacy on openness and information-sharing rather than on blind allegiance to formal authority allows everyone to have a say, which is why the best ideas, more often than not, rise to the top. In other words, tech embraces what works, as opposed to what's procedurally approved to work. Like the hacker culture described in chapter 1, the tech ethos holds that skill matters, not credentials. Authority comes not from what fancy college you went to or title you might hold, but from who can make tech work better, period. According to the tech ethos, a 17-year-old hacker living in his mom's basement or a cybersecurity expert at the Pentagon are equals, if they can code equally effectively.

One of the reasons that the military fails to "understand the idea" (to echo Nagata's wording) behind the tech ethos rests with its reliance on experts. National security circles are filled with experts: policy experts, cybersecurity experts, military strategy experts, grand strategy experts. Of course, experts are indispensable: we absolutely need to draw on the deep wisdom and practical know-how of people who understand every nook and cranny of their field.

The problem with experts in national security, however, is the governmental practice of relying *only* on experts. In Philip Tetlock's 2006 book, *Expert Political Judgment: How Good Is It? How Can We Know?*, he showed, after decades of data collection, that experts run into two problems when it comes to making predictions about upcoming threats.[22] First, expertise can lead to overconfidence. Since experts can rely on a deep well of existing knowledge, they might overlook the importance of gathering information in real time or double-checking their assumptions. Second, while experts tend to specialize in one thing, our increasingly complex world is also in need of generalists with multidisciplinary knowledge—knowledge that helps them connect seemingly unrelated things. Tetlock likens these two groups—experts and generalists—to hedgehogs and foxes, modeling his analogy on the philosopher Isaiah Berlin's essay "The Hedgehog and the Fox." Berlin adopted the phrase from the Greek poet Archilochus. "A fox knows many things," Archilochus wrote, "but the hedgehog one important thing."

In a 2007 lecture, Tetlock argued that foxes are, in fact, better at predicting what's going to happen in the future precisely *because of* their lack of expertise.[23] Generalists make predictions by considering a wide array of real-time information, double-checking their assumptions, and reading up on areas they may not be familiar with. Generalists don't have a deep knowledge base to rely on. As a consequence, they're more willing to relinquish ideological loyalty and adjust their hypothesis when they uncover new facts. On the other hand, "partisans across the opinion spectrum," Tetlock noted, "are vulnerable to occasional bouts of ideologically induced insanity."

Expertise is associated with confidence. "Blinded by beautiful theory," as Tetlock describes it, experts have a tendency to overly rely on their knowledge base and are, in general, less willing to adjust their predictions based on changing circumstances or new information. The political forecaster, as Tetlock puts it, who "bores you with a cloud of 'howevers' is probably right about what's going to happen. The charismatic expert who exudes confidence and has a great story to tell is probably wrong."

The Ivy-trained Tetlock didn't rely on his own expertise to come up with this theory: he tested it.[24] First, he studied those in the know, examining the aggregate predictions of 284 experts making 28,000 forecasts.[25] Then he studied nonexperts, launching a decade-long social science experiment called the "Good Judgment Project."[26] Sponsored by the Intelligence Advanced Research Projects Activity (IARPA, the US government's think tank for intelligence project moonshots), Tetlock invited volunteers to participate in online political forecasting exercises, in which teams of strangers tried to predict the outcome of a political election or a world event. (The project is ongoing—you can sign up to be a forecaster yourself on www.gjopen.com.)

Among the 2,000 participants in the initial group, Tetlock identified a small group of "superforecasters": foxlike generalists who outforecasted the experts.[27] In a tournament that pitted five research teams of generalists against professional intelligence analysts with access to classified data, Tetlock's superforecasters outperformed the intelligence analysts in accuracy at every turn.[28] According to Steven Rieber, program manager at IARPA, "Team Good Judgment . . . beat the control group by more than 50%." What's more, he noted, "This is the largest improvement in judgmental forecasting accuracy observed in the literature."

Despite evidence like that generated from the Good Judgment Project, unfortunately generalists don't make it very high up the promotion ladder in national security agencies. This is because such agencies hire and promote based on credentialed expertise. And therein lies the problem.

Again, there's a good reason to value expertise in the intelligence community. In a refreshingly sober assessment of what it takes to work in the intelligence field, "So You Want to Be an Intelligence Analyst," former analyst Matthew Burton writes, "The primary point of intelligence analysis is not to defeat current threats. Instead, it is to foresee, prepare for, and prevent tomorrow's threats. . . . When the next unforeseen threat arrives, it's very important that the policymakers have immediate access to expertise—expertise attained through years of quiet research."[29] A slew of experts in the intelligence community churn out reports on Ukrainian ethnography, to use Burton's example, *just in case* this topic later proves to be valuable in assessing an emerging threat.

However, as Tetlock's investigations show, we *also* absolutely need generalists: foxlike adaptable thinkers who can scan the horizon and make assessments based on the facts of the day, not a lifetime of research. The tech ethos, which prizes in-the-moment capabilities over long-earned credentials, is thus grossly mismatched with the military ethos, which prizes earned credentials above all else.

While achievements during time in service matter, the primacy placed by the military on needing to simply do time in your job is out of joint not only with the tech ethos, but also with the threats the military faces today, which are unpredictable and increasingly tech-based. Combined, the military's time-in-service requirements and the up-or-out promotion schedule (if you don't get promoted on schedule, you're out of a job) skew the national security workforce toward people who have been in their field forever. As veteran military journalist Leo Shane put it, these policies prevent bringing in people qualified "to tackle specific challenges—issues like cybersecurity—because [those] individuals don't have the requisite time in service to step into senior leadership roles."[30]

ONE MILITARY EXPERIMENT, HOWEVER, TRIED TO INCORPORATE THE tech ethos into a military operation. In 2004, General Stanley McChrystal, then commander of the Joint Special Operations Command (JSOC), was eager to change the United States' strategy in Afghanistan and Iraq. Desperate to turn around two poorly faring wars, the career military man unexpectedly adopted critical elements of the tech ethos into his battle plans. As McChrystal describes in his book *Team of Teams: New Rules of Engagement for a Complex World*, the military in 2004 wasn't set up to win—in the most literal way possible.[31] Its bureaucratic chain of command and siloed sources of information were simply too cumbersome to compete with the networked command structure of al-Qaeda. McChrystal recognized that there was one exception: small teams within the armed forces—such as the Special Operations forces—also managed to be agile, fluid, and networked. But he couldn't figure out, in the parlance of Silicon Valley, how to scale the qualities that made Special Ops so effective to the entire armed forces.

"It was impossible to foresee what elements of our organization would and would not need to know a given piece of information," McChrystal noted. "So instead of trying in vain to control information, we reversed direction and shared it as broadly as possible. . . . Our motto became 'share until it hurts, then share some more.'"[32]

While keeping intact the command structure that has dominated the US fighting forces for centuries, McChrystal instituted three key changes to operationalize this radical form of information-sharing. First, he modified that hierarchical command structure to include "boundary spanners," or people who could "provide critical links between two groups of people that are defined by functional affiliation, physical location, or hierarchical level."[33] Just because the military relied on hierarchy, McChrystal realized, didn't mean that everyone on the same tier was equally qualified for every task. Some people are more social than others; some are connected to lots of people above and below their official rank. These are the "boundary spanners"—the "culture carriers," as McChrystal's former aide-de-camp Chris Fussell termed them in his book *One Mission*. This is similar to Malcolm Gladwell's conceptualization of the people he calls "connectors" in *The Tipping Point*.[34]

Whatever we call them—boundary spanners, culture carriers, or connectors—we intuitively "get" who these people are: people who just seem to know everyone else. McChrystal recognized the value of people who could span boundaries and quickly granted them permission to consult with members of the military above and below their rank. This empowered the boundary spanners to share information throughout the chain of command via informal networks, not unlike how gossip spreads through neighborhoods.[35]

McChrystal then instituted a daily 90-minute video conference call, the "Operations & Intelligence Forum," or O&I, which allowed thousands of soldiers of different ranks around the globe to participate in the briefing. The purpose of the call was, on one level, to go over the previous day's events, establish what was on the docket for the hours ahead, and ensure that everyone understood their respective assignments. But, as Fussell describes it, the O&I was more forum than traditional meeting. McChrystal encouraged participation, setting an example in how he presented his own reports: "McChrystal had to lead from the front," Fussell explained. "First, he himself had to share all news, whether positive or negative. Second, he had to show during the meetings that he was nonjudgmental about the news, whether positive or negative."[36] Doing so set the tone that enabled others to come forth with their potentially less than stellar news or partially fleshed-out ideas.

The O&I simultaneously served two critical purposes: (1) to enable boundary spanners to emerge on any level of any team; and (2) to let McChrystal reiterate daily what he called the "aligning narrative," a reminder that everyone on the call, regardless of rank, was part of one mission, something bigger than any of them as individuals.

McChrystal's O&I shows that the even the military, the country's most rigid hierarchy, can bring elements of the tech ethos into their operations. True, the O&I forum was not exactly an example of "rough consensus and running code," as Clark had described the tech ethos, nor was it an example of rejecting "kings, presidents, and voting." But it was a hybrid model that embraced the emphasis of the tech ethos on collaboration, inclusion, and openness to ideas, all within the firm scaffolding of the military hierarchy that would run the call, deliver the narrative, and, once the call was concluded, issue orders throughout the remainder of the day.

The need for our security forces to "get" the tech ethos grows by the day. While physical threats to our collective security remain, more and more threat actors lob their opening salvos digitally.

On October 25, 2018, approximately 50,000 military and civilian support personnel from 31 NATO member states and partner countries assembled in Brydalen, Norway, a tiny town known primarily for a church constructed in 1884 by the entirety of the town: 20 residents. Brydalen was one of several sites on land and just off the Norwegian shoreline hosting "Trident Juncture," the largest NATO military exercise in northern Europe since the end of the Cold War.

"It is happening in the air, on land, at sea and in cyberspace," NATO's website explained. Quoting Admiral James G. Foggo, commander of Allied Joint Force Command Naples, the site read, "Trident Juncture will show the world that NATO is relevant, unified, and ready to defend itself in this Article 5 scenario, testing our collective defense."[37] With temperatures dipping down to the low 20s overnight, the operation was clearly intended to test not only coordination of land, sea, and cyberspace response, but the physical rigors of cold weather warfare. It is likely no coincidence that the region selected for Trident Juncture was at roughly the same latitudinal coordinates as those between St. Petersburg and Moscow, Russia.

With NATO's shift from Afghanistan and the Middle East to activity in Europe and North America, cold weather warfare is coming into focus as a necessary skill to develop. "NATO has standing forces in Poland and the Baltics, and these are also very cold areas. The Alliance needs winter knowledge," notes Major Knut Hummelvoll, chief for the Norwegian School of Winter Warfare.[38] "Cold knowledge," he reports, is a growing interest among militaries, especially in the United States and the UK. In 2017, the school held 1 course for 30 participants. In 2018, they saw a 1,900 percent increase in enrollment, forcing them to expand from 1 to 10 courses, now accommodating 60 participants apiece.

Despite careful planning and precision execution, just a few days after Trident Juncture launched, officials detected a serious problem: NATO's GPS system had been jammed.[39] Scrambled GPS signals threw off communications among the 50,000 participants and threatened aerial maneuvers for the military aircraft as well as commercial airlines, which transported upwards of a million passengers for the duration of the exercise—from October 25 to November 7.[40]

Norwegian and Finnish defense agencies traced the jamming to Kola Peninsula, site of a heavily fortified Russian base just over 200 miles southeast of the Norway-Russia border. The US denied that the GPS jamming was a problem, telling CNN it had "little to no effect on US military assets" during the NATO exercise. NATO spokesperson Oana Lungescu was slightly more transparent, telling reporters, "Jamming of this sort is dangerous, disruptive, and irresponsible."[41]

This wasn't the first time the Russians had been caught deploying GPS-jamming technologies against the US military. In April 2018, NBC News reported that four American officials had confirmed Russia had been jamming US military drones operating in the airspace over Syria, causing them to malfunction or crash.[42] Todd Humphreys, the director of the Radionavigation Laboratory at the University of Texas at Austin, said the impact of these tactics was significant. "They are a little less hostile looking than a kinetic bullet," he said, "but sometimes the effect can be just as damaging. It's like shooting at them with radio waves instead of bullets."

Here's where net states could be helpful partners to the military: working in concert to develop effective anti-GPS-jamming systems. But net states have yet to step up to the table. Despite all the monetary incentives for nabbing a military contract, most net states simply don't seem to be interested, as Google's Project Maven (described in the introduction) exemplified. The question remains, then, How can the military change tech companies' minds?

THE US GOVERNMENT / NET STATE HEAD-BUTTING DESCRIBED IN THE introduction—both Microsoft's battle with the Department of Justice and Google's retreat from the Pentagon—may be celebrated by net state employees who lean more toward the hacker ethos than the military one. But where does this leave the nation-state, which is responsible for pursuing justice for and physical protection of its citizens? The military's frustration at being left behind by net states was eloquently expressed by First Lieutenant Walker D. Mills, a US Marine Corps infantry officer, who wrote an open letter to Google: "Dear Google, please help your country defend itself."[43]

In the letter, Mills argued that while Google's staffers may feel morally superior by cutting ties with the defense industry, they're not necessarily making the country—or even themselves—safer. "Neutrality, even well-intentioned, is not possible," wrote Mills. "We live in a world where Facebook's platform has been corrupted to influence our elections and terrorists use the secure messaging of WhatsApp and Telegram to coordinate their crimes." He added, "In the pleading words of Merry from the 'Lord of the Rings' trilogy: 'You're part of this world, aren't you?'"

Mills's letter embodies a barely contained freak-out within the American military establishment—why can't we get American tech firms to work with us?—on a very human level. The American defense community has been scrambling to convince tech companies such as Google and Apple to work with them for years now. Having discovered that net states aren't offering themselves up as willing partners, the military has begun reaching out to cash-starved startups directly, with the promise of fast-tracking prototypes with plenty of funding.

For instance, in 2018, the Army opened its newest unit, the Army Futures Command—part of the Army's first major reorganization effort since 1973—in the notoriously liberal city of Austin, Texas, which has a burgeoning startup scene.[44] One of the first actions the Army Futures Command pursued was to set up two cross-functional teams, one of which is the Assured Position, Navigation and Timing Cross-Functional Team (APNTCFT)—or, in lay terms, the deal-with-the-GPS-problem team.[45] William Nelson, director of the APNTCFT, said they're working simultaneously on two tracks: first, developing methods to deploy pseudolite (from "pseudosatellite") transmitters—sort of like ground-based, hyperlocal satellites—that can deliver a high-powered signal that's difficult to interfere with; and second, pursuing "software solutions to prevent jamming." GPS and the technologies that are reliant on it—drones, for example—are priority one for the Futures Command: the very first event they hosted in Austin was called "Hack of the Drones."[46]

The other military branches are also trying to address GPS and drone vulnerabilities. In 2016, the US Naval Academy reintroduced courses on celestial navigation—how to plot a course by the stars—which it had discontinued over a decade ago.[47] And not to be outdone by its counterparts in the Army, in March 2019 the US Air Force announced it would hold its own "Pitch Day," a New York City–based event designed to attract startups and investors. As an Air Force official told me in an email describing the event, "It's a high priority for the Secretary of the USAF [United States Air Force] and she wants it to be known the USAF is 'open for business.'"

Despite these outreach efforts to the cash-hungry startup community, the defense sector is finding a chillier welcome than hoped for. Their messaging probably hasn't helped. For instance, the Army Futures

Command mission statement leads with the goal of being better able to kill people. It states, "The Army's modernization strategy has one focus: make soldiers and units more lethal." Given Google's mutiny over the very hint that its AI technology might be used to harm humans, the Army's choice to lead with lethality language underscores just how much the military fails to understand its tech audience.

To be fair, it's no mystery that this is what the military is designed to do: war isn't like boxing in a ring, with rules and penalties. It *is* about killing people. But if the goal was to engage the tech sector in modernization of the military, plenty of other approaches for partnership would have been more effective. For instance, there's a long history of military-technological innovation they could have pointed to: past partnerships led to the creation of the internet itself (first as ARPANET), GPS, encrypted communication, and scores of other hugely useful inventions.

The military's choice to woo tech companies with words of aggression rather than of collaborative innovation reveals its failure to understand that the tech ethos isn't just a mindset; it serves as a sort of moral compass for many in the tech industry. Sometimes, money alone isn't enough. Certainly, the moral aspect doesn't deter every net state from working with security forces altogether. Amazon, for instance, holds the contract for the CIA's cloud services. Microsoft provides the cloud services for the Immigration and Customs Enforcement agency. But the idea that a code of ethics influences the broader industry at all—and that it's ubiquitous enough that the military, the best-funded military in the world, is caught flat-footed in their quest to get the tech industry to work with them—reveals that many tech devotees want to do more than just code. They want to *build*: to contribute to some idealized, upgraded version of the world they're in. And formal authority—whether it be the military, the president, or the king—is less important in this quest than whether those who create tech generally believe in the worth of what they're creating.

In addition to highlighting past partnerships in innovation, there is yet another route the military could take to working with net states on securing the nation-state. This is to pursue partnerships on the flip side of defense: diplomacy. That would be a wise place to start, especially given that net states are already making historic moves diplomatically that would have been unthinkable just a decade ago. Let's start with the "Digital Geneva Convention."

When Microsoft's Brad Smith announced that the cyberfield was a battlefield back in March of 2017, he also argued for the world's first diplomatic treaty on behalf of the tech industry—one that would help regulate internet-based conduct. The Cybersecurity Tech Accord, often referred to as the Digital Geneva Convention, was officially announced just a year later, on April 17, 2018. Thirty-four of the biggest global tech firms, including Microsoft, Facebook, Cisco, Dell, Oracle, and LinkedIn, pledged their support, though Google didn't join until the following year (and Apple and Amazon have yet to join as of this writing). Within the next six months, they brought an additional 37 tech companies on board, internationalizing the group with the addition of the Japanese manufacturers Hitachi, Nokia, and Panasonic.

I met Brad Smith for an interview at Microsoft's Washington, DC, outpost not many months after the Tech Accord was signed. It happened to be the day after President Donald Trump's infamous speech at the 2018 United Nations General Assembly. As a reminder, in response to Trump's boasting of his administration's accomplishments, the audience erupted into an unprecedented outburst of laughter. Unfazed, Trump continued. "We reject the ideology of globalism," he told the more than 140 presidents, prime ministers, and kings gathered before him. "We embrace the doctrine of patriotism."[48]

The nation-state—America, specifically, ostensibly the most powerful nation on earth—is backing away from multilateralism. By contrast, the tech industry leaders—Microsoft, Facebook, and the other signatories to the Cybersecurity Tech Accord—are embracing it, binding themselves together under a set of principles aimed not at protecting their business, but rather, at the global good: the rights of citizens.

Despite having just flown in from an Orlando, Florida, event with the comedian Trevor Noah, when we met at 2:30 p.m. on a Wednesday afternoon in September, Smith appeared remarkably alert, his eyes beaming with curiosity. With a congenial air that made me feel like we were just two buddies having a chat, Smith walked me through why Microsoft had taken the step to create the digital equivalent of the Geneva Conventions: "It's definitely our view that a world with rules is better than a world without them," he said. "It doesn't mean that there won't be people who violate them, but at least you have rules to point to when the violations take place."

That a collection of the world's most influential tech companies likened the Tech Accord to the Geneva Conventions—a series of diplomatic treaties signed by all 193 countries in the United Nations, pledging to protect civilians during times of war—signifies that this was a big deal. But one particular piece is the real reason this treaty is of moon-landing significance: principle 2, point 2, which states, "We will not help governments launch cyberattacks against innocent citizens and enterprises from anywhere."[49]

Again: the keepers of the digital world told the governments of the physical world, including their *own* governments, that they will not use their platforms, software, or any other tool or service for the purpose of waging cyberattacks. They declared themselves neutral—a "digital Switzerland," as Smith termed it in his 2017 speech.[50]

Significantly, the signatories of the Tech Accord are not distinguishing among *citizens* of any particular *country* in their declaration of protection. Instead, they're pledging to protect any and all tech users—the people who employ their products and services—*regardless of the nation-state in which users physically reside.* In establishing this accord, net states cemented the notion that a new class of citizen had sprung into existence: the previously discussed citizen-user. Though the accord didn't use this specific term, it implied that individuals have rights that exist independent of whatever other rights they may have by virtue of the particular nation in which they live. Perhaps most significantly, in signing this treaty, Microsoft and Facebook and their allies have acknowledged their own power as players alongside countries. As the providers of both the battlefield and the weapons for the next generation of warfare—cyberwar—they have declared that they play a role in influencing diplomacy.

Smith reflected on the irony of this during our meeting. "When we sit down with the other companies that are headquartered in other countries [outside the US], we find ourselves perhaps more the spear-carriers for multilateralism at the moment than governments themselves." He paused. "I don't know that it's healthy. We'd much rather be in a different place, but it is where we are."

Smith explained how Microsoft and Facebook and their tech partners hoped to expand the Tech Accord's impact later that year. "We've been working with a number of different stakeholders and our hope is . . . that governments can sign as well. . . . How governments, in particular, step up to embrace those [principles of the accord] could be very important."

By any measure, they succeeded. In November 2018, the principles underlying the Digital Geneva Convention were folded into an international initiative sponsored by French president Emmanuel Macron. Named the "Paris Call for Trust and Security in Cyberspace," this was the first-ever diplomatic accord between nation-states, net states, private-sector businesses, and not-for-profits. Fifty-seven countries, 70 tech firms, 188 private-sector businesses, and 112 not-for-profit organizations signed the Paris Call. This historic diplomatic agreement not only metaphorically but literally signified a new kind of power-sharing among different groups of global players, with company names affixed beside those of countries, on the same level, as coequals.

DIPLOMACY IS MORE THAN MULTILATERAL TREATIES AT INTERNATIONAL conferences. The day-to-day work of diplomacy resides in issuing official statements.

That may sound like a snoozer of a topic: "issuing official statements."

But it represents a seismic shift in how companies (and similar entities) communicate to government, about government, and with government.

Take, for example, the matter of augmented reality (AR). It's pretty new tech. As such, it makes sense that it's yet to be regulated. But precisely because it's so new, we can use AR as a test case for why regulation is crucial. With AR technology, as with all digital technologies, the big question comes down to what happens with the data the devices collect. If net states already know us through our digital lives—Amazon, through what we browse and buy; Google, through what we look up; Facebook, by whom we connect with—then AR-powered net states will add to our profiles those currently missing elements about our *physical* lives: where we go, what we see, and which people we interact with. To ensure that we don't end up in some dystopian future where personally tailored advertisements pop up in our AR-mediated physical environment (as in the film *Minority Report*), we need to pay attention to how AR develops and, what's more, how it's regulated.

US regulation of net state technologies has been largely absent up to now, as we saw in the previous chapter. All we really see is a growing list of official statements. Amazingly enough, it's not government issuing those statements; it's the net states.

In July 2018, Microsoft released a statement calling on lawmakers to regulate facial recognition technologies. Now, the fact that Microsoft, whose cloud platform, Azure, is used by the Immigration and Customs Enforcement agency (ICE), came out against the practice of using facial recognition tech may have been a PR move. While it may have been fueled by genuine moral outrage, perhaps it was an attempt to defuse any implication of Microsoft's involvement with ICE, which in 2018 was wildly unpopular due to the brief but terrible practice of separately detaining immigrant parents and their children. Whatever the motive, the issuing of statements on this topic by net states was widely adopted: other tech companies that might have been considered remotely associated with ICE's family separations—including Apple, Google, and Facebook—issued statements of condemnation as well.

The significance of net states issuing official condemnations against a government can hardly be overstated. Official condemnations are generally issued *by* governments criticizing activities by *other* governments. For instance, the US State Department issued 93 statements of condemnation from 2017 to 2018 alone, on topics ranging from hostilities in South Sudan, Russia's violations of Georgian sovereignty, and the criminalization of protests in Venezuela, to attacks on security forces in Egypt.[51] Official statements are *governments'* way of talking to each other. They're bureaucratic turf. To see tech companies embracing official statements of condemnation sure makes it seem like they're adopting governmental practices.

Here's why statements matter in diplomacy: they are the first move one country takes against another in a chain of potentially escalating events. By themselves, they may seem toothless to the casual observer: How far does saying "This is bad" or "We call on So-and-So to change" actually go toward fixing a bad situation? one might wonder. But issuing an official statement is like moving the first piece in a game of chess; it declares intent against the other party, who, through this act, becomes the opponent. That statement may stand alone, the game never further engaged in. On the other hand, it may turn out to be the *catalytic* step, the next move being official talks. Official talks may then be the precursor to official actions. This is where things go up a level. Because official actions invariably hurt: they include economic sanctions, military strikes, war.

For tech companies to issue official statements—in this case, to condemn the United States government—was almost unheard of a few years ago. Indeed, Microsoft had issued only three statements against government entities prior to calling on lawmakers to regulate facial recognition technologies: one condemning family separations in 2018, one on the CLOUD Act governing overseas data access, also in 2018, and one in response to the European Union in 2009. But the fact that it and other net states are starting to take up the practice is a sign that it's something we should be paying attention to.

Take, for example, the Trump administration's first travel ban, which barred visitors from seven Muslim-majority countries from entering the United States. Apple, Google, Facebook, Intel, and Expedia (owned by Microsoft) all came out with public statements against the policy.[52] Apple's Tim Cook, Amazon's Jeff Bezos, Google's Sundar Pichai, and others also signed a group letter calling on Congress to institute protections for so-called DREAMers—immigrants born abroad but brought to the United States as children.[53] "As entrepreneurs and business leaders, we are concerned about new developments in immigration policy that threaten the future of young undocumented immigrants brought to America as children," reads the joint statement. "We call on Congress to pass the bipartisan DREAM Act or legislation that provides these young people raised in our country the permanent solution they deserve."

Now, again, these official statements of condemnation could cynically be construed as PR moves: both the travel ban and the family separations were hugely unpopular, making the companies that condemned them look good in the eyes of the public. But the more significant issue is the very use of the official statement of condemnation against a government—not by another government, but by an industry. That practice further underscores the quasi-nation-state status that net states are growing into and is likely a sign of more activity in this area in the future.

AT THE HEIGHT OF ITS REIGN, ISIS CONTROLLED SO MUCH PHYSICAL territory in Syria and Iraq that it rivaled Great Britain in size.[54] ISIS wielded power over a population of approximately 8 million people, and through the banks, oil fields, mines, and other resources on its land it raked in $2 billion in 2014 alone. It was, as one reporter put it, "the most powerful, wealthiest, best-equipped jihadi force ever seen."[55]

Three years later, ISIS's physical footprint was gone. Its nominal capital in Raqqa, Syria, fell in October 2017. From its peak fighting force of 20,000 to 30,000 in 2014, ISIS recruits have largely scattered to the wind, their primary connection with one another being on digital platforms.[56]

ISIS first gained power through those same digital platforms. Just because their physical presence has been defeated doesn't mean we should believe that terrorism itself has been defeated. ISIS's attraction to recruits was never solely about land. ISIS, and other terrorist groups, fed on recruits' fear, targeting the vulnerable, the disaffected, the marginalized; encouraging them to take action on their own—regardless of whether they were on ISIS-controlled land.

And they did. From when ISIS declared its caliphate in June 2014 through February 2018, ISIS or ISIS-inspired terrorists carried out 143 attacks in 29 countries, killing 2,043 people[57]—and 14 of these attacks occurred *after* ISIS had been officially defeated. Indeed, ISIS hasn't disappeared; they've simply relocated. Since the caliphate's fall in October 2017, the US Department of State has identified four new ISIS-linked terrorist organizations: ISIS-Bangladesh, ISIS-Philippines, ISIS–West Africa, and ISIS–Greater Sahara.[58] And ISIS continues to find connections and recruits via net state tools and services—most notably, Twitter and Facebook.

But perhaps what's more concerning for the average American isn't the threat of terrorists who are officially defined as such, but rather the so-called lone wolf attackers. There's no one profile that fits all lone wolf attackers, but they do tend to be motivated by personal grievances rather than political goals.[59] And they're on the rise, especially in the US: in the six-month period between January and July 2018, there were 205 mass shootings in the United States that together killed 214 people and injured 832, all perpetrated by lone shooters.[60]

Research suggests that social media may be contributing to the rise in mass shootings. According to a study conducted by Harvard University researchers, the rate of mass shootings in the US has tripled since 2011.[61] While no single reason was identified in that study, other researchers have noted what's called the "contagion effect": in an average of 13 days following a mass shooting, a subsequent mass shooting is more likely to occur—indeed, between 20 and 30 percent of mass shootings appear to stem from the contagion effect.[62] Wall-to-wall media exposure inspires would-be copycats. Studies further suggest that social media exacerbates this effect: in 2000, pre–social media, there were about three mass shootings per year. By 2017, this had escalated to 346 per year, or an average of one mass shooting every 12.5 days.[63] And it's only gotten worse since then: according to statistics from the Gun Violence Archive, between January and August 2019, there were 261 mass shootings, an average of one mass shooting every single day.[64]

As social media gained widespread popularity—2011 being widely regarded as the tipping point for that—would-be mass shooters both heard more about other shootings and were more able to envision their own rising fame should they take similar action, thanks to social media. While various motivations drive mass shootings, Jennifer B. Johnston, a researcher at Western New Mexico University, noted, "Unfortunately, we find that a cross-cutting trait among many profiles of mass shooters is desire for fame." The quest for fame, she said, is "in correspondence to the emergence of widespread 24-hour news coverage on cable news programs, and the rise of the internet during the same period."[65]

Social media outlets don't just spread the idea of mass shootings; many shooters post about their intentions *on* social media before acting. Following the Parkland, Florida, shooting in 2018, 600 copycats posted similar threats across social media targeting other schools.[66] The problem is, law enforcement doesn't have the capacity to investigate every single threat they encounter. According to one report, the FBI received 50,000 tips in the month of January alone.[67] Even if, as the report suggested, they get only about 100 "actionable" tips per day, that's still a staggering amount of potential evidence to wade through.

This is where "defeating the idea"—again, Nagata's words—becomes so important. Chasing threat-makers is a Whac-A-Mole exercise, making it virtually impossible to fully explore every potential threat.

Here's where net states could help: they have more than just data; they have user profiles. With these profiles, law enforcement could potentially narrow down who among the threat-makers is likely to take action.

This is not to suggest that net states hand over user data to law enforcement. Rather, net states have the opportunity to put their user data to use for public safety and security—an opportunity that they don't seem to be taking advantage of at this point. While Facebook has a significant counterterrorism (CT) team—with 200 staffers at last count—it is unclear whether these same individuals also seek out what could be called domestic terrorists: mass shooters. What is even less clear is what Facebook's CT team would do with the information if they *did* find potential bad actors. It's unclear what kind of relationship, if any, they have with the FBI, the Department of Defense, or local law enforcement. As citizens whose tax dollars fund law enforcement and national security agencies, and as citizen-users whose activities online make it possible for platforms like Google and Facebook to exist, we should expect transparency about the relationships between our net states and our nation-states.

Living as we do in the physical world, we must seek transparency from any and all entities that could do more to help ensure our physical safety from violence: whether that violence be from foreign or domestic terrorists, mass shooters, lone wolves, or however else malevolent individuals might be described. Net states might not be able to "defeat the idea" that inspires bad actors alone. But they are certainly better equipped to identify both the ideas that inspire and the actors inspired by those ideas than are governments working on their own on the outside.

Keeping a country safe from violent offenders in the information age means understanding that the goalposts are constantly moving and adapting in real time as the actors and targets change. Governments need net states; and net states—as keepers of the ecosystem in which terrorism plots are hatched, planned, and reported on—need to consider sharing their insights with governments. They should do so safely, securely, to be sure—but in some kind of partnership. The health of the internet ecosystem and all its inhabitants would be better for it.

FIVE

A GREAT WALL OF WATCHERS

Via social media profiles, we already have many opportunities to curate idealized versions of ourselves. This chapter explores visions of a possible future in which nation-states mandate curation of the ideal self via a system of punishments and rewards—making moral judgments about the appropriateness of our information—as viewed through China's experimental Social Credit Score system. It examines our own messaging practices in the West, starting with our earliest profile-building practices on Facebook, asking what our networks of connection and our information-sharing practices say about our needs as both individuals and societies.

We may not admit it, but most of us likely get a tiny thrill when we notice that someone has "liked" a comment of ours on Facebook, retweeted a Twitter post, or "hearted" an Instagram photo. These small acts of validation can go a long way toward not only showing our friends and acquaintances what we approve of, but—in China, at least—showing the government who the "good citizens" are. In what sounds suspiciously like an episode out of the British science fiction television series *Black Mirror*, China is in the process of unrolling what's colloquially called the "Social Credit Score" (SCS).

The premise is simple: the more likes, retweets, and hearts you get on your social media profile from "good citizens," the more rewards you will gain in the form of desirable services—everything from expedited travel permits to bank loans. Unveiled in 2014, the Chinese government's proposal, titled "Planning Outline for the Construction of a Social Credit System," called for the Social Credit Score to be fully implemented nationwide by 2020.

It's not hard to imagine how it would work. While in the West we are relatively free from governmental oversight of our social media activities, net states routinely collect data on who likes what content on our profiles (how many likes or retweets or hearts we get), what we spend our money on (through our purchase history), our whereabouts (using cell tower data, regardless of whether we "check in" or even whether we opt out—as recently admitted by Android),[1] our health (for Fitbit or other wellness app users), and so on.

While net states' goals for this monitoring include, among other things, to better target users with ads, China's goal with the SCS is explained by the Chinese government as serving loftier ends. According to official documents, the SCS—already in use in parts of China—will enhance "trust" nationwide and build a culture of "sincerity." The plan goes on to say that the SCS will "forge a public opinion environment where keeping trust is"—and this is the government's word, not mine—"glorious." Finally, the SCS will purportedly "strengthen sincerity in government affairs, commercial sincerity, social sincerity and the construction of judicial credibility." By punishing citizens who get a low score and rewarding those who score high, the SCS will make everyone *want* to be better citizens. Thus it will be easier to trust everyone, as people will simply *be* better citizens—never mind that they've been *forced* to become better citizens.

Or are they? Are Chinese citizens *compelled* to use social media? Are any of us?

There's a saying in Chinese: Rén pà chūmíng zhū pà zhuàng (人怕出名猪怕壮), which means, "People fear fame like the pig fears being fattened for the slaughter." In other words, greater visibility comes at a cost; fame has its price.

Nowhere is this more apparent than in China's implementation of the Social Credit Score. As a person's online profile grows in visibility, so too does the data that feeds his or her Social Credit Score: a ranking of how "good" a citizen that person is.

The SCS works much like a financial credit score. Having no credit history is almost as problematic as having a bad credit history, as banks have no information by which to assess your reliability as a debtor. With the SCS, rather than assessing your financial reliability, they're assessing your reputational reliability.

In the United States, if you have a "good" financial credit score, you're more likely to be approved for a loan by a bank. In China, if you have a "good" SCS, you're more likely to be approved for a myriad of services; yes, economic services like loans, but much more than that, as we'll see.

So what exactly are the actions that will improve or destroy a person's SCS? Given that the SCS is being implemented among the largest online population on Earth, it's surprisingly difficult to obtain a precise list of offenses that will damage one's SCS or smiled-upon activities that might give it a boost. When the initiative was announced in 2014, the Chinese Communist Party (CCP) cited four categories for evaluation: "commercial sincerity," "social security," "trust breaking" and "judicial credibility."[2] But what these categories translate to in action is much harder to discern. As noted on the previous page, social media activity is crucial (both what one posts and how others respond). Beyond that, though, all we have to go on is anecdotal evidence. And from what that tells us, the SCS reaches into every nook and cranny of a person's digital existence.

We have a few recent examples of what'll get a person's SCS dinged: In March 2018, 9 million people with "bad" credit scores were banned from purchasing plane tickets. An additional 3 million travelers were denied business-class train tickets, despite the fact that they had the money for them.[3] Their offenses? In some cases, there had been pretty clear-cut infractions, such as attempting to board the train without a ticket or smoking in a nonsmoking area. But deeds that appeared to be borderline benign or, at the very least, difficult to clearly define—loitering, for example—also contributed to a bad SCS. In one case, a Chinese lawyer was barred from buying a plane ticket because a court-ordered apology—which he had in fact turned in, in writing—was deemed "insincere."[4]

A person's spending habits are subject to rating by the SCS as well. Pay bills late? Lose points. "Waste" money on "frivolous" purchases? Lose points.

How wisely a person spends time is also subject to evaluation. Spend too much time playing video games? Lose points. Spread "rumors" on social media? Definitely lose points.

But does all this really matter? What impact might the SCS have on a person's day-to-day life? Well, a low score might mean that your kids could be denied access to certain schools, regardless of their grades or other qualifications, or, as noted, you might be banned from travel.[5] If you have a bad score long enough, you might be put on a government blacklist, which could keep you from getting a job.[6]

Conversely, with a high score you might get enhanced book-borrowing privileges at the local library and, on a rainy day, a loaner umbrella.[7] You might improve your chances of securing a mortgage, enhancing your likelihood of financial stability, which in turn could make you a more desirable mate. This is reflected in the practice of those with a good SCS being featured more prominently on China's largest dating site, Baihe. "A person's appearance is very important," said Baihe's vice-president, Zhuan Yirong, to the BBC. "But it's more important to be able make a living. Your partner's fortune guarantees a comfortable life."[8] You also get to enjoy the incalculable gains of knowing that *other* people know you are considered "good" in the eyes of the government.

Clearly, there are several problems here, the overwhelmingly obvious one being this: What criteria are being used to decide what's "frivolous"? What's a "rumor"? What makes an apology (a court-ordered apology, remember) "insincere"? There doesn't appear to be a nationwide set of rules or standards for how the SCS is to be implemented. In other words, some officials within a certain branch of the Chinese government decide what's frivolous or a rumor or what makes an apology legit, but nowhere does anyone share these decision-making criteria with citizens, nor are the criteria uniform across the country. You just have to *hope* that your actions don't irk the SCS monitors, learn from your neighbors, and be "frequently posting but careful about content" in your social media activity.

In some places in China, the incentive is especially high: citizens with a good SCS get a special shout-out from certain local governments. In one of the three dozen cities where the SCS system was first piloted en masse, a midsize city called Rongcheng, photographs of citizens with a good SCS were displayed on public bulletin boards in the town square under a banner that read: "Civilized Families." (Check *Foreign Policy* magazine's excellent feature on the SCS for pics of this.)[9] And in Rongcheng, the oversight authority for the SCS has devised a mathematical approach for managing the SCS of its 740,000 adult citizens, potentially lessening uncertainty. According to interviews with officials conducted by reporter Simina Mistreanu, each Rongcheng adult citizen starts with 1,000 points. You gain or lose points depending on your behavior. Traffic ticket? Minus 5 points. Do a "heroic deed," help your family "in unusually tough circumstances," or conduct "exemplary business"? Plus 30 points. Win an award at work? Another plus 5. Whether this rubric is shared publicly with the citizens of Rongcheng is unclear, nor is it apparent whether other cities are using the same criteria. But at least there's a discernible pattern here: Do good stuff and you will be rewarded. Do bad stuff and you will be caught—and because your every online action is tracked, chances are you *will* be caught . . . and penalized.

Given what seem to be rather draconian oversight measures, it's surprising to learn that the SCS does actually seem to be increasing trust among citizens, just as designed.

"I feel like in the past six months, people's behavior has gotten better and better," reported a resident of Rongcheng to *Foreign Policy*. "For example, when we drive, now we always stop in front of crosswalks. If you don't stop, you will lose your points."

As counterintuitive as this may seem to Westerners, whether it's out of fear of losing points or the desire to gain points, the SCS seems to indeed be welcomed by at least some Chinese citizens. According to an interview with Rachel Botsman, who recently wrote a book on the SCS, "When I interviewed people from China on this, they don't necessarily see this as a bad thing."[10]

To understand why this might be, it's helpful to know a little about the history of China. In the 1950s a national development plan was launched, led by Mao Zedong, called the "Great Leap Forward." It brought massive upheaval to the country and caused a famine that killed an estimated 15 to 30 million people.[11] This was followed by the "Cultural Revolution" in the 1960s and '70s, when Chairman Mao led the country through a violent social scrub, vilifying members of "Five Black Categories"—namely, rich farmers, landlords, "counterrevolutionaries" who didn't support the communist wave, "rightists," as well as a catch-all class called "bad elements" that included those with high levels of education, such as teachers and "intellectuals."

This era was defined by rampant mistrust: families were torn apart as students—who led the revolution through the youth army, the Red Guards—turned in parents and grandparents for "reeducation" at forced labor camps in the countryside. The Red Army destroyed thousands of temples and antiquities in the name of ridding the country of the trappings of the unjust hoarding of wealth. The country remained in chaos for almost 10 years. Schools and universities were completely shut down from 1966 until 1972. No one studied. No one went about business as usual. Everyone in the country just labored—children and parents alike.

While the Chinese Communist Party has come a long way since the

Cultural Revolution, China's history of surveillance, morality-based punishment, and widespread fear of being "turned in" by even one's own family casts a shadow over at least the elder population to this day. As Botsman reported, "[People] say in their grandparents' generation, [people] knew that the Communist Party had a file on them, but they had no idea what was in that file. This is actually the same system. Digitized, but it's more transparent."

CHINA HAS JUSTIFIED ITS SOCIAL CREDIT SCORE SYSTEM BY POINTING out a very real need: to know who you're doing business with. Brand reputation works that way. For instance, you may decide to spend the extra money on a backpack from L.L.Bean rather than buying some knock-off brand because you know it will likely hold up for a while; the company has a reputation for quality products. In China, as described earlier, mistrust has been a societal issue for decades. And in the absence of a strong legal system, it's difficult to hold people with shady business practices to account. With the SCS system, in theory everyone knows how reliable everyone they're engaging in business with is, and therefore consumers have the means to protect themselves against less than honorable purveyors of goods and services.

Such a system is easier to enforce in China, with its population engaging in the most mobile transactions in the world, than it would be in many other countries. In a ten-month period of 2017, mobile payments amounted to $12.8 trillion in China, outstripping the United States' $42 billion in mobile transactions by "lightyears."[12] While this was largely attributable to city dwellers, the diffusion of smartphones among the massive rural population contributed as well; rural folks are leapfrogging straight from cash to mobile payments, without ever having had a checkbook or credit card. The cash-based population with no bank account—estimated by the World Bank to include around 21 million people in China—went from being under-the-mattress savers to players in the digital economy almost overnight.[13]

The problem is, studies show that newcomers to cashless transactions tend to underestimate the damage late payments can do to one's credit. Thus they're more likely to make late payments.[14] Given this, it seems that the 21 million financially inexperienced Chinese now using mobile payments are extremely vulnerable to accidentally damaging their SCS, simply by virtue of being new to the very notion of credit.

Praise of the SCS system by ordinary Chinese for how it "establishes the idea of a sincerity culture," as the government planning documents put it, is almost impossible to find, with the Western media understandably focused on the Orwellian overtones. But, as noted above, it's important to recall the rampant mistrust that permeated Chinese society for decades. Furthermore, prior to the SCS system, the average Chinese citizen had few means of redress to deal with fraud, corruption, or other bad business practices.[15]

The SCS system may be terrifying to Westerners, and certainly a sound constitution and legal system for the Chinese people would be more just than the less-than-transparent SCS. But to the average Chinese person used to dealing with hucksters and swindlers—*pianzi*, in Chinese—the SCS may make life slightly more manageable—so long as they stay on the government's good side, of course.[16] As China scholar Jonathan Margolis wrote in a *Financial Times* op-ed, "I am almost embarrassed to say, I get it. Bearing in mind China's violent history, I understand its preoccupation with order and harmony."[17] Society would indeed be "glorious" if we could all trust and be trusted.

FACE-TO-FACE, WE RELY ON A HOST OF VARIABLES TO DETERMINE whether or not to trust someone: their smile; how well they maintain eye contact; their body language; and, of course, their words. Our social networking platforms have far fewer data points to go on. Facebook, for instance, can't (yet) gauge whether someone is exhibiting suspicious cues such as shifty eye contact or excessive fidgeting. Perhaps it's not surprising to find that they've identified other ways to assess whether we might want to "trust" someone—in other words, to make a person part of our social network. One such way is to determine whether we're already in contact with that person *outside* of Facebook.

And such a determination *is* made.

"Creepy. Phoned a guy on mobile after he gave me his number in the weekend at the retro caravan show (first time I have met him). Today @Facebook serves him up as People you may know. We have no mutual friends so Facebook is looking at my phone call history. Intrusive?"[18]

This tweet reflects one of the many practices that Facebook has taken heat for in the past few years: the "People You May Know" feature. Just who gets served up reveals how extensive and detailed Facebook's profiles are on its users. First off, your profile includes anyone stored in your phone's contacts list.[19] In addition, Facebook also stores your connections to "shadow profiles"[20]—detailed accounts about people you interact with who've never signed up for Facebook.[21]

The public reaction to learning about these practices was, predictably, outrage. This was followed by a small wave of folks fleeing to Instagram (which, perhaps they didn't know, Facebook also owns).[22] Yet despite the public pie-in-the-face that Facebook has taken in the last few years for these practices, along with scandals like Cambridge Analytica, the platform remains extremely popular, especially outside the US.

Almost 2.2 billion people use Facebook across the globe. US engagement rates on Facebook have declined, but worldwide adoption climbs steadily each year.[23] Even among those who are less engaged with Facebook now than in the past, the site remains a staple of the digital experience. It's so popular that in many countries, Facebook is almost synonymous with simply going online. "It seemed that in their minds, the Internet did not exist; only Facebook," recalls one researcher who studied internet use in Indonesia.[24] Indeed, 11 percent of the 94 million Indonesians using Facebook reported that they did not use the internet but *did* use Facebook. In their minds, Facebook *is* the internet.

To understand why Facebook remains so popular despite its flaws, it's worth remembering what Facebook gave us—and continues to give us. Facebook gave us the first major digital space—the first space anywhere, for that matter—where absolutely anyone could create an idealized version of themselves. Once satisfied, they could then share that with a handpicked collection of other people.

Prior to Facebook, only the very famous could curate their public persona, via expensive public relations firms; and even they couldn't control bad news about themselves. Facebook gave rise to the curated self: a place where you could amass and flaunt the most complimentary photos, issue the most modestly worded self-congratulatory posts, share the most aren't-I-smart-and-socially-conscious news stories. And you got to pick who would see all of this—in other words, create your very own collection of fans. Any random thought, quip, joke, observation about your day, the world, the nature of existence was fair game. What's more, people would reward you for sharing these tidbits—with their likes and comments, and now, with a whole range of emoticons designed to validate your musings.

Even given this somewhat cynical take on what Facebook does to one personally, it's impossible to discount the incredible influence it's had (and has) on the world socially. True, grassroots social protests have always spread from person to person—from individual friends to extended family members to neighborhoods to communities. But with Facebook's network effects—where the more people who are part of a network, the more powerful it becomes—movements have the capacity to hit the public consciousness with the force and speed of a hurricane.[25]

Social movements, energized on Facebook, become simply inevitable. You may not join the movement yourself, but you will see something about it; know of its existence. Think #MeToo: more than 12 million people posted with this hashtag on Facebook within the movement's first 24 hours.[26] Think #BlackLivesMatter: over 3 million people watched the shooting of Philando Castile on Facebook Live the day it happened.[27] Think #ArabSpring: the five days of protests in Cairo that toppled the almost-30-year reign of President Hosni Mubarak were organized on a Facebook page.[28] Even if you don't know what these hashtags represent, chances are you've at least come across them. And even greater are the chances that you did so on Facebook.[29] I teach a class called "Technology, National Security, and the Citizen" at Columbia University, and I asked my students in February 2019 how many of them were on Facebook. Everyone raised a hand, if somewhat sheepishly. As one student commented, "Even if we *wanted* to stage a mass exodus from Facebook, where would we organize such a thing? Yep—on Facebook."

Whether you're on it or not, for good or for ill, Facebook helped define the past decade of the digital age. While Facebook the net state may someday perish, its innovations have already influenced the next generation of programmers and engineers. Politicians too. And countries with one-party political systems, like China. There would be no Social Credit Score without Facebook, because there was no such thing as a digital social network for the masses *before* Facebook. Though by 2019 many Americans have lost their faith in Facebook, it's impossible to deny the influence that Facebook has had over the past fifteen years, shaping a generation of social interactions, political movements, and personal acts of identity-curation.

JUST BECAUSE AN AMERICAN TECH FIRM CREATED THE SOCIAL NETworking phenomenon doesn't mean that it could give rise to an American version of the Social Credit Score system, right? We do have the US Constitution, the bedrock for law and order in the United States, which seems to stand in the way of a China-like Social Credit Score taking hold here. But aspects of the SCS system and its related technologies have gotten the attention of Western governments and net state leaders alike in ways that we should pay careful attention to.

Let's start with facial recognition. For the SCS, associating a person's face with a name is in many ways just another data point: Can we access person X's buying habits? Check. Do we know people's social media handles so that we can track their behavior online? Check. Can we identify who's jaywalking and instantly send them a fine via text message?

That's a check, too. In Shenzhen, an industrialized city bordering Hong Kong, China rolled out a facial recognition program to do just that back in March of 2018.[30] With 170 million surveillance cameras already in place countrywide and plans to up that total to 570 million by 2020, the CCP will soon be able to devote one camera for every two citizens to track when they board and leave the subway, verify student identity for school tests, even dispense toilet paper to ensure that no individual takes more than their allotted two sheets.[31]

But again—that's just in China, right, where citizens' rights are somewhat nebulous, right? If only we could claim this with confidence. Other countries and net states are paying close attention to how this mass surveillance pans out in China, with an eye toward importing it for use closer to home.

Germany, for instance, is purported to have modeled its facial recognition pilot program at Berlin's Südkreuz train station after the Chinese system—and it was recently extended for further data collection, despite protests by German citizens over privacy concerns.[32] China outright exported its facial ID technology to Zimbabwe, although CloudWalk, the company that manages the facial recognition technology, insists that it was sort of an add-on: "The Zimbabwean government did not come to Guangzhou purely for AI or facial ID technology; rather it had a comprehensive package plan for such areas as infrastructure, technology and biology," according to CloudWalk CEO Yao Zhiqiang.[33]

American net states are keeping an eye on developments in China as well. For the first time ever, China's World Internet Conference—formerly described in the *New York Times* as a rather dinky affair attracting "local tech execs and leaders of impoverished states"—in 2017 managed to secure the attendance of some of the major net state players, including Apple CEO Tim Cook and Google CEO Sundar Pichai.[34]

The focus of the conference? China's surveillance state.

At the conference, Face++, one of the fastest-growing Chinese facial recognition companies, displayed a demo in which they surveilled the conference attendees, then broadcast from their booth how they could identify individual attendees' gender, hair length, color, and even clothing. Another exhibit showed off a map of Beijing that pinpointed every single foreigner in the city, using surveillance footage and GPS data from their cell phones.

The American tech industry isn't stopping at mere observation of China's surveillance state: it's betting on its success with investments. A report by Reuters identified three primary companies in China's mass facial recognition surveillance triumvirate: SenseTime, DeepGlint, and Face++.[35] All three companies, according to that report, are recipients of significant funding from Silicon Valley venture capital firms. Sequoia Capital, one of the largest venture capital funds in the Valley, has made 206 investments in China since 2000, including an $18 million investment in DeepGlint in June of 2018. Another major Silicon Valley venture capital firm, Silver Lake, was one of eight lead investors in SenseTime.[36] And the artificial intelligence venture capital fund Comet Labs' very first investment back in 2013 was in Megvii—better known as Face++, China's biggest facial recognition software firm, which is now valued at roughly $1 billion.[37]

It matters that the American tech industry is aiding China's surveillance program, because by any account the program is anything but benign. It's being used to target specific populations—not because of criminal activity, but because of suspicion that these populations *might* someday engage in such activity. The most extreme example of facial recognition targeting is in China's westernmost region, Xinjiang—one of the so-called autonomous regions. The ethnic minority native to Xinjiang, the Uighurs, is undergoing what might be the world's most intensive surveillance-state experiment to date.

Using some Uighurs' involvement in past protests as an excuse, China was able to justify a comprehensive tracking system for the minority group as a national security measure. But regardless of the justification, the measures are extreme: according to a grim op-ed by historian James Millward,[38] every 100 feet Uighurs pass police checkpoints at which police scan their ID cards, irises, and cell phones. Using sophisticated facial recognition software—which works well to distinguish Uighurs, who have different physical features than Han Chinese—the government has built a vast database of the roughly 11 million Uighurs, a feat that would have been impossible before the advent of AI surveillance.[39] Even routine purchases by Uighurs that might be considered threatening, such as buying a kitchen knife, are logged in databases. In the city of Aksu, shop owners are now required to etch the buyer's name and ID number into the blade of any knife a Uighur buys with a QR code tied to that buyer's name, a policy "intended to trace a knife back to its owner in the event it's used to commit acts of violence," as reported by Radio Free Asia.[40]

While no state in the US has yet to roll out a comprehensive suite of tracking and facial recognition tech such as that in Xinjiang, there aren't any laws on the books prohibiting it. As noted in the previous chapter, Microsoft recently issued a statement calling on Congress to institute some sort of regulation for facial recognition technology.[41] "Advanced technology no longer stands apart from society; it is becoming deeply infused in our personal and professional lives," wrote Microsoft president Brad Smith. "[The] potential applications are . . . sobering. Imagine a government tracking everywhere you walked over the past month without your permission or knowledge. Imagine a database of everyone who attended a political rally that constitutes the very essence of free speech. Imagine the stores of a shopping mall using facial recognition to share information with each other about each shelf that you browse and product you buy, without asking you first."

Tech companies aren't known for asking permission before they develop a life-altering product—that would be almost like a teenager asking his parents for *more* rules. Their call for regulation should give us all pause; it's an acknowledgment that the makers of this technology recognize how possible it would be to use it to restrict user liberties.

At the same time, however, many elements of a system to assess social credit are already in place in the West—largely because of efforts by tech companies and emerging net states. In the United States, almost every adult has a financial credit score, for example. There are other scores out there too—scores assessing intangibles such as reliability and friendliness: anyone who ride-shares with Uber or Lyft, for instance, gets rated by their driver (and vice versa). In addition, everyone with a social media presence is *connected to* people with either good or bad credit scores and good or bad Uber or Lyft ratings—connections that would count in China. But these numbers aren't readily available to the public except on a site-by-site basis.

Which is ironic. In the West, private corporations like data brokers can see our *compiled profile* of rankings. But we, as individuals, can't. In China, on the other hand, the government itself compiles the profiles, and while citizens don't necessarily know what will improve or degrade their ranking, at least they can *see* their own profile.

If you took the time, however, you could make a little compilation on yourself. Get your credit score from each credit agency (which you can do for free) and collect your ratings from all your sharing-economy apps—home-sharing services like Airbnb, ride-sharing services like Uber and Lyft. Then assess your public digital footprint by Googling yourself. This wouldn't be just a vanity project—you'd be assessing how others size you up. For instance, if the search turns up a flattering Facebook profile picture, your chances of getting a job interview increase by as much as 40 percent.[42] But don't stop there. Your Fitbit data may reveal how healthy you are; your Amazon purchase history, how lavish you are. Your Google search history may uncover a pattern of queries that are impressive—queries that make you seem socially conscious and inquisitive, for example—or less than flattering—like Google-stalking a new acquaintance or asking about things you're embarrassed that you didn't already know. Indeed, with enough patience and perseverance, you too could create your own version of a social credit score. But even then, you'd still "know" less about yourself than one of the country's five major data brokers, which typically have about 3,000 "data segments" on every American.[43]

There's another saying in Chinese: Er tīng wéi xǖ, yǎn jiàn wéi shí (耳听为虚, 眼见为实). It means, in effect, "Seeing is believing." We apply that saying every day: we trust people we've had a chance to "see," to get to know—or at least we did until the internet made us question what we see. Our ability to craft the perfect version of ourselves with our online profile makes it difficult to know if we can trust the *other* digital personas we encounter. And in an age where we have increasingly frequent interactions with people we've not met in person—work colleagues we communicate with only via email, social media "friends" who don't even live in our country—any proxy for trust could be a big help in discerning who's reliable and who's a sham.

But it's unlikely we'll find a replacement for our own personal assessment. We've yet to find a substitute for good old-fashioned instinct. Our "gut" can help us sniff out a criminal,[44] pick the right romantic partner,[45] or prevent us from making mistakes.[46] We barely understand how the human brain actually works, so we can't replicate it. "We know about connections, but we don't know how information is processed," reflected Stanford University neurobiologist Lu Chen at a lecture on the challenges of neuroscience. "We are still in need of an understanding of the fundaments," echoed another neuroscientist in attendance, Nobel laureate Tom Südhof.[47]

Until we've got the brain figured out, it's unlikely we'll be able to recreate a copy with the fidelity to replace our deeply felt instincts about whether or not to trust someone. What we *can* rely on net states for are the troves of data that can aid in either confirming or denying whether our instincts are spot on. But so long as we're pheromone-emitting, subconscious information–processing creatures, tech (hard as it may try) can't yet replace face-to-face interactions when it comes to establishing trust. China may be creating a centralized database of "trust breakers," but coworkers, neighbors, friends, and family will—one hopes—still establish their own relationships, regardless of what their government or net state rankings say.

FOR ALL OUR CONCERNS ABOUT PRIVACY, WE SEEM TO PLACE A LOT OF trust in our technology, seemingly without a second thought. Take search engines, for instance. It probably doesn't occur to most of us to think of typing in search terms as an expression of trust. And in some ways, it's not—we don't need to have the usual hallmarks of trust (feelings of respect and admiration, say) for a search engine to proceed, after all. It's a tool; we use it, it gives us answers.

But we trust in the search engine's ability to give us answers—indeed, *right* answers. It's implied in the very fact that we use the tool. In exchange for our trust, search engines do more than give us answers. They give us power. Not just the *feeling* of power, either; actual power—power we could never possess without them. Perhaps this is why we're so willing to trust our personal data, including the most intimate questions that cross our minds, to search engines—to Google, specifically. Because the power we get in return is worth it.

Consider what goes into a search. Each and every query we type into Google Search travels an average of 1,500 miles to a data center somewhere in the world and engages over 1,000 computers, generating a result in 0.2 seconds.[48] This is possible only because Google Search is constantly crawling over 100 billion web pages, then storing and organizing the information found. "It's like the index in the back of a book—with an entry for every word seen on every web page we index," reads Google's blog post on "How Search Works."[49] It's a staggering amount of data, processed extraordinarily fast. And Google Search feeds us remarkably accurate results even if we use poorly worded search terms that the search engine has never encountered before. (Each day, 15 percent of all queries have never been asked of Google before.)

And we ask our questions in increasingly personal spaces: our phones. Ninety-six percent of cell phone owners in the US used their phones for searches in 2019 (with Google handling almost all of those searches).[50] But while Google may help us satiate our curiosity, it's largely thanks to Apple's invention of what we now think of as the smartphone that we can do so at any time and anywhere we happen to want to know something.

This brings us to the question, So what? We find stuff out, fast. And the data we create in the process of finding stuff out is available for us to review, if we want. If we go to MyActivity.Google.com, we can see every search we've ever entered, on any device, since we created an account however many years ago. We can even download the entire archive so that we can peruse our curiosity profile in our spare time. But if we bothered, what would we find out about ourselves that we don't already know, somewhere in the deep recesses of our minds?

In the book *Everybody Lies: Big Data, New Data, and What the Internet Can Tell Us About Who We Really Are*, data scientist Seth Stephens-Davidowitz traces the fascinating history of how we tend to fool ourselves, especially when reporting on our behavior to others. Surveys—one of the most common social measurement tools in use—are particularly problematic. People tend to overestimate their positive attributes and downplay the negative when reporting on their own actions (even when doing so anonymously), an effect called the "social desirability bias." Studies suggest that people self-report eating less, being more physically active,[51] voting at higher rates,[52] drinking less alcohol,[53] being more (for men) or less (for women) sexually active,[54] and giving to charity and volunteering,[55] to name just a few.

And we don't just pretend to be better people to other people; we fool *ourselves* into thinking we're generally better and, specifically, smarter than we actually are. In what's known as the "Dunning-Kruger effect," everyone from students to long-serving professionals overestimates their intelligence and abilities. In a *Forbes* profile on this effect, the author cites studies showing that 32 to 42 percent of software engineers rated themselves as being in the top 5 percent of performers in their companies.[56] A nationwide poll conducted by CNBC revealed that 20 percent of Americans thought it was "fairly likely" they'd be millionaires within the next decade[57] (while only 10 percent of all Americans actually *are* millionaires). Take the fact that we think we're smarter than we are in general and combine it with the fact that search engines make us feel smarter still, and we've got a fairly unrealistic—and inflated—sense of self on our hands.

What makes us so bad at estimating our own abilities and predicting a realistic future? The answer lies, in part, in confidence—or rather, overconfidence. It's not something we can help; overconfidence in our abilities is an innate bias, built in and hardwired. It makes sense on some intuitive level: if we believe in ourselves, we're more likely to try difficult things. Imagine the opposite bias: underconfidence. We'd never have tried anything hard and would likely never have made it very far as a species. Indeed, researchers believe that there's an evolutionary advantage to overconfidence: it rewards people for stretching themselves to the limits of their abilities. Perhaps they won't achieve their original goal, but they'll certainly be better off than if they hadn't bothered trying. Overconfidence is also linked to increased morale, to persistence, to the ability to project credibility, and to creating self-fulfilling prophecies.[58]

The problem is, overconfidence also leads us to wildly underestimate how difficult tasks are and how powerful our foes or challenges are. Overconfidence bias has been linked with historically cataclysmic events ranging from World War I, the Vietnam War, the first Gulf War, the 2008 financial crash, and the federal government's underpreparedness for natural disasters such as Hurricane Katrina.[59] Ironically, the more expertise we have in a given area, the more susceptible we are to the overconfidence bias. Cultural norms also come into play. It's probably not surprising to learn that men exhibit overconfidence more often than women, who are culturally encouraged from girlhood on to be accommodating rather than assertive with their opinions.

This matters because if we aren't honest with ourselves about our own abilities, we pursue goals that are simply not realistic. This both sets us up for failure and causes us to waste time that could be better spent on endeavors more closely matched with our actual abilities.

Getting around this is pretty tricky. Because to know what competence looks like—how good you have to be to achieve a goal—tends to require actually *being* competent already. As a result, being unable to do something at a level of excellence means that you can't really imagine what it would take to *be* excellent at it. Think about something basic, like riding a bike. To look at the process from the outside, it's straightforward. It looks like you just get on the bike, pedal, and you're off! But as any of us who've learned to ride a bike might recall, it's nowhere near that simple. There's the issue of how to balance before you pick up speed, how to pick up speed from a standstill, how to steer while also picking up speed, and how to do all these things in concert without crashing into anything. Like most easy-looking things, bike-riding involves multiple steps to achieve even basic competency, let alone mastery. In other words, we tend to look at things that *look* easy as being, well, easy to do. And this is another reason why we overestimate our ability to be good at things. Until we *experience* mastery, we simply can't fathom how hard it is to get there.

This relates to search engines in two ways. First, online searches can make us feel like geniuses; with Google, we're all information-finding ninjas. We've already established that we tend to feel more powerful and more intelligent thanks to Google Search, even when we're unsuccessful at finding what we're look for—after all, we have this incredible processing power working to augment our brain's ability to find the answer to something. Second, online searching feels anonymous, perpetuating our ability to lie to ourselves about what we really want and thus who we really are.

Here's where data collection once again comes into play. As you do general web browsing and searching, your computer is collecting data on what websites you visit, and the content on specific pages, and sending that information back to the host website. Ever notice how you might be shopping for shoes one morning and then, later that day, you get an ad for the same shoes on Facebook or Gmail? This is a reminder that our activities, while they might feel anonymous, are in fact tracked. Even when we enter into "incognito mode"—Google Chrome's "private" web-browsing option—while the web browser doesn't track your history or the content of the sites you visit, your ISP (internet service provider) certainly does. In short, you can't access the internet without letting at least the entity that's providing you with a connection *to* the internet see what you're searching for.

As we use search engines, we're fueling our innate sense of overconfidence, empowered by the sense of intelligence we get by getting answers to our queries so damn fast. Imagine what this does to us, as a people, on a global scale. If on average, 57 percent of the population of Earth does a search on Google at least once a day, then once a day 4.2 billion people get a little cognitive boost. This boost makes us slightly more overconfident in our intellectual abilities—a problem we already struggle with, thanks to our innate overconfidence bias. We feel empowered, smarter, and as a consequence, take greater risks; we overestimate our abilities to succeed.

If one of the goals of being human is to know oneself, search engines both help and hurt us. They allow us to find out all kinds of things we might never know otherwise. But at the same time, they fuel what is ultimately an unrealistic view of ourselves and our abilities. Search empowers us, but in so doing it further obscures our sense of reality. To know ourselves, then, maybe we need to look further inward for answers and rely on search engines a little less.

ULTIMATELY, ALL OF THIS SHOWS WHY RELYING ON NET STATE DATA AS the primary means to assess a person's worth is a problem. Net states—and nation-states, such as China, engaged in digital monitoring—know only *versions* of us: the idealized self that we curate for Facebook; the frustrated self that we reveal in what questions we seek answers to on Google; the hyperconfident self that posts clever reviews on Amazon; the isolated self that unlocks our iPhone 63 times a day, hoping for something, anything, to distract us from our loneliness.

We're infinitely complicated creatures, inhabiting multiple and often contradictory versions of ourselves all at once. A national scoring system (like China's Social Credit Score) that assumes it can know people based only on their digital footprint is dangerously naive; it's like taking a person's photograph and then claiming knowledge of their soul.

Governments will never be able to fully know their citizenry—not even China's rapidly developing surveillance state. The danger with systems like the SCS is that they *assume* they can—and with a high enough degree of confidence that they act on that assumption, sometimes harming people's physical and interior lives.

One last proverb. It goes something like this: To know a man, look only to his friends. To know a king, look only to whom he employs. "The wise man," it concludes, "is exceedingly careful about the company he keeps" (君子慎所藏).

Here's the thing: the company *I* keep, personally—and I suspect many of us—isn't, for the most part, kept online. For instance, I don't think I've ever exchanged Facebook messages, comments, or likes with my husband; not my brother or my best friend; *certainly* not with my kids (they'd be mortified). I rarely even send any of those people emails. That's because, despite all the evidence of my life online, I still live the majority of my life offline. Most of us do. Even with our eyeballs on screens hours upon hours a day, we still have to reside someplace physically and interact with neighbors. We have to go to work and converse in person with colleagues, or go to school and engage with our teachers. We see friends. We have experiences—all offline.

Thus, any government mapping of our worth based on our digital footprint alone overlooks the most important data points about us: the vast majority of time we spend living life when *not* on tech. Our digital reflections may be highly sophisticated renderings of the self we want the world to see, but they're not trustworthy stand-ins for the real person—not yet.

SIX

THE ALL-KNOWING INTERNET OF THINGS

Tech is increasingly personal, collecting data on our individual behaviors whether we're aware of it or not. This chapter explores our daily life with technology, examining the tech we use in our homes and our public spaces with the so-called Internet of Things (IOT) and how the IOT currently influences and is likely to affect our daily existence. It looks at developments in user profiling, starting with Amazon's recommendation system and "smart" technologies such as Amazon's Echo that gather data on our health, our environment, the information we seek, the music we play. It tackles the question of whether these practices let our net states "know" us, and whether, ultimately, users have any say in the matter.

If you Google the phrase "What is the Internet of Things?" in 2019, you'll still find explainers and primers: "What Is the Internet of Things? A Complete Beginner's Guide" (*WIRED*); "6 Things Everyone Needs to Know About the Internet of Things" (Business Insider website); "The Internet of Things: Could It Really Change the Way We Live?" (*The Telegraph*). That's how new it still is. In 2014, a much-cited Accenture poll reported that 87 percent of people had never heard of the internet of things.[1] Five years later, that's transformed from "never heard of" to—apparently—"don't really understand."

Incredibly convenient, the Internet of Things is arguably also the greatest threat to individual privacy to date. It includes lots of things. Internet-enabled household devices, like coffee makers, refrigerators, and televisions. Internet-linked wearables, like smart watches and Fitbit-type health trackers; voice-enabled assistants, such as Apple's HomePod, Amazon's Echo, and Google's Home; as well as smartphone apps we keep tethered to our physical health, like heart rate monitors and step counters. Computer-assisted cars. The smart thermostats that regulate our air conditioning and heating. Our cloud-based home security systems.

Soon, the Internet of Things will include anything you can stick a sensor on: in 2018, scientists developed a thin-film electronic sensor—an internet-enabled sticker, essentially—that could conceivably turn *any* object into an IOT device.[2] "We could customize a sensor, stick it onto a drone, and send the drone to dangerous areas to detect gas leaks, for example," said Chi Hwan Lee, Purdue assistant professor of biomedical engineering and mechanical engineering, one of the scientists who created the device.[3] Or, for example, you could stick this sensor on a flower pot, suddenly enabling the pot to detect temperature changes that might affect how the plant grows.

The phone giant Ericsson forecasted that there would be more IOT devices in 2018 than the world's 7.4 billion mobile phones.[4] Other industry experts predict that this number will increase to 30.7 billion by 2020 and 75.4 billion by 2025. Here's another way to look at it: by 2020, there will be an average of four IOT devices per person on the planet; by 2025, that will increase to nine devices per person.

Considering that most of us aren't even aware of how many IOT devices we have in our vicinity, this is a disconcerting thought. Because smart devices talk to each other, an IOT doesn't necessarily require our intervention to work—which is exactly the point.

This raises the question of what the IOT industry means for our ability to exert control over our data and our identity. Opting out of the IOT doesn't appear to be an option moving forward. When polled by Pew Research in 2017, industry experts almost universally concluded that despite a certain creepiness factor inherent in the IOT, it would indeed be adopted by almost everyone, either because "the magical behaviors that the new devices will provide will be too strong for people to resist" or because "a significant number will have the illusion of being disconnected [when they actually are not]."[5]

Since an IOT device doesn't necessarily require user intervention to function, once we set it up it's entirely possible to forget that it exists. The technology, like so many other technologies, may seem magical at first, but it quickly becomes invisible to users—both the tech itself and the implications of what net states do with the data they collect from IOT devices.

The marketing term for Internet of Things products—"smart" devices—is not entirely accurate: they're less smart than just chatty. They're in near-constant communication with their manufacturers, checking for updates and sending notifications that they're still functioning.[6] Smart TVs, for instance, track and report back to their manufacturers on what you watch, for how long, and how often. But they also connect with the other internet-enabled devices in your home—your phone, your laptop, your smart thermostat, your smart door-locking system—anything using the same internet connection as your television.[7] Given that 45 percent of American homes currently house smart TVs, this generates a tremendous amount of data about a public that remains largely unaware it's happening.

In a chilling experiment to discover what it's like to live in an IOT-connected home, Gizmodo reporter Kashmir Hill converted as many of her devices as possible into internet-enabled ones: this included her Amazon Echo, light switches, coffee maker, baby monitor, kids' toys, vacuum, television, toothbrush, and (to monitor and report on sleep patterns) even her bed. "This must be what it's like to be in a documentary or in a reality TV show," she wrote.[8]

Hill reflected on two disturbing trends. First, she quickly stopped noticing the devices in her home. Being surrounded by internet-connected gadgets became normal—just part of the background. When she stopped to think about it, this lack of awareness troubled her: "If homes become sentient, and it becomes the norm that activity in them is captured, measured, and used to profile us, all of the anxiety you currently feel about being tracked online is going to move into your living room."

The second disturbing trend Hill noted in IOT-connected homes is the data collection power of the ISP, the internet service provider. Not only do individual internet-enabled devices send data to their manufacturers, but a person's ISP—Comcast or Verizon or whatever company is paid for internet access—scoops up all that data, too. While in some ways, the ISP is just a utility, like the electrical station that supplies power to a home, it's also a data sieve. Think of it this way: any data that comes into your home via the internet (your TV shows, your email, the websites you visit) and any data that gets sent out from your home via the internet (your search history, your web-browsing patterns, your purchase requests) goes through your ISP, which acts like a giant net through which all data traffic flows and gets captured; like a toll collector between your home devices and your destination, whether it be Amazon or Google or Facebook, with your data being the payment that your ISP collects.

And ISPs are now permitted to sell that data. In 2017, Congress passed a bill repealing restrictions on ISPs selling customer browsing and app usage data without seeking individual users' consent.[9] Whom can they sell this information to? Anyone who wants to pay for it, really. At the very least, though, this likely includes data brokers.

Also, it's worth remembering that all digital data is leakable and hackable. In June 2018, for instance, the marketing firm Exactis was called out for having publicly exposed its database of 340 million records containing people's personal information. The records included details ranging from phone numbers and email addresses to the number, age, and gender of a person's children. "It seems like this is a database with pretty much every U.S. citizen in it," reported security researcher Vinny Troia in an interview with *WIRED*.[10] "I don't know where the data is coming from, but it's one of the most comprehensive collections I've ever seen."

That a company like Exactis has this much data on us is noteworthy for two reasons. First, Exactis isn't exactly a household name. Likely, most of us have never heard of it. The only reason it even merited a news story is that it screwed up. This, though, leads to an inevitable question: How many other marketing firms out there have data profiles on us that we *don't* know about? NSA consultant Edward Snowden shocked the world with his 2013 revelation that the US government ran a program, PRISM, to collect data on American citizens. Yet Americans seem somehow less shocked at the prospect that unknown multitudes of market research firms likely have far *more* information on us.

We are, as a people, increasingly distrustful of government. Our confidence in our government has been declining since the 1950s and now rests at rock-bottom historic lows: only 18 percent of Americans report trusting government, down from 77 percent in the 1960s.[11] We don't trust the people who run our country with our private information—though these are people we have the power to elect or reject—but seem resigned to having no power over net states' possession of our private information. Even Facebook, whose credibility took a historic walloping in 2018, still enjoys a 49 percent trust rating from its users. While this is pretty low on an objective scale, it's practically off the charts when compared with our government: the last time 49 percent of Americans trusted their government was 1972.[12] The US Congress, with its 11 percent trust rating, could only dream of Facebook's 49 percent.[13]

Even with our middling trust rating in Facebook, we're still on it, en masse. Sure, we've directed outrage at this particular net state, hauling its founder, Mark Zuckerberg, before Congress for a good, old-fashioned grilling. But we haven't given up our allegiance to it or the other net states. (Remember: it's not just Facebook—*all* our net states engage in data sharing; it's a core element of their business models. Facebook just happened to be the one in the hot seat in 2018.)

The data that shapes our individual profiles will only increase as we move from wearing internet-enabled devices, as with phones, to installing them in our homes, as with smart TVs, to having them baked into the very infrastructure of our environment, as with smart cities. Smart city technologies are already being deployed, and in plain sight, but likely to our utter obliviousness.

It's easiest to observe the implementation and grasp the possible implications of smart city tech in an extreme example. Let's take a look at a smart city that's being built from scratch in our neighbor to the north.

CANADA DOESN'T TEND TO COURT CONTROVERSY. IT'S A PLEASANT, polite place that plays by established norms and rules.

Quayside could change all that.

In the summer of 2018, Google's sister company Sidewalk Labs (also owned by Alphabet)—which declared its mission to be "reimagining cities to improve quality of life"[14]—embarked on its first actual foray into city-building. In partnership with a government-funded corporation called Waterfront Toronto, Sidewalk Labs will convert Quayside, a 12-acre stretch of waterfront property, from what locals describe as "blighted, underutilized, and contaminated" into a fully smart city.[15] According to Sidewalk Toronto, as the joint initiative is called, they'll "combine forward-thinking urban design and new digital technology to create people-centred neighbourhoods that achieve precedent-setting levels of sustainability, affordability, mobility, and economic opportunity."[16]

The ultimate vision is something like a city version of your iPhone. Sidewalk will provide the infrastructure, similar to how Apple provides the iPhone itself. The iPhone apps—what makes the phone interesting—are all designed by third-party developers: Facebook and AccuWeather and the like. Sidewalk's Quayside will have features that are likewise customizable: "If you think of the city as a platform and design in the ability for people to change it as quickly as you and I can customize our iPhones, you make it authentic because it doesn't just reflect a central plan," Rit Aggarwala, the executive in charge of Sidewalk Labs' urban-systems planning, explained to *MIT Technology Review.* "It also reflects the people who live and work there."

This is where Sidewalk Toronto emerges as an intriguing test case. The 196-page "vision document" that Sidewalk Labs presented to Toronto proposed razing the 12-acre property and creating their city of the future from scratch. The plans include a self-contained geothermal grid to recycle energy, which can then be used to both heat and cool buildings. The city's carbon emissions will be negligible, since cars won't be permitted—only emergency vehicles and driverless transit shuttles will be. Sidewalks will be kept clear of snow with built-in melters—a must, considering that walking and cycling paths will form the primary arteries for people to get around. Buildings will be constructed from recycled materials, the city will employ smart waste disposal systems, and so on.

Quayside will also be populated with sensors—many, many sensors. They're billed in the vision document as a "digital layer" that covers all aspects of the city. Built into every facet of the neighborhood, the proposed digital layer—read, "IOT, everywhere"—will provide "an unprecedented degree of insight into the physical environment."[17]

That such technology will permit an unprecedented degree of insight into the inner workings of city life is perhaps inevitable. But what's still up for debate is who owns that technology and who has access to those insights. Public infrastructure—protecting and maintaining the roads, bridges, electrical grid, and other spaces for use by any and all citizens—has typically been the domain of nation-state governments. That this city will be constructed and operated by two corporations—Sidewalk Labs and Waterfront Toronto—raises the question of what remains "public" in this scenario. If Quayside's public sidewalks and parks are permeated by sensors, are they really "public"?

It's unclear what, exactly, Sidewalk Labs and Waterfront Toronto plan to do with the data they collect from Quayside's "digital layer." As of this writing, the questions of data governance (who has access to the data collected at Quayside), data residency (what country the data will be stored in), and how citizens' privacy will be protected are all still being debated. As Anthony Townsend, a former consultant for Sidewalk Labs, noted in a *Politico* profile on Quayside, the fact that these issues have yet to be worked out spells trouble. "This is the city-level question of our time," he said. "And [the governmental leaders of Toronto] haven't taken the slightest step in trying to even create a vision there."[18]

Quayside is tiny—you could likely walk from end to end in just over half an hour—but it will be a test case for what a truly smart city, one not just *built with* but *integrated into* the IOT, could actually look like. "This is really a grand experiment, in many respects, that is going to teach not just Toronto but really cities all across the world what the future city is going to look like," Bruce Katz, a former Brookings Institution official, reported in that same *Politico* profile. "Of course, if you're Toronto, you're the lab."

In some ways, Sidewalk's Quayside is a technical and ideological moonshot: a future-leaning pilot project designed to test the limits of what cities could be, but also an experiment into how people's lives could be lived. This is not a bad thing. Indeed, it invites a kind of "first principles" thinking that net state leaders like Elon Musk have credited for their many successes.[19]

First principles thinking is a concept borrowed from physics. "The normal way we conduct our lives is we reason by analogy," Musk explained in an interview. "[With analogy,] we are doing this because it's like something else that was done, or it is like what other people are doing. [With first principles,] you boil things down to the most fundamental truths . . . and then reason up from there."

This means that instead of looking at how to solve a problem using existing road maps, you start with the very essence of the problem itself, questioning every assumption and every previously attempted solution. Cities are an excellent target for first principles thinking, because they're quite literally built on assumptions.

Case in point: city roads. Most cities' road systems evolved through use. Dirt roads at one point became paved, not necessarily because they were the most direct route from point A to point B but because those were the routes people were accustomed to. Boston is a famous example of this, with its tangled web of streets evolved from the practice of what's called "paving the cow path."[20] A look at a map of Boston reveals a tangled, swerving network of roads, seemingly devoid of forethought and planning—precisely because it *is* largely devoid of forethought and planning, reflecting instead the concretization of human routines.

How roads are planned affects how you live your life in a city. In a place like Boston, you have to know where you are and where you're going—there's no predictable logic you can use to plot your path from, say, Faneuil Hall to Chinatown. Compare this with the experience of getting out of a subway station at 14th Street in Manhattan. Even if you don't know the city, you can use logic to determine that heading in a southerly direction means counting down to 13th Street, 12th Street, and so on.

First principles thinking also matters because it represents a chasm between how technologists approach a problem and how government officials approach a problem. As the stewards of taxpayer money, governments are guided—in theory, at least—by finding the most cost-effective solution that will cause the least amount of disruption in people's lives. Technologists, on the other hand, are all about disruption—so much so that the very term "disruption" has become a cliché in the field. The aim of technologists is to find the *best* solution, not the most inexpensive one that causes the least amount of upheaval. And what better place to explore the very foundation of a problem than where people live and how they get around? In other words, in cities.

The problem with technologists increasingly investing energy, time, and money in cities and their infrastructure comes down to two key points: access and the public interest.

The access issue relates to whether what was "public" transportation (for instance) in the past continues to be equally accessible to everyone under net state supervision. Government, after all, is a servant of "the people." It's required by law to serve *all* the people: those who can easily adapt to new services and spaces as well as the most vulnerable among us who cannot. "It's easy to be good," Mayor Duarte Eustáquio Júnior, who governs the Brazilian city of Mariana, once told me. "It's much harder to be fair."

Technologists may indeed be *thinking* about doing good; fairness, however, is not necessarily the highest item on their priority list. This is not out of malice, but simply because a fairness orientation goes against first principles thinking, which would scrap all assumptions and just build the best possible tech, period, not the perhaps-mediocre-but-hey-it-works-for-everyone tech. Fairness means ensuring that *everyone* has equal access—that the disabled, the disadvantaged, and the infirm have the same access as the able-bodied, the young, and the wealthy.

From this perspective, could we consider the iPhone, for instance, to be "fair"? It did introduce the accessibility feature VoiceOver—a function that lets you double-tap anything on the screen to have it read aloud—in 2009.[21] That's just two years after the iPhone hit the market, so Apple is apparently doing pretty well at introducing features to make their product accessible to the visually impaired.

But what about to the disadvantaged? iPhones are the most expensive smartphones out there, topping $1,000 back in 2017.[22] By 2019, a top-of-the-line iPhone XS cost $1,449,[23] exceeding the average American's monthly mortgage payment of $1,030[24] by hundreds of dollars.

Here's where net states and nation-states sharply diverge: the nation-state is responsible for *all* of its people, not just those who can afford services or don't need special accommodations. When net states or other private-sector entities begin taking over what used to be public infrastructure, who will ensure that those roads, squares, trains—or entire cities, in the case of Quayside—and other services are fair? Just as we saw in the example of Tesla and Google in Puerto Rico, net states might *choose* to do something to benefit the public, but they're not *required* to—nor are they required to be *fair*, as governments are. And this is where governments handing over public services to net states becomes worrisome.

If the first question is, Who's looking out for public *access*? the next is, Who's looking out for the public *interest*? This is perhaps best illustrated through an example.

Take Waze, for instance, the crowdsourced navigation system acquired by Google. Waze used to rely primarily on data that users input directly, such as instances of traffic jams, accidents, or other obstacles on the roads. Then, in 2014, Waze launched the Connected Citizens program, partnering with 10 locales around the world—including Boston, the states of Florida and Utah, and the New York City Police Department—so that they could include in their app government-owned traffic data, including that from traffic sensors and street cameras.

What exactly does this platform include? Many cities have already begun to adopt smart city technologies with respect to traffic lights. At most recent count, 363 cities across the United States have swapped out traditional traffic lights for modern adaptive traffic control systems (ATCSs).[25]

ATCSs are pretty much what they sound like: traffic systems that adapt the timing of signals based on what they can detect from current traffic conditions.[26] These systems help to reduce traffic jams, which cause substantial problems in cities of all sizes. According to the INRIX Global Traffic Scorecard, in 2019 the average commuter will lose 97 hours just sitting idle in traffic. This costs Americans $87 billion in lost time—an average of $1,348 per driver.[27]

Even more important, smart traffic systems save lives. "If you have a smooth traffic flow, that potentially will reduce the number of accidents. Every time you make a stop, there's a chance of a rear-end collision," said Aleksandar Stevanovic, director of the Laboratory for Adaptive Traffic Operations and Management at Florida Atlantic University. With roughly 1.7 million incidents of this type in the United States each year, rear-end collisions make up 40 percent of all traffic accidents, killing 1,700 people annually and injuring another half million.[28]

With this context in mind, it's understandable that so many cities are scrambling to implement smart traffic technologies. As with Quayside, the trick is figuring out who controls these technologies and who's in charge of the data they produce.

It seems like it should be citizens. After all, taxpayer money pays for traffic sensors and street cameras. And we as citizens tolerate sensors and street cameras because they provide a public good: in addition to reducing traffic congestion, sensors and street cameras make our streets safer. However, Waze used data from sensors and street cameras to create a feature on its app that serves as a police locator. This is to alert drivers about the presence of police, most likely so they know to slow down to avoid getting speeding tickets. The police locator, however, could also theoretically be used by criminals to pinpoint where cops are stationed throughout the city for the purpose of avoiding the police when committing crimes. This precise complaint was lodged by Charlie Beck of the Los Angeles Police, who wrote in a letter to Waze that the police locator could be "misused by those with criminal intent to endanger police officers and the community."[29]

This is purportedly what happened when two New York City police officers were gunned down and killed in their patrol car on December 20, 2014. The gunman's phone contained screenshots of Waze's police locator, indicating that he may have used the Waze app to pinpoint the location of the police officers he targeted in the days leading up to his lethal attack.[30]

Clearly, the technologists at Waze did not include the police-tracking feature to facilitate crime. Indeed, Waze's spokesperson reported that the company's relationship with police departments "keep[s] citizens safe, promote[s] faster emergency response and help[s] alleviate traffic congestion."

Yet this example shows how technology designed with first principles thinking can be manipulated in unexpected and unfortunate ways. It turns out that sometimes there are good reasons—public safety and access for all citizens among them—for assumptions and previously implemented approaches. Waze's programmers were most likely just trying to make the best navigator app they could. The more features they could include, the better the product would be. But coders and software engineers aren't policy experts; they don't necessarily think about the potential impact of their wares on the public in the ways that government officials might. While net states should absolutely continue to innovate ways to make our cities smarter, more connected, and better for citizen-users to live in, governments cannot simply step to the sidelines and watch them do so. They must be integral partners to ensure that innovation doesn't leapfrog over boring yet necessary considerations such as public safety and fair and universal access.

The example above shows public taxpayer dollars being spent on a tech system that may, in fact, have made the public less safe by putting law enforcement in harm's way. This raises the question of whether the people of New York City actually approved of a partnership between Waze and the NYPD. Did they even know about it? It wasn't exactly advertised: in the city's press release announcing the Connected Citizens partnership, Waze was never mentioned, and even Google was mentioned only as one "partner" among many in an announcement on a series of tech partnerships.[31]

While dragging up a vaguely worded press release may seem nitpicky, it underscores a critical point: the people of New York City—*citizens* to the police and *users* to Waze—were taken out of this endeavor's equation. Citizens should have the right to weigh in on how their tax dollars are spent, and citizen-users should have the right to weigh in on whether they approve of new features in their tech. Just as cities like New York should solicit input from citizens on partnerships that may affect public safety, net states like Google, Waze's parent company, should recognize that the people who use their apps are not just users, but citizen-users whose physical welfare is affected by their technological products and services.

If net states start to recognize citizen-users as such, they may be more likely to invest in adding policy experts to their teams of coders and designers. In this way, they would be able to meet the comprehensive needs of citizen-users, including public safety, and not just the needs obviously related to their app (say, traffic navigation).

One last risk inherent in the IOT and smart cities is, as with all things digital, their vulnerability to attack. In 2013, Iranian hackers launched a cyberattack on a small dam in Rye Brook, New York, a town 30 miles north of Grand Central Station in midtown Manhattan, successfully accessing its command and control system. A nuclear power plant near Burlington, Kansas, was targeted by Russian hackers in May 2017. Unsurprisingly, dams and nuclear power plants count among the 16 sectors that the US Department of Homeland Security deems "critical infrastructure": defined as any sector "so vital to the United States that their incapacitation or destruction would have a debilitating effect on security, national economic security, national public health or safety."[32]

Security agencies rarely report details on such attacks—the Rye Brook dam attack was reported on only three years after the breach—making it difficult to discern how vulnerable the nation's critical infrastructure really is. But attacks on what are known as SCADA systems—supervisory control and data acquisition systems, which are increasingly used to manage infrastructure—have skyrocketed globally, from 90,000 in 2012 to 675,000 in 2014, a staggering 650 percent increase.[33]

As with most innovations, there's a trade-off with the IOT. The more technologically connected our infrastructure, the more we can learn about the human condition: how we live, where we go, what we do, en masse. At the same time, we become more exposed and more vulnerable. The question is whether the trade-off will be worth it.

THERE ARE SOME REASONS TO TAKE HEART THAT NET STATES SUCH AS Google and Tesla are increasingly investing in, and thus in charge of, future-leaning infrastructure projects. In addition to expanding into cities' electric grids, as described in chapter 2, Tesla is investing in transportation projects: the city of Chicago has approved Tesla affiliate Boring Company's proposal to build a high-speed transit system from its O'Hare airport to its downtown area.[34] The reason? Given that technology is their core business, net states may be in a better position than governments to manage high-tech projects.

Often bogged down by politics and bureaucracy, government doesn't have a great track record when it comes to implementing and managing tech projects, especially in the United States. Take Social Security, for example. In 2000, recognizing that a massive backlog of claims had amassed, Congress approved an overhaul of the claims-processing technology used by the Social Security Administration (SSA). By 2007, the SSA had spent upwards of $381 million trying to integrate its 54 separate information technology (IT) systems, followed by another $200 million in 2011.[35] Six years after they'd launched the project, "they found out that they really didn't have anything. In fact," commented reporter Bob Charette, "the initial system has consistently been projected to be 24 to 32 months away. So for five years, it was always kind of like Wimpy in Popeye: 'I'll gladly pay you on Tuesday for a hamburger today.'" The rest of the US government doesn't fare much better, and it's no wonder: 70 percent of government IT budget dollars go to maintaining legacy—read "really, really old"—systems.[36]

Again, however, even though net states may be in better positions to successfully pull off large-scale technology infrastructure projects, they are ultimately not legally required to act in the public interest. Net states have shareholders. They have bottom lines. Democratic governments may be delinquent in how they execute their policies, but at least they work to ensure that those policies serve all their people. In the absence of government regulation of technology, especially as it pertains to data sharing, there's no guarantee that net states will choose to protect people's rights as new data sets become available in a suddenly digitized physical environment. While attempts to protect user data exist here and there, we still lack a cohesive guide that countries and net states can follow to protect user data. As data scientist Venkat

Motupalli wrote, "There is a bewildering array of acts, guidance, and agreements across dozens of countries that attempt to address the new size and shareability of citizen data." Most important, he noted, "While almost everyone agrees that personal data should be protected, there is little consensus as to how and who should be responsible."[37]

In the absence of consensus, net states are left to take up the responsibility for themselves. And, in the absence of protests from their populations—the citizen-users who use their products and services—they will most likely continue along the path they've taken thus far: sharing data with their partners, not only to fuel innovation, but also to maximize profit.

BEFORE WE DELVE INTO THE NEXT PHASE OF FUTURE TECH, IT'S USEFUL to reflect for a moment on how our relationship with technology flipped from active to passive—and how fast that happened. With the IOT, our physical environments—from our smart homes to our neighborhoods, with devices embedded in appliances, roads, traffic lights, and sidewalks—are the active agents, the data collectors. We, the users, are passive; we're the data creators, the things being observed. The IOT requires no special action from us, just that we go about our business—nothing to see here.

This is a tectonic shift in our relationship with tech. Not two decades ago, our physical environments were dumb, inert. Even our *digital* environments weren't all that capable. It wasn't until websites became interactive that we entered a truly dynamic relationship with our technology. Up until then, information flowed along a one-way street, from human to machine. Even with that, we—people—remained in the driver's seat; we *chose* what information to give our tech.

And we chose carefully—in the early days, at least. We were hesitant, distrustful even. We didn't even reveal our true names. Indeed, most early websites not only permitted anonymous interaction with their sites; it was the norm.

This was by design, a manifestation of the early web's influence by hacker culture. As web scholar Derek Powazek noted, "Back then, it was normal to have a pseudonym or 'handle' that you went by. Most communication online was hidden behind handles, which reinforced the idea that the internet was not 'real' in the same way real life was. People treated everything online like a game. It was 'cyberspace,' not reality."[38] Our handles both identified and in some ways defined us to others online.

Then came e-commerce, which effectively killed anonymity as common practice. After all, you couldn't use a credit card to order a book on Amazon without giving up your real name. Soon even free websites began requesting "registration"—inviting you to explore their sites for free, so long as you created an account using, at the very least, a name (which could of course be faked) and an email address (which could also be created using a fake name).

For a long while, this seemed like an eminently reasonable tradeoff. After all, websites were entirely free back then—as they remain today, for the most part. While the *devices* that grant access to what's online can involve substantial costs, once you're on your phone or computer, you can peruse the world's knowledge—from maps to media to movies—largely without spending a dime.

Even if you did have to give up some identifying information about yourself in the early years, doing so yielded perks. In 1998, Amazon launched "item-based collaborative filtering"—or, in plain English, the recommendation system.[39] This marked a turning point in how we shop. "Amazon.com has been building a store for every customer," noted Amazon's Brent Smith in a study on its recommendation system. "Each person who comes to Amazon.com sees it differently, because it's individually personalized based on their interests. It's as if you walked into a store and the shelves started rearranging themselves, with what you might want moving to the front, and what you're unlikely to be interested in shuffling further away."

Here's how this translates to a person's actual life. In 1998, e-commerce accounted for only 0.8 percent of all sales in the United States.[40] This means that nearly the entire population had to go to a physical store if they wanted to buy something. That store looked the same for you and me and everyone else who entered. You had to browse the aisles, maybe ask a salesperson for help, and wait in line to pay; then, eventually, your clothes or books or groceries were bagged and packaged, and off you went.

Enter Amazon. Let's say I go to the site. Because I've bought a few books there in the past, I'm now shown books similar to those purchases. But when *you* go to the site, you see entirely different books, based on the books *you've* bought. With just a few clicks, you go on to buy from the books recommended to you—all without interacting with a single human.

Note: it's not a salesperson who has sized you up and proposed a product to you based on personal judgment and experience.[41] It's an algorithm. And the algorithm probably did a better job of assessing your tastes than that human salesperson could. Already by the 2000s, the algorithm could do a better job of recommending the books or movies you liked than even your best friend could. It now knows you better than anyone because it's mirroring you. It does such a good job that Amazon won an Emmy for its personalized recommendation system for video discovery in 2013.[42]

Recommendation systems have become so sophisticated that they can tailor offerings to you from your first visit to a site—what's referred to as a "cold start," where you've yet to input any information that the recommendation system can draw from.[43] Even in cold start cases, Amazon can draw on "implicit interactions": the date, time of day, and your geographic location. Based only on these three bits of information, it can then recommend items that others who live in your zip code and browse Amazon at, say, 3:00 a.m. have also purchased.[44]

If you've shared your specific preferences—or, as noted above, just by virtue of the location of your IP address—recommendation systems customize the web for you whether you want them to or not. We live in a digital world that is tailored to our likes and tastes. *My* web experience doesn't look like *your* web experience—or anyone else's, for that matter. Based on the fact that I'm researching this book, read sci-fi for fun, and have kids, Amazon offers me titles on the history of Google and Facebook, biographies of Isaac Asimov, and manuals on mindful parenting. It's a combination of products designed for me and me alone.

But this raises the inevitable question of what it means that the web "knows" me—and if it's too late to do anything about it.

The comic W. C. Fields once wrote, "It ain't what they call you, it's what you answer to." Recommendation systems know us not because of what we *tell* them we want, but because of what we're willing to put money on wanting. In other words, since we buy only things we want or need, recommendation systems call out to us, offering precisely the kinds of products we've given them evidence of wanting and needing. They display products as do salespeople who guide us through a store. The difference is that they take us directly to the wares we do, in fact, really like, even if we may not want to admit to it. While Facebook may permit us to create idealized versions of ourselves on our profiles, recommendation systems know the real us. They mirror our desires. In this way, they are perhaps the most honest reflections of ourselves in the digital world.

There's a lot to be said for recommendation systems. They may help us make wonderful discoveries of new books, movies, and products that we never would have found otherwise. But there's a hidden cost to being so well known. We become less a simple consumer and more a profile of consumer behaviors—a data set that maps our browsing patterns (by how long we spend per product page), purchase temptations (by what we put in our shopping carts and later delete), and commitments (by the products we actually purchase).

If this profile were merely stored in Amazon's database, waiting until our next visit to the website to be launched into action, it would seem a clear-cut victory for consumers: tailored, customized sales support, plain and simple.

But it's not, of course. Amazon and Netflix and YouTube and likely any other website with a recommendation system spell out clearly in their terms of service that your information may be shared with "third parties." Amazon attempts to deflect this in its privacy notice. To the question "Does Amazon.com Share the Information It Receives?" the response reads, "Information about our customers is an important part of our business, and we are not in the business of selling it to others."[45] Great, if that were the whole story. But it's not. The privacy notice then details all the ways in which they *do* share your information with others, including "Marketplace Sellers,'" or "businesses [that] operate stores at Amazon.com or sell offerings to you at Amazon.com." This would account for a whopping 52 percent of all sales on Amazon.[46]

Amazon also shares your personal data with third parties for the purposes of "fulfilling orders, delivering packages, sending postal mail and e-mail, removing repetitive information from customer lists, analyzing data, providing marketing assistance, providing search results and links (including paid listings and links), processing credit card payments, and providing customer service." In short, just about every step of the transaction process at Amazon involves third parties. So, while Amazon may not "sell" your data, it certainly does share it with numerous third parties.

This matters because we no longer transmit data only when we're aware that we're online. Our data from conscious actions, such as making purchases on Amazon, as well as the data that streams almost constantly in the background, gets gobbled up by various third-party data brokers to create ever more detailed profiles on us.

AMAZON HAS SINCE EVOLVED. IT MADE ITS MARK BY "KNOWING" US: identifying our wants and needs and providing recommendations for other things we might want based on those wants and needs. But it wants to know us even better—not just our preferences, but our physical profiles as well. With its new recognition software, Amazon is now wading into the IOT.

Foundationally, recognition software isn't so different from recommendation systems. Both know you—an individual user—in a way. Both can identify you based on various attributes: your likes and preferences, for recommendation agents; your physical characteristics, for recognition software.

Yet, while the foundations are similar, the aims are very different. Amazon's recommendation systems identify you for the purpose of showing you stuff you like. Rekognition, as Amazon's facial recognition system is called, identifies you for the purpose of pinpointing you for law enforcement, with the aim of getting you translated into a profile in a police database.

The ultimate aim, of course, is public safety. If law enforcement officers can pick a known criminal out of a crowd on a busy street, they can move to protect other citizens from that criminal. But questions have emerged regarding how the system works and who handles the data collected.

Rekognition is already being tested in real life—police departments scattered across the country use the system, though few advertise that fact. Amazon boasted in a 2017 blog post that Rekognition can detect "up to 100 faces in challenging crowded photos"[47] and proudly lists the Washington County, Oregon, sheriff's department on its Rekognition website as a satisfied customer. The department reciprocated, testifying on the Amazon Web Services (AWS) website that they "were able to index more than 300,000 photo records within 1–2 days, and the identification time of suspects went from 2–3 days down to minutes."[48]

While the US surveillance system is far less developed than that of countries like the United Kingdom, which has over 6 million closed-circuit television (CCTV) cameras,[49] or China, with 570 million cameras,[50] we fundamentally diverge with those camera-heavy countries with respect to owner and operator. In the UK, their CCTV network is operated by the government of the UK; in China, it's the government of China. Yet it is *not* US law enforcement agencies that developed and deploy surveillance technology in this country; it's Amazon. And it is *not* US law enforcement agencies that wholly own the data they input into their surveillance technology: it's Amazon.[51] Recall, law enforcement agencies, being an arm of the nation-state government, are legally prohibited from monetizing, making public, or sharing personally identifiable information (PII) with anyone.

But Amazon is a private corporation. No laws yet regulate what it can do with its data. As of this writing, it and any other net state that cares to get into the surveillance game can share with as many third parties as they see fit.

Doing so has not been without controversy. The American Civil Liberties Union (ACLU) publicly criticized Amazon for marketing directly to police departments.[52] Independent privacy scholars have also expressed concern: "The idea that a massive and highly resourced company like Amazon has moved decisively into this space"—meaning, directly marketing to law enforcement entities—"could mark a sea change for this technology," said Alvaro Bedoya, executive director of the Center on Privacy and Technology at Georgetown University Law Center.[53]

In some ways, the fact that net states are moving into law enforcement support is not exactly shocking, or even all that new. Security agencies from the NSA down to local sheriff departments have been using computer systems to track and monitor criminal activity for decades. In a way, implementing facial recognition software might be seen as a natural evolution to what beat cops have been doing for years: picking out faces from a crowd for the purpose of getting criminals off the street.

The problem is, law enforcement doesn't exactly have a squeaky-clean record in terms of *who* it picks out of a crowd. When Congress heard about Rekognition's use in law enforcement offices, several members of Congress, led by Rep. Jimmy Gomez (D-CA) and Rep. John Lewis (D-GA), as well as the Congressional Black Caucus, sent letters of concern to Amazon CEO Jeff Bezos, citing studies showing that facial recognition software consistently misidentifies minorities—specifically, persons of color and especially female persons of color.[54]

The software misidentified white men 0.6 percent of the time but misidentified women of color 35 percent of the time. According to an MIT study of the technology, this was due to the fact that the data set training the program was 77 percent male and 83 percent white.[55]

"Surveillance of perfectly legitimate and constitutionally protected activity will only further erode the public's trust in law enforcement," Chairman Cedric L. Richmond (D-LA) wrote in his letter to Bezos. "We urge you to be thoughtful, deliberate, and assiduous as development of this technology advances."

This is the key question: Are these technological advances being thought through—and, if so, how and by whom? Or are advances simply made because they're the next logical step in tech's evolution?

Scott Carpenter, Google Jigsaw's managing director, isn't optimistic. "A lot of companies, they're just trying to get off the mat and into the game," he said. Right now, the implications of how their tech will be used in the future is less a consideration than figuring out how to get their tech used at all. Tech users—you and I and all of us—are responsible to hold our net states to account for the ways that their new technology is being implemented.

But we can't do that if their use is secret. This is the crux of the ACLU's argument against law enforcement's use of Rekognition. As the *Washington Post* reported, "Not even the FBI is using its facial recognition software to track people in real time."[56] Privacy scholar Bedoya was quoted in that article as saying that the ACLU's findings are the "clearest example yet that these dragnet, real-time face recognition systems are real."

AMAZON KNOWS MUCH ABOUT US, AS WE'VE SEEN. THROUGH OUR browsing patterns, it knows what products tempt us. Through our purchase history, it knows what temptations triumph over us. Through these, it can infer our drives and impulses and create a profile of our consumption behavior—a mirror of who we are as people. With Rekognition, it will be able to assign a face to the behavior. With Amazon's home assistant, the Echo, it can also assign a voice to the face and the behavior. The question then becomes: If tech can know what we do, what we want, what we sound like, and what we look like, what's left? Are we truly and fully known by our net states? And, if so, what happens with that knowledge?

It would be naive to think that net states won't make use of this knowledge for some purpose. Evgeny Morozov, who has taken on the mantle of technology distruster, recently dubbed this practice "data extractivism"; he argues that its purpose is to train artificial intelligence systems.[57] "Once built," Morozov argues, "this AI capacity can be lucratively rented out to governments and companies."

With the IOT and smart cities, citizen-users will feed the net states with databases that profile us. Will we be aware that it's happening? And, if so, will we care?

As it turns out, people report worrying about what happens to their data, but they're not willing to give up their tech to protect it. "The trust train has left the station," as one professor reported to the Pew Research Center for Internet and Technology in a survey on technology and trust.[58] A software engineer wrote in the same survey, "People's trust will have zero correlation with reality. It is not appropriate to expect their feelings of trust to correlate with actual technological details."

Trust in technology may indeed be an issue of the past. It's certainly not the driving factor people rely on when they decide whether to use technology. Convenience, connection with people we care about, and plain old curiosity are much better indicators of whether we'll sign various terms of service and give away our data to parties unknown.

In this way, Amazon knew us best from the very beginning: we don't look to tech for trust, we look to *other people* for trust. The customer reviews on Amazon—the first of any site on the web—got it right. We want to hear from other people what they think about the world. We want to get in on the experience they describe, whether it be a book or film or product we're told they just can't live without. And, in this way, we simply want to be connected with others; to be part of the conversation, even if it's just as an anonymous, passive observer peeking in on an online comment. For by reading others' thoughts, we are less alone with our own thoughts. And perhaps we feel just a little less alone.

We may not trust tech, but somewhere deep down, we still trust people—even strangers online. Technology is simply the conduit that gives us access to others, especially like-minded others who love the same books we love, the same movies and gadgets. For all the ways that tech isolates us, it also creates avenues for connection to a quasi-experience of kinship—ways to find others like us, who like what we like and want what we want. It may not be as satisfying as a friendship with a person we interact with face-to-face, but maybe it's better than no connection at all.

The IOT won't connect us to other people—yet. It connects tech to other tech. But the data collected by IOT devices will be applied to some purpose at some point. And, if history provides any indication, increased connection to others like us is a fair bet for future applications. For all our solitary endeavors, we are social creatures. And we love finding others who love what we love. Net states would be wise to capitalize, and likely already are capitalizing, on that for their next generation of products and services: providing yet more ways for us to find a feeling of serendipity—even if we know deep down it's just an algorithmically generated facsimile.

SEVEN

THE MIND, IMMERSED

This chapter moves from our habits to our head, exploring what happens to our brain, for good and ill, when engaging with net state technology. It describes how immersive technologies, starting with our most closely held technology—the smartphone—both facilitate learning and interfere with cognition. It also examines what we stand to gain from new forms of immersive tech and asks how our desire for mediated reality that's especially designed for us as individuals might be fulfilled by these new technologies (and what that fragmenting of experiences might mean for us as a people).

The modern person drives a lot. And much of the technology installed in our cars has made transportation radically safer than in the past. For instance, the National Highway Traffic Safety Administration estimated in 2017 that airbags alone had saved 50,457 lives in the US since they were first introduced in 1987—the number of people it would take to fill a major-league baseball stadium.[1]

Yet some of the technology we *use* in our cars may endanger us and the people around us even more dramatically. This section explores how.

On an average day in 2018, American car-owners spent 55 minutes apiece behind the wheel, taking, as a group, approximately 1.1 billion car trips.[2] That's 1.1 billion chances, every day, that someone—managing a complicated piece of machinery hurtling through space at high velocity with thousands of other vehicles doing the same in the face of countless potential distractions—could crash.

But, for the most part, people don't. Despite distractions from the weather, the road, the radio, other drivers, passengers, food, and beverages, we manage—by and large—not to drive our cars off the road in a fiery blaze.

This holds true even when we space out—and apparently we do that a lot. According to a 2016 report published by the National Institutes of Health (NIH), every single respondent in the study admitted to "mind wandering" every time he or she drove.[3] "Mind wandering" behind the wheel of a car sounds like a recipe for disaster. But statistics suggest that people overwhelmingly manage to stay on the road, despite being checked out at least part of the time.

Consider the numbers: in 2018, nine of every 1.1 billion car trips ended in a fatal car crash due to distracted driving.[4] That sounds like a lot, and even one fatal crash is, of course, too many. But let's change the perspective for a second and consider this statistic another way: out of the 1.1 billion car trips taken every day, fewer than 0.000001 percent of them resulted in a fatality related to distracted driving. Statistically speaking, that's a bizarrely small percentage.

To understand what really happens in our brain when we're operating a car, a team of research scientists at the University of Texas decided to study drivers during episodes of various types of distraction. In a study published in *Nature* in 2016, they found that drivers were surprisingly capable of staying on the road, not only in the presence of garden-variety mind wandering, but even when experiencing emotionally distressing thoughts and cognitively demanding tasks.[5]

What they found is that even when we don't pay active attention, some part of our brain keeps watch. As Ioannis Pavlidis, who led the study, remarked, "The driver's mind can wander and his or her feelings may boil, but a sixth sense keeps a person safe, at least in terms of veering off course."[6]

This "sixth sense" refers to the part of the brain—the anterior cingulate cortex, or ACC—that automatically intervenes when it detects conflict from cognitive, emotional, or sensorimotor stressors. In other words, we can space out or think angry, sad, disturbing, or distressing thoughts and still not crash because there's a part of our brain standing guard. The brain's ACC minds the sensory-information shop, so to speak, and ensures that whatever we smell or taste or hear or see or feel that is somehow not right triggers a response.

There is, however, one exception that prevents this sixth sense from working—one activity that fails to trigger the ACC's intervention response.

I'm sure this won't surprise you: it's texting.

Based on the findings in Pavlidis's study, the only time they found that the ACC failed was when people engaged with their phones. The subjects' neurological watchdog simply went offline. "The moral of the story," Pavlidis explained, "is that humans have their own auto systems that work wonders . . . until they break." Ultimately, what is causing the systems to break is tech.

This holds for anytime we're using our tech. Take walking. Emergency room visits for pedestrian injuries increased a sobering 500 percent from 2005 and 2010.[7] And the most tech-obsessed demographic is also, depressingly, the hardest hit: in 2016, out of almost 6,000 pedestrian fatalities by cars, 61 percent were 15 to 19 years old, with most deaths due to what the National Safety Council described as "distracted walking"—that is, walking while immersed in one's tech.[8]

We'll delve later into what it is *about* tech that causes the break in our ability to save ourselves. Suffice it to say here that something about using tech wreaks havoc on our neurological protection system; tech is too immersive, too distracting—too *something*—to let our brain step in and save us from the consequences of inattention. In other words, something about tech draws us in—and not just our active mind, but our subconscious—even to the point where it can put us in danger.

"TECHNOLOGY MAKES US SMARTER."

"Technology makes us stupid."

"Technology is an extension of our brain."

"Technology impedes our ability to think."

And so goes the debate about the effects of technology on our thinking. Rather than hash out every aspect of the tech/brain debate, let's focus on the most personal aspects of our relationship with net states: how our technology use impacts our ability to learn, focus, and function. Let's look at the first of these abilities next. We'll circle around to the latter two later in the chapter, when considering internet negatives.

Of that bunch—learning, focusing, and functioning—learning is the given for positive impact. If our capacity to learn is tied in part to having access to materials from which to learn, then technology provides a slam-dunk win for the human race. Technology provides more *opportunities* to learn—not necessarily to focus or recall, but to find information that can teach us stuff. As technology has evolved into smaller, more personalized devices, we can take advantage of opportunities to learn faster and more easily than ever before. Consider that just two decades ago, if you didn't know the capital of Mongolia or how to calculate standard deviation, you'd have to look it up in a book in your home, if you happened to have an encyclopedia or a statistics textbook, or physically go to a library or bookstore. Now, thanks to Google and Bing and other search engines, the answers are literally seconds away. (That would be Ulaanbaatar and, well, you can look up how to calculate standard deviation yourself.)

Microsoft may have brought computing to the masses, but it was Apple that made computing an individualized experience. From the iMac to the iPod to the iPhone, Apple made computing increasingly personal throughout its product evolution. As noted previously, the key is in the marketing label "i": tech products for you and you alone. For decades, Apple has promoted computing as a relationship between human and machine, and a very intimate one, at that. True, encouraging each individual to buy their own Apple computer or phone rather than sharing one with their family means that Apple sells more Apple products. But there's more to it than that.

Throughout the history of Apple's computer products, they have been the drivers behind making tech smaller and more portable. Back in the early 1990s, it was eminently reasonable that each household would have only one computer: desktops were big; you needed a desk to house the tower (the computer itself) plus the monitor, mouse, keyboard, and modem—all of which were separate, space-consuming devices. Apple led the way toward shrinking tech products, starting with its earliest desktops and leaping into the modern age in 1999 with the unveiling of the iMac—available in five candy-pretty colors.[9]

Despite the size—it was still a hefty machine, weighing in at 40 pounds—it begged for personalization: one's teenage daughter might opt for tangerine, while the grown-ups in the house might choose the slightly more muted blueberry.[10] Then with the iPod music player in 2001, Apple's offerings graduated from customizable machines to customizable music collections. With the iPod, users could not only upload just the music *they* liked, but make multiple playlists. With this move, users liberated themselves from the studio album format and began creating their own versions of albums with as many artists as they wanted.

Finally, the iPhone arrived in 2007, taking Apple's personalization/shrinkage quest to a whole new level. In a single, pocketable device, web browsing and picture-taking and texting and game-playing were suddenly available to you and you alone, anywhere you happened to be. Users were now liberated from the desk and from the house. With the iPhone, the internet left the building.

Here's how the iPhone relates to our intelligence—or, specifically, our learning.

According to what's known as "the Flynn effect," IQs—intelligence quotients—have been steadily rising over the past hundred years, gaining approximately 30 points worldwide across that century.[11] This means that the average person in 2012 (when the study was conducted) was smarter than 95 percent of the population from 1900.

While several factors likely contributed to this effect, including better nutrition and medicine, the global revolution in schooling remains the most influential contributor. In 1900, for instance, only 6 percent of Americans had a high school diploma.[12] Compare this with 88 percent of Americans in 2015.[13] Higher education was even more elusive a century ago. In 1900, only 3 percent of the population had finished college. By 2014, more than 40 percent of Americans had college degrees.[14] A meta-analysis of studies suggests a direct correlation between how many years one attends school and his or her IQ score, with a global composite average of an additional 1.9 IQ points gained for every year of schooling.[15] The general takeaway here is that, on the whole, greater education yields a slight increase in cognitive abilities.

Learning is certainly associated with formal schooling, but teaching oneself counts, too. And thanks to the internet, opportunities for self-education have exploded. According to a study by the Pew Research Center for Internet and Technology, Americans with technology assets that include a smartphone *and* a broadband connection are much more likely to engage in personal learning than those without tech.[16] And we don't feel forced to learn; we dig it—72 percent of respondents report enjoying having access to so much information online, with only 26 percent reporting feeling "overloaded."[17]

Perhaps in some ways, the reason most of us don't feel overloaded is simply that we've adapted to our current reality. As James R. Flynn, the scientist behind the rising IQs study and author of the book *Are We Getting Smarter?*, noted in an interview, our increased IQs and global commitment to educating our youth don't mean we're somehow better than our counterparts of a century ago; we've simply adapted to a more complex information environment. We *needed* to become smarter, in other words, to be able to manage more complex information. Indeed, it's not that the entire year-1900 population wrestled through a daily struggle to survive, dragged down by their "low" IQ scores. For 1900, their scores weren't low at all. They were just era-appropriate—exactly the kind of IQ necessary to cope with *their* information environment. The world that people encountered simply wasn't as complicated back then; 30 fewer IQ points was no hardship, it was the norm. So, while we can view the global elevation of IQ points as a sign that we are becoming smarter as a people, we didn't really have a choice in the matter—improved cognitive abilities were, perhaps, just a natural adaptation for us to survive in the complex landscape that defines our time.[18]

This landscape is both fueled and shaped by technology. As noted, the internet makes self-education easier than ever. Indeed, the tech *industry* itself is the paragon of self-teaching: 69 percent of coders report being self-taught, as opposed to having learned their skills in a classroom.[19] Even beyond coders, regular people learn things online fairly frequently: 59 percent of Americans use search engines on a daily basis, which equates to roughly 191,750,000 people searching for information daily, totaling at minimum around 321,000,000 queries a day.[20]

Clearly, not all queries are about substantive or educational topics. The five most popular search terms in 2017 were "Hurricane Irma," "Matt Lauer," "Tom Petty," "the Super Bowl," and "the Las Vegas shooting."[21] While two of the five concern socially impactful events—the hurricane and the mass shooting in Las Vegas—the others appear to be more focused on entertainment and scandal-of-the-moment curiosity.

But the broader point is actually not *what* people search for; it's *that* they're searching, period. People are curious, and tech provides them the means to satisfy that curiosity. Whether they're inquiring about how to make slime or solar eclipse glasses (those would be the top two "how-to" searches in 2017), or about national tragedies like the largest domestic-terrorist mass shooting in the history of the United States, almost doesn't matter.[22] What matters is that, thanks to tech, people have the power to find things out. And, given the high number of searches that come from mobile devices, we're often finding out things from our tech the very moment we wish to know them.[23]

Curiosity matters. It's one indicator of cognition—different from IQ, but important in its own right. In fact, in these increasingly complex times, some researchers have suggested that in addition to our IQ, we should take stock of our EQ, our emotional intelligence quotient (or how good we are at recognizing the emotions of ourselves and others), and our CQ, or curiosity quotient (how hungry our mind is to find things out).[24] The reason our curiosity quotient matters so much is that people who are more curious tend to be more comfortable with ambiguity.[25] And comfort with ambiguity is an indicator of tolerance.[26] Tolerating ambiguity—the unknown or the unresolved—is difficult, but it's also really good for us: people who can tolerate unknowns for sustained periods have been shown to be more innovative, more optimistic, less vulnerable to depression, and more comfortable with taking risks.[27]

Clearly, there are downsides to checking our phone 80 times a day on average.[28] It's highly likely that at least some of these checks are compulsive, almost unthinking. We've already established that phone-checking can be a ruse to avoid interacting with people nearby, at least some of the time. But it's also an indicator that we're curious—to learn new things, for new stimulation, for updated information. Regardless of the subject of our curiosity, just being curious is cause (albeit small) to celebrate.

NO ONE IN THE UNITED STATES IS FORCED TO CREATE A SOCIAL MEDIA account, to buy or sell anything online, or even to maintain a digital footprint of any kind. But almost three decades after the advent of the World Wide Web, is it really possible to be a fully functioning citizen without some sort of digital presence? What would digital invisibility cost you?

To answer this requires first understanding the current context. Americans are lucky, digitally speaking. As I'm writing, 90 percent of us have internet access, and that's *high-speed* access for 68 percent of us.[29] We take advantage of that access too: 90 percent of Americans use the internet on a regular basis.[30]

Which means, of course, that 10 percent of Americans don't. The reasons *for* being online are fairly well understood by now—fast and convenient access to information, services, shopping, social interaction (of a sort), and so on. With all these benefits, several questions arise: Why would anyone stay away? And, in our increasingly digital economy, how would they manage? Finally, is there really any downside to staying offline?

For all the free information online, accessing the internet isn't actually all that cheap. Computers and smartphones cost money—even a cheap laptop runs around $300, and the going rate on smartphones is $400 to $500. Data plans or broadband connections will also cost you: somewhere in the neighborhood of $75 to $125 a month for either one. Add this up over the course of a year and you're spending $1,500 or more just to go online.

But money's not actually the issue. Of the 10 percent of Americans who are offline, only 19 percent cited lack of money as the prohibitive reason. Over a third reported that they simply weren't interested in the internet or didn't think it was relevant to their lives. Age, income, and educational attainment were all factors associated with being offline—those over 65, those earning less than $30,000 a year, and those with less than a high school diploma were the most likely groups to stay offline.

These all seem predictable. For example, 47 percent of the group who are offline are over the age of 50, with decreased use more prevalent the older the person. But what's more surprising is the cohort of younger, wealthier, and more educated people who aren't online. *This* group isn't keeping offline because they can't afford internet access or don't understand it. They simply choose not to use it—or don't permit their children to use it.

Chief among this group are the net state developers: in 2018, Silicon Valley parents were among the cohort most likely to keep their children from using the very same technology they created. This group, through the years, has included Apple's late Steve Jobs, Microsoft's Bill Gates, *WIRED*'s Chris Anderson, and Twitter's Evan Williams, all of whom practiced what's called "low-tech" or "no-tech" parenting.[31]

The rationale behind low-tech parenting is fairly sobering. Tech, in the view of many (including a number of its creators), can do bad things to kids. As Napster cofounder and former Facebook president Sean Parker put it, "I don't know if I really understood . . . the unintended consequences of a network when it grows to a billion or 2 billion people. . . . It literally changes your relationship with society, with each other. . . . It probably interferes with productivity in weird ways. God only knows what it's doing to our children's brains."[32]

While low-income families struggle to ensure that their children don't miss out on the chance to be "digital natives," the trend among the wealthiest and most tech-savvy Americans is to keep their children *away* from digital devices for as long as they can. Former Facebook executive Chamath Palihapitiya made headlines when he said publicly that he felt "tremendous guilt" for what he'd helped to create.[33] "The short-term, dopamine-driven feedback loops that we have created are destroying how society works," he said. And he doesn't let himself off the hook for it. "In the back, deep, deep recesses of our mind, we kind of knew something bad could happen."

Keeping children away from social media sites like Facebook or limiting time spent on online games is one thing. There's plenty of research that now supports the notion that excessive use of social media leads to negative health effects, especially among the young, such as anxiety and depression.[34]

But there remains an unexplored question: Beyond parents who choose to keep their children offline, what about adults who can afford internet access and are educated enough to navigate the web, but still choose themselves to remain offline? Why might *they* do it? And more important, what does it cost them, if anything?

One of the first reasons to stay offline is to avoid a digital footprint, a publicly accessible online profile. Strangers can get a pretty detailed impression of us from reading about us online—and many people aren't comfortable with that.

Even if we don't admit it, most of us have Googled ourselves at some point, if only to ensure that nothing damning or inaccurate comes up in search results. It's time well spent. Employment recruiters attest that one of the first things they do when assessing candidates is to look at their digital footprint. And not having one can be just as problematic as having a negative one. It can signal the absence of tech smarts, as one recruiter reflected: in the new millennium, "lack of technical skills makes you an employment pariah."[35] Simply put, in our computer-dominated work world, most positions require some degree of technical know-how. The total absence of an online persona may look more suspicious than even a slightly embarrassing one. That's a significant cost to avoiding the internet.

Another reason some adults choose to remain offline is the threat of reduced focus. While we *can* learn just about anything using the tech at our fingertips, tech is killing our concentration, a necessary precondition of learning.

Problem number one, focus-wise, is our technological attention-grabbers. Thanks to haptics—the physical buzzes and tics our phones give off when we get new messages, new calls, new texts, and new alerts—we're physically nudged by our phones, gently yet unfailingly, on a fairly regular basis. This effectively shatters our ability to focus. Tech may make it possible to learn whatever we want, but whether we can pay attention long enough to *process* what we've learned is a different matter entirely.

Most of us have apps on our smartphones, and most of those apps send us alerts. Getting a single notification on our phones may not seem like a big deal. As it turns out, though, we're not the best judge of what constitutes a distraction.

In a recent study, participants were asked how often they got push notifications on their phones. They reported that they got around 20 a day. That's not nothing, but it's a seemingly manageable amount, averaging about 1.2 notifications per waking hour, spread out over 16 hours (assuming most people sleep eight hours). In reality, however, these participants actually received an average of 63 push notifications per day—more like four per waking hour. What's more, participants in the study didn't just *get* the alerts; they usually took the time to unlock their phones and view the notifications within minutes of receipt.[36]

That push notifications might negatively impact the brain isn't hard to imagine. In an NIH study on the specific brain waves associated with concentration and cognition, researchers found that push notifications resulted in impaired cognition in the period immediately following receipt of the notification.[37] What's more, the subset of participants who were deemed at risk for overusing their smartphones also exhibited a diminished ability to complete tasks after receiving notifications. In other words, just getting notifications hindered people's ability to focus. For some people, it hindered their ability to do things as well—in other words, hindered functioning.

Even when we're not getting alerts, our bodies have become conditioned to think we are. In a 2012 study reported in the journal *Computers in Human Behavior*, 90 percent of respondents reported feeling "phantom vibrations"—the eerie sense that their phone was buzzing, even when no alert had come through. And this was fairly frequent: at least once every two weeks.[38]

To track the ill effects of our forever-on technology, a group of former net state engineers who founded the Center for Humane Technology (CHT) created what they called the "Ledger of Harms," a roster of studies into just how damaging our tech can be to our attention.[39] "This Ledger collects those negative impacts of social media and mobile tech that do not show up on the balance sheets of companies, but on the balance sheet of society," the website says. That tech creators themselves—the founders of the CHT come from Facebook, Google, and other major net states—have mounted a campaign *against* their own products is a powerful indicator that the current quantity of time we spend with our tech is likely far from benign.

But let's break down what the evidence really means. As we just read, our phone physically prods us 63 times a day, or once every 15 minutes, with a sound or vibration that temporarily annihilates our focus and messes with our ability to function—to do the thing we were doing before we got the notification. Worse yet, checking our phone because we got a notification is just a subset of the overall number of times we check it: that would be an average of 80 times a day, or once every 12 minutes.[40]

There are a number of obvious problems with this intermittent activity. Our train of thought is routinely derailed. Our ability to be present in the moment with the humans standing physically in front of us is greatly diminished. Our neck gets cramped from craning over our phone for long periods. But what's really interesting about this isn't how tech affects our attention, our cognition, or even our ability to function. What's truly intriguing about all of this is the possibility that we're seeing a new kind of information environment that demands its own form of adaptation—a new kind of survival mechanism. It's called the "adaptability quotient."[41]

Imagine this scenario. You're at work. Your phone buzzes. You check it; then you go back to the email you were in the middle of writing.

Three things have to happen here. First, you have to break away from a train of thought. Second, you have to engage in a new thought process to deal with whatever the notification was about. And third, you have to gather the thoughts that got scattered about when the notification distracted you. If you have a high adaptability quotient, you may be able to train yourself *not* to lose your train of thought, but rather to hold it aside, in a suspension of sorts, so it's still there when you come back to it.

Before we look in more detail at what the adaptability quotient is, let's consider what it is *not*. It may sound a bit like multitasking, but there are key distinctions. First, no one can actually perform multiple tasks at the same time. What we think of as multitasking is actually performing tasks in batches and quickly switching between them.[42] As it turns out, doing so comes at a cognitive cost. We're ultimately less productive and even less smart: our brains operate at a lower intellectual capacity—our IQ literally drops downward about 10 points—when multitasking. This is because we experience what's called "attention residue," a sort of mental jetlag that ties us to whatever we were just paying attention to. And it lingers, especially when we've switched to a new task without first completing the previous one.[43]

Attention residue isn't just annoying; it gums up our mental works, stripping us of the ability to cleanly switch cognitive gears. In other words, task-switching before we've finished what we were working on hurts us in multiple ways. First, our original task is left incomplete. Second, we have less brainpower to focus on our new task, making us approach it a little less effectively than if we had just finished the first thing we were doing.

We know from the earlier example about distracted driving that tech shuts down the brain's sensory information monitor, the ACC, whose job it is to detect conflict from cognitive, emotional, or sensorimotor stressors, ensuring that even if we're sort of spacey, we notice when something dangerous comes our way. With the ACC shut down, when something dangerous *does* comes our way, we may not notice until it's too late.

Now, consider the number of distractions we get per day via tech: again, an average of 63 notifications. We don't know when we'll get them, but we might anticipate that they're coming. The anticipation alone takes a toll on our cognition. A study in the UK showed that just knowing an unread email was waiting in one's inbox made workers less cognitively capable and resulted in an average loss of 10 to 15 IQ points.[44] As we saw earlier in this chapter, research predicts a 1.9-point increase in IQ for every year of schooling. Turning that finding on its head, we could conclude, based on the IQ loss found in the UK study, that just by virtue of knowing you have an unread email waiting in your inbox, you lose 7.5 years of school, going from being a college graduate to a freshman in high school.

This all points to a big problem. The fact that mere *anticipation* of unread messages obliterates our cognition, and that we anticipate that we're going to get alerts on our phone every 15 minutes, seems to suggest we are, as a people, probably a bit dumber than we would be without our phones.

Today's constant notification environment is quite new. It started with the pager, a pocket-size device popular in the early 1990s that would alert you to call someone. It may be difficult to imagine more than two decades into the internet era, but pre-pager, once you left your home or office, you were simply unreachable. Unreachable—or free, depending on how you looked at it. Pagers changed all that. You could be pinged (or "beeped"; the other popular moniker for the device was a "beeper") when someone was trying to reach you, no matter where you were.

Pagers had their moment. Back in 1994, when Motorola dominated the market, more than 61 million people clipped the little things to their belts or tucked them into their pockets.[45] Though physically insubstantial, they changed everything. They were the first mass-produced, reach-me-everywhere device and the first mass distract-me-anywhere device.

In the 20 years since the pager's heyday, technology has only become more immersive—first with cell phones, then with smartphones. Pre-pager, we had to make a choice to connect with individuals who were not in our geographic vicinity—we had to physically go to them, or go to our home or office or a payphone to call them. Now we've lost the element of choice; we're on the receiving end of alerts telling us that other people want to connect with us—people we know, who have our number and send us texts or emails—and also telling us that certain apps want our attention: weather apps warning of incoming storms, game apps inviting us to pick up where we left off, news apps with "breaking" headlines, social media notifications that someone has tagged us in a tweet or a post. We're in an era of relentless incoming attention-grabbing.

Imagine the IRL equivalent: someone tapping you on the shoulder, no matter where you were or what you were doing, 63 times a day. The difference, of course, is that you can choose to ignore your tech-fueled shoulder-tapping. Yet, as the studies cited above show, simply getting the notification (even absent reading it) is enough to disrupt your focus. That's because the act of choosing to ignore a notification means first cognitively registering that you got it—all of which temporarily takes you out of the moment, away from doing whatever it is you were doing.

To survive this new threat to our mental functioning, we need to get smarter, but in a new way. We need to be able to adapt, quickly and frequently, to preserve our brainpower. We need an adaptability mechanism that can manage both the *possibility* of distraction and the *actual* distraction from the technology we carry and wear.

Which brings us, at long last, back to the adaptability quotient. Harvard professors Robert Kegan and Lisa Laskow Lahey argue in their book *Immunity to Change* that to stay relevant and competitive in our distraction-fueled environment, we need to get comfortable with making complicated, complex decisions regularly—so comfortable that we become, in his words, immune to even frequent change.[46] In other words, to survive this era, we need not just intelligence, emotional intelligence, and curiosity, but adaptability as well.

Training ourselves to become more adaptable, all while coping with distraction and the related diminished brainpower, is exhausting. It's no surprise, then, that distractions that create a sense of escape, like television, are as popular as ever. The good news is that there's a silver lining for our brains in our quest to escape our exhausting information environment. Thankfully, it turns out that certain kinds of TV may be just a little bit good for you, albeit in a sneaky way.

To figure out how, we can take a lesson from a high school chemistry teacher who transformed himself into New Mexico's crystal meth king.

Walter White's dying. His wife's pregnant. His teenage son has cerebral palsy. He's so broke that in addition to his job as a high school chemistry teacher, he works a second job at the local car wash. To leave his family something they can use to build a better life after his death,

Walter comes to the less-than-obvious conclusion that he should use his chemistry know-how to make crystal meth.

This is the premise behind the Emmy Award–winning show *Breaking Bad*, which aired from 2008 to 2013. It's one example of a new genre of show that exhibits what academics call "narrative complexity." Shows with deep narrative complexity—*The Sopranos*, *The Wire*, and *Game of Thrones*, to name a few—use a "historically distinct set of norms of narrational construction and comprehension" that distinguishes them from procedural shows like standard sitcoms, cop dramas, or reality shows.[47] They're what they sound like: complex. They're not just stories; they're fictional worlds with massive casts of characters. It is precisely because of their complexity that they're somewhat good for our brains.

Infusing our entertainment with complex narratives is important because it may very well counterbalance, to some small extent, the cognitive damage done by tech in other ways. To be clear, this is not to suggest that we should seek out *more* screen time or that doing so would be somehow good for us. Rather, it's a way of seeing one form of screen time we already escape to—TV—from a new perspective.

Ironically, it's the same tech that distracts us that's responsible for enabling this new genre of narratively complex shows. Thanks to the proliferation of tech platforms producing television shows—as when Amazon and Netflix entered the market—there was suddenly an appetite for content, opening up space for more shows that experiment with increasingly complex narrative structures.

And narrative complexity does do good things for our brain. It activates areas of the prefrontal cortex associated with episodic memory retrieval, thus enhancing overall memory retrieval abilities.[48] The more complex the characters in the story, the better: character-driven stories stimulate production of oxytocin, the hormone we feel when we experience love.[49] Narratives of all kinds also activate multiple areas of our brain—the language-processing areas, of course, but also, somewhat surprisingly, any area associated with the activity happening in the story.[50] This means watching a story about a person running through a forest activates the area of your brain that deals with your fight-or-flight response, simply because you can imagine running away from someone.[51]

The real reason this matters is that we continue to spend a lot of time with narratives. We're still watching a lot of TV, upwards of four hours a day.[52] Interestingly, the younger generation—18- to 24-year-olds—are actually watching TV far less: 2.7 hours a day, or about three-fourths that of their older counterparts. Games, social media, and short-form videos on YouTube account for a substantial portion of youth entertainment time. Surprisingly, so does reading.

Yes, reading is making a comeback. According to a report by the National Endowment of the Arts, reading rates have been on the rise since 2002.[53] And not just any reading: literature. People—especially younger people—are returning to stories: the 18- to 24-year-old cohort made the largest gains in reading, increasing in readership by 9 percent between 2002 and 2008, and the trend has only continued since then. Millennials are the most widely read age group, with 87 percent of them reporting having read a book in some format—print or digital—in the past 12 months, compared with 73 percent of the general population.[54]

Whether we read on our phone, an e-reader, or in print, reading gives us an invaluable path to elsewhere. It's called "transportation": stories "transport" our minds out of our present environment and into something else. Since our earliest ancestors first discovered the joy of swapping stories around the campfire, there's been only one surefire way to transport ourselves: to hear a really, really good story.

Technology may be changing that. Through truly immersive technology—augmented reality and virtual reality—we may be able to achieve this sense of being transported not only without a gripping story, but without any story at all.[55]

"NOBODY IN HERE . . . THINK[S] IT'S ACCEPTABLE TO BE TETHERED TO a computer, . . . because we're all social people at heart. I think that a significant portion of the population of developed countries, and eventually all countries, will have AR experiences every day, almost like eating three meals a day, it will become that much a part of you."[56]

This is Apple CEO Tim Cook in a conversation with Senator Orrin Hatch at a technology conference in Utah. In Apple's endless quest to personalize technology for its users, it's now placing bets on a next-generation technology interface: augmented reality, or AR.

Most of us will likely have heard of Google Glass, the short-lived, groundbreaking gadget that put computers on tiny displays you could see out of the corner of your eye. It was revolutionary. It was the best tech development since the iPhone.

It was also a total flop—in its first commercial iteration, at least. One unfortunate side effect of the Google Glass nosedive is that it gave *all* AR a bad rap. To explore what AR really is, it's first necessary to distinguish it from the product we most associate it with.

Google Glass sounded like science fiction come to life. Everyone could wear glasses that displayed features similar to those on your smartphone—just projected onto your glasses for you and you alone to see, with the real world perfectly in view in the background. The information you saw would depend on where you were in the real world. For instance, when you stepped outside in the morning, you'd see a weather notification with the current temperature. Walking down the street near your local laundromat, you might get a pop-up reminder that your dry cleaning was ready to be picked up. As you headed into the subway station, you could get a travel advisory warning you that the number-6 train was running with delays. Once you arrived at your subway stop, you could be shown turn-by-turn directions to navigate an unfamiliar neighborhood. And if you liked what you saw, you could take a photo with just a wink. Should you want to share what you were seeing with others, you could livestream your point of view directly from the glasses to the internet.[57]

Google Glass failed for several reasons, many of them marketing-related. When first rolled out, they weren't available for sale to the general public, only to "Glass Explorers," who had to pay $1,500 for the privilege. They also creeped people out: the notion that someone standing in front of you might be looking at a display—or, worse, taking your picture without your knowledge—rather than making eye contact with you was deeply unsettling.

AR hasn't gone away, though. Instead, it's migrated to games. And here is where the kinks are being worked out to pave the way for mass adoption. The first wave of mass-personal-AR use arrived with Pokémon Go. Yes, it's technically a kids' game. But it was also groundbreaking, raising public awareness about what AR is and can do in an easy-to-understand way.

The premise is simple: Download the Pokémon Go app on your Apple iPhone or Android device. Load the game. Then point your phone's camera at the path in front of you. By tracking your physical location using your phone's GPS, accelerometer, and compass, the game pops cute little animated creatures—those would be Pokémon—into your field of vision. You then have to physically approach them to "catch" them, which of course wins you points.

Pokémon Go marked the world's first mass-AR game success. Around the world, more than 45 million users downloaded the game within the first days of its release in July of 2016. In the intervening year, the game was downloaded 750 million times[58]—equating to a stunning 10 percent of the global population. It's been heralded as the breakthrough tech for AR, just as the iPhone was for smartphones and Fitbit was for wearable health trackers.[59] Pokémon Go succeeded largely because users already had the device necessary to access it: a smartphone. You didn't have to carry around any additional gadgetry; you just had to download an app.

Apple's Tim Cook is betting on AR, as we heard in the earlier quote—not just thanks to the success of Pokémon Go, though it probably didn't hurt that iPhones were among the primary reasons for the game's massive success. Apple has partnered with several big brands, including Ikea and American Airlines, to promote the iPhone's AR capabilities[60] and has released a kit for app developers, the ARKit, containing tools to help them create new apps using AR.

While the AR market is still small, recall that the website market was small back in the 1990s too. It's only a matter of time before AR use reaches a tipping point where any serious company will have to develop an AR-enabled presence. It may not be overnight. After all, the first handheld e-reader, the Rocket, was introduced in the late '90s, but it wasn't until Amazon launched the Kindle in 2007 that e-books really took off.[61] But all signs point to AR being at the forefront of the next phase of our relationship with technology products.

Immersion technologies such as AR matter because they represent a shift in how we experience reality. Pre-AR, we're already presented with different versions of reality when we go online: as noted earlier, the products Amazon displays to you are different from what they display to me; our news, Facebook, Twitter, and Instagram feeds are similarly customized. Even our search results are tailored: you and I could both Google the same keywords and end up with different results depending on "personalization" factors that include geographic location, search history, and the customized profiles created from our interaction with any and all Google products.[62]

With AR, we'll be able to see different images right in front of us, in the real world—though mediated through our phones. What's more, soon we'll be able to share what AR shows us with other users: in June 2018, Apple announced a new feature that enables users to share AR imagery between two phones.[63] We'll move from personalizing our online experiences with our handpicked network of "friends" to personalizing our real-world experiences with the same. While the present capability permits only two people to share an AR experience, there's no reason to believe this won't eventually scale to include groups as well.

C. S. Lewis wrote, "What you see and hear depends a good deal on where you are standing; it also depends on what sort of person you are."[64] As human beings, we already experience reality differently. Depending on our background, our values, our mood, and our state of physical wellness, you and I and someone else entirely may interpret the scene of a parade as exciting or overwhelming, as beautiful or depressing.

Technology won't be able to level the perceptive playing field for us; indeed, it will only further fragment what we see. In other words, we already don't share the same *perception* about a uniform set of things in the world. Soon, with tech, we won't even *see* the same set of things in the world. We'll immerse ourselves in customized, personalized versions of reality that are unique to us and, perhaps, a select few we invite along for the ride.

Like it or not, immersion technologies such as AR will likely be at the center of the next stage in our relationship with tech. Not only will we wear our devices, but we will use them to filter what we see. Apple CEO Tim Cook doesn't see this as a bad thing. "AR has the ability to amplify human performance instead of isolating humans," he remarked.[65] AR's biggest success story, Pokémon Go, certainly seems to support this hypothesis. According to one study on the game's effects on users, the AR experience on one's phone is associated with "increased positive affect, nostalgic reverie, friendship formation, friendship intensification, and walking," all of which are associated with well-being.[66] All of this is to say that AR may well alter our sense of reality, but that new reality may equate to a really good time.

IN NEAL STEPHENSON'S AWARD-WINNING 1995 SCIENCE FICTION NOVEL *The Diamond Age: Or, a Young Girl's Illustrated Primer*, a destitute young girl accidentally gets hold of an interactive virtual reality (VR) "book"—a primer—intended to train an aristocrat's daughter on how to comport herself in the upper echelons of their world. The book "grows"—becomes more complex—as the girl ages, adapting to her intelligence, her curiosity, her emotions, and her experience. Lots of other stuff happens, but for our purposes, let's just focus on the primer itself.

Immersive technology that adapts to our every need is not a new concept (think of the Holodeck on *Star Trek* and the training modules in *The Matrix*), but the fact that the idea keeps coming back in story after story reveals something about human desire. We seem to long for our tech to not only entertain us, but train us. But there's a catch: it must be in exactly the right modality for each of us, personally. This tool would feed us the precise experience and challenge to advance us to the next level of expertise, of ability, of comprehension—a "gamification" of learning that allows us to level up all the way to enlightenment.

Perhaps no other technology outside AR and VR has ever held as much potential to meet this desire. As VR filmmaker Eugene Chung noted, "We've effectively had the same flat screen medium since 1896. VR/AR uniquely provides a sense of presence and immersion; it's a brand new art form and brand new form of experiencing."[67]

Immersive tech is not being developed just for gamers and niche markets. Tim Cook describes AR as potentially game-changing for everyone's everyday experience of technology, just as the iPhone was. "The smartphone is for everyone," he reported in a 2017 interview with the *Independent*.[68] "We don't have to think the iPhone is about a certain demographic, or country or vertical market: it's for everyone. I think AR is that big, it's huge." Apple is putting significant money behind this sentiment. In the spring of 2018, reports leaked of an Apple headset in development that would facilitate both VR and AR. With Apple's characteristic commitment to visual aesthetics, the leaked plans describe an 8K resolution display per eye—for context, this would be a higher resolution than the best TVs on the market today.[69]

Whether Apple's headset, slated to come out in 2020, will allow us to truly immerse ourselves in another reality—one that's vivid and complex and that *feels* real—has yet to be determined. But given how life-altering our current form of immersive technology is—introduced to us through Apple's iPhone—it's possible that this company has what it takes to produce yet another mass-magical tech moment with its AR/VR device: the ability to become fully engaged not in one's physical surroundings, but in one's mental experience. Whether this will ultimately be good for us is almost impossible to predict; what we *can* predict is that such technologies are coming. We should be prepared for what could be the next great unknown tech-mediated life-changer.

EIGHT

A DECLARATION OF CITIZEN-USER RIGHTS

It's time to take a hard look at where we've been in recent history and where we are now, with staggering rates of depression, addiction, and—unique to America—acts of gun violence. This chapter gives us options for how to reconcile our feelings of disempowerment via tech with the feelings of being empowered by it at the same time; it explores what we as citizen-users must do to ensure that we remain actively engaged in the development of net states, monitoring their relationship with our nation-states and with our own lives; and it offers a citizen-user pact with net states.

In 1948, only 82 independent sovereign nations ruled the world—113 fewer than there are today—thanks in large part to colonialism. And the countries that did exist were unstable and weakened: the Second World War killed over 55 million people through conflict, disease, famine, and other war-borne causes. It took a devastating toll the world over.

While the war formally ended in 1945, the years following were far from peaceful. The Soviet Union, which had lost a staggering 27 million people to the war, quickly consolidated power by occupying the neighboring countries that had fallen to shambles following the German invasion: Poland, Czechoslovakia, Hungary, Bulgaria, Romania. In 1946, British prime minister Winston Churchill called these Soviet postwar occupations the lowering of the "Iron Curtain," thus describing what would become a 45-year-long Cold War between the Soviets and the Western world—former allies that had fought alongside one another just one year prior.

Over the next few years, the world only grew more complicated. In January of 1948, Mahatma Gandhi was assassinated. In May, the United Nations recognized Israel, at which point Egypt, Jordan, Syria, Lebanon, Iraq, Saudi Arabia, and Yemen attacked, setting the tone for 70 years of hostilities in the region. In June the same year, the Soviet army blockaded Berlin, preventing the delivery of supplies through all roads and rails and fanning the flames of the nascent Cold War. The government of South Africa, also that year, launched a policy of apartheid, creating an institutional system of discrimination that would plague the country for the next almost 50 years. In short, the world was a disorderly and dangerous place.

Despite this, the majority of countries did agree on one thing: the atrocities of World War II, a war in which more civilians were killed than uniformed soldiers, had to be prevented from recurring at all costs. The world needed rules for war to safeguard the innocent and to protect fundamental rights. This gave rise to the United Nations, the Universal Declaration of Human Rights in 1948, and the expansion (from multiple earlier forms) of the Geneva Conventions in 1949.

Unlike the Geneva Conventions, which set out the rules and conditions for "humane" conduct during wartime, the Universal Declaration of Human Rights wasn't legally binding, making its success all the more astonishing. It was, at the time, almost like a formal suggestion, albeit a powerful one. It enumerated minimal rights that should apply to every living person on Earth.

On December 10, 1948, for the first time in history, dozens of nations committed themselves to a pact not to other nations, but to humanity itself: acknowledging that regardless of nation of origin or citizenship or religion or any other factor, humans anywhere and everywhere, from the moment of birth through their last breath on Earth, possessed intrinsic rights. And with this document, the idea of people having rights that supersede and exist independent of any particular nation-state bloomed into life.

There is no net state equivalent to the United Nations—no body where net state leadership can come together and question whether the application of their products and services will ultimately improve the world and its citizens. Absent that body, there's no pact defining standards and protections that members adhere to.

This chapter argues that there should be: our world needs a pact that establishes citizen-user protections from net states; a set of principles that guide net state developers in becoming better stewards of the digital ecosystem (more later on just what that ecosystem is).

Protective pacts generally reside in the realm of the nation-state—either individually, through a constitution; or multilaterally, through documents such as the Geneva Conventions. Before turning to what a pact between net states and their citizen-users might look like, it's worth taking a look at the pact already familiar to most of us: the US Constitution with its Bill of Rights.

TEDDY ROOSEVELT IS ONE OF AMERICA'S MOST FAMILIAR AND MOST beloved presidents, coming in second only to Abraham Lincoln in one ranking of past presidents.[1] Roosevelt was one of the most literary of all presidents, writing 35 books in his lifetime. He was also one of the most idealistic, devoting his considerable writing talent to urging his fellow Americans to live principled, "strenuous" lives.

On government, he held particularly strong views. In a 1902 speech to thousands of citizens gathered in what is now Pack Square Park, a tree-lined, open-air meeting grounds in Asheville, North Carolina, Roosevelt spoke about the need to heal the wounds from the Civil War.[2] He called upon Americans to do so by becoming active participants in their democracy, saying, "We get in the habit of speaking about government as if it were something apart from us. Now, the government is us—we are the government, you and I. And the government is going to do well or ill accordingly."[3]

If we, the people, are indeed government, who does that put in charge? It turns out that both conceptually and actually, the American government does have a boss—and it's not the president. Government is both empowered and restrained by the US Constitution: it can do only what's been outlined in this document. Of course, Congress has passed many laws since the ratification of the Constitution back in 1789, but none of them can override the Constitution itself. The American government serves the American people, according to the rules detailed in the Constitution, not the commander in chief: even he or she is beholden to the Constitution.

The president and everyone else who works for the American government has, then, five basic responsibilities that are detailed in the Constitution's preamble: to "establish justice, insure domestic tranquility, provide for the common defense, promote the general welfare, and secure the blessings of liberty to ourselves and our posterity."

The first three responsibilities don't tend to be terribly controversial. As a people, Americans largely support the idea of the rule of law: 68 percent of Americans trust the judicial branch of government, a far higher amount than those who trust the executive (45 percent) or the legislative (11 percent).[4] Americans generally prefer a peaceful and orderly society over, say, anarchy. Americans also really like the military, with 74 percent trusting that it will keep them safe. But the final two—to "promote the general welfare" and "to secure the blessings of liberty"—are harder to define and much more difficult to find consensus on.

There are exceptions: as regards "liberty," 85 percent of Americans believe in the freedom of speech. In fact, so important is this particular freedom to Americans that 73 percent report they would be willing to die to defend it.[5] In the liberty department, Americans are generally content: 87 percent report having a lot of freedom in their lives.[6]

While we may be happy with the degree of freedom we enjoy, this doesn't tell the whole story of American happiness. Because Americans aren't actually doing all that well, as a people.

We're in greater danger than we used to be, for instance. Life in the United States is less predictably safe than it was twenty years ago. As discussed previously, part of the reason for this is the epidemic of gun violence: from 1995 to 2005, there were 112 mass shootings in the US.[7] In the decade from 2006 to 2015, there were 310 mass shootings: a 176 percent increase. In 2018, there were 340 mass shootings: an average of 28 a month.

Drug overdoses are up, too. Overdose fatalities from opioids were five times higher in 2016 than in 1999.[8] It's a widespread problem: in 2016, 26 states saw a statistically significant increase in fatal drug overdoses. And it's getting worse: by 2019, 130 Americans died each day from opioid overdoses.[9]

Suicides are also up. Death by suicide is now the 10th leading cause of death in America.[10] And there's been an especially troubling recent spike among young people. From 1950 to 2014, deaths by suicide among 15- to 24-year-olds increased by 255 percent.[11] Among even younger children—those 5 to 14 years old—suicides increased by a heartbreaking 500 percent.[12]

Americans are not doing all that well as a democracy, either. As noted earlier, we don't trust our government. Very few of us vote. And increasingly, we can't even tell facts from falsehoods in the news. We *feel* free, at least, and we're proud of that. But on balance, we as a people and as a citizenry are less *well* than we used to be.

This isn't just an American problem. The world is, in many ways, unwell. Democracy, once a seemingly inevitable form of governance spreading across the globe, is suddenly in decline.[13] Attacks on a free press are on the rise in authoritarian states and major democracies alike.[14] We've seen an overall loss in global freedoms for eleven consecutive years, thanks to what the human rights watchdog organization Freedom House calls a "crisis in confidence in democracies," along with a rise of populist and nationalist movements fueled by antiestablishment politicians.[15]

In 2017, survey respondents from 145 countries reported feeling more worried, stressed, anxious, angry, sad, and physically in pain than at any point in the previous decade.[16] Depression is on the rise: the World Health Organization (WHO) reported global depression rates spiking starting in 2005—not coincidentally around the same time that Facebook (2004), YouTube (2005), and Twitter (2006) hit the scene. And it's only gotten worse: after years of stagnation, from 2005 to 2017 rates of depression leaped up 18 percent. Depression has become so widespread that in March 2017, WHO declared it to be the number one cause of illness worldwide.[17]

As developed as we are technologically, we as a people are on the edge of an epidemic of bad feeling. There's been plenty written about excessive social media usage exacerbating depression—a phenomenon dubbed "Facebook depression."[18] But I argue that there's more to what's happening than just heightened FOMO ("fear of missing out") or the negative impacts of cyberbullying, though those are problems in their own right.

I contend that we are experiencing a mass sense of cognitive dissonance. We feel both empowered and powerless by the same tools, at the same time.

Net states matter here, because they provide us with the tools to enhance our sense of freedom—the one thing we've really got going for us. But as we've seen, just because they make us feel free doesn't mean that they don't come with a cost. We are empowered by the devices that give us the world's answers at the touch of our fingertips. Yet we feel powerless to do anything about what personal information is collected in the process. We are empowered by our incredible increase in awareness about what's going on in the world, thanks to the massive amounts of media we consume each day. But we feel powerless to do anything to stem the tide of terrible things we read and hear about. We are, in feeling and in fact, more cognitively powerful—we can see more, find more, learn more, connect more—than we've ever been at any point in our history. Yet we're paralyzed, unable to engage in meaningful action; our attention is drained, our bandwidth exceeded; our emotions and cognition are overloaded and exhausted.

We can't persist like this. We risk being—or worse, risk our children being—drugged, depressed, or suicidal. While the causes for our drug use, depression, and suicidal urges are complex and multifaceted, there's no escaping that one of the most obvious correlations in the rise of all three is the simultaneous rise in our technology use. The more we use tech to excess, the more drugged, depressed, and suicidal we've become. The more media we expose ourselves to through technology, the more we feel beaten down by the incivility we see in partisan politics—in Washington, DC, yes, but also among the acquaintances in our social networks.

Take, as just one example, Russian misinformation warfare during the 2016 election. We know more about it than we ever could have, pre–social media. Yet we don't see action on the part of our elected leaders to protect the 2020 election from similar warfare. And we don't feel as if we can do anything about it on our own. Given so much awareness and the feeling of so little power to effect change, it's no wonder we feel terrible.

Our world is nowhere as dismal and chaotic as it was just after World War II. But we've been experiencing our own version of hell ever since the Global War on Terror was unleashed almost two decades ago—that is, when we're not too distracted, exhausted, or depressed to think about it. Our particular fight, here in the United States, is diffuse, protracted, and unpredictable; it is both land-based and cyber; it is also distant and surreal. We Americans may even feel guilty about the fact that the war feels so far removed, if we even remember that it's still happening. And if we do remember, that's a sure path to feeling awful, because terrorism is incredibly scary. Here, as with the Russian misinformation campaign, we feel powerless to do anything to change its course. We're in what feels like an endless war—multiple wars, if you count the onslaught of cyberattacks on our country: 4,000 a day, according to the FBI.[19] It's almost impossible to conceptualize it all, let alone feel as if we can be part of the fight.

There's no silver bullet for our problems. We can't fix all the world's ills. But some component of our sense of powerlessness lies with our (simultaneously empowering) technology, and I do have a suggestion to address that.

We do not need to feel powerless. We are not, in fact, powerless. We need to remind ourselves and the keepers of our digital realm of that. We need, as once in a while the world seems to need, a declaration: a set of suggestions that establish ground-floor, foundational conditions that should apply to every human being on Earth.

We need to establish a new relationship with our net states—one that reclaims power that users can exert over their own data, their sense of privacy, and their experiences of the digital ecosystem. We need a pact that puts citizen-users at the center—a digital descendent of the Universal Declaration of Human Rights, a tech-savvy version of the Constitution's Bill of Rights. We don't need any more terms of service. We don't need Bible-length provisions to counter the unreadable terms net states provide us. We need just a few basic principles—brief but unassailable fundamentals that everyone should be assured. We need a Declaration of Citizen-User Rights.

The goal of such a pact—not an agreement that *net states* write and users must accept, but rather an agreement that *citizen-users* craft, enumerating rights we expect to be respected—is to create a set of ground rules that net states must agree to in exchange for our using their products: specifically, in exchange for our data and our attention.

To figure out how to create such a pact, we can look to several precursors, road maps for what a meaningful pact between net states and users could look like. The Cybersecurity Tech Accord (aka the Digital Geneva Convention) described earlier may seem like a likely candidate. But for all its laudable qualities, it's actually not the best model for what citizen-users need for their protections among net states. The Digital

Geneva Convention gets a lot right. But the main issue is that it's not really *for* users. It's a pact *by* net states *for* net states. Users are among the objects of the pact, but they're not really a party to it; only the net states are.

Before we can craft a user-centric declaration, let's quickly review the current state of affairs with what we've already signed on to.

Everyone who uses technology is already party to multiple versions of terms of service (TOS). But these, too, aren't *for* us; they're for tech companies to protect themselves from liabilities. They're the gate through which we must pass to use a company's products and services. So we agree to them, time and again, without reading them. Even if we did read them, we'd discover that they don't put us—the users—and our protections front and center. The focus is on the tech company itself, ensuring that it doesn't get sued when it eventually shares your data with third parties.

There are various websites that attempt to help users navigate which platforms and services have TOS that are more or less risky for the user. For example, the website Terms of Service; Didn't Read, or TOSDR (a play on "too long; didn't read," generally shortened to TLDR, which refers to online content that people caution they didn't bother reading), was launched in 2012 in collaboration with the Electronic Frontier Foundation (EFF). It classifies sites depending on their risk to users, from A to E. (The search engine DuckDuckGo gets an A since it doesn't keep track of search queries; YouTube gets a D, mostly because it keeps deleted videos for its internal purposes, unknown to the user who created the content.) The rating of TOS is a collaborative effort: users from all over the web note changes to TOS on GitHub, one of the most widely used code-sharing platforms. But this also means it's uneven—without armies of online volunteers, it's difficult to assess how "good" TOS are for users. TOSDR does helpfully compile various TOS into a single location with plain-language summaries. Even if out of date and far from exhaustive, its list is still quite instructive (as are those of other sites).

To see if I could flesh out what was missing on TOSDR, in August of

2018, I compiled and read all the terms of service for Amazon, Apple, Facebook, Google, Microsoft, and Tesla. Here's what I found.

First, there's a wide range of "average" when it comes to the sheer length of terms of service.[20] (Note: for the purposes of this exercise, I examined only language explicitly called "terms of service"—not ancillary policies, such as those mentioned in FAQ or other legal policies listed; in other words, I stuck to the agreement between the user and the platform or service.)

Apple's iTunes TOS section is a handful, adding up to around 6,804 words.[21] Google embraces brevity (and ambiguity—more on that shortly), with its TOS clocking in at about 1,869 words.[22] Facebook is somewhere in the middle, at 3,243 words.[23] Amazon is about the same, at 3,387.[24] Microsoft blasts the others out of the water for length, at 15,290 words.[25] Tesla covers their business in 5,736 words. All this averages out to terms of service composed of 7,265.8 words for your standard net state. Even if these terms were written in plain English, as many of them now are (thanks to the need to comply with the EU's General Data Protection Regulation), if the average person reads 200 words per minute, it would take 36 minutes to read one single platform's terms of service; 3 hours and 20 minutes for all the ones I calculated—just six of zillions.

In practice, we not only don't want to spend this much time reading TOS, we come across too many to realistically do so. Researchers at Carnegie Mellon University estimated that we encounter 76 sets of terms of service in a year, based on our 2008 browsing habits[26]—back when we spent only 13 hours a week, or 1.8 hours a day online.[27] Terms of service are shorter now—again, thanks to the EU's GDPR. But we're still encountering them quite regularly. In 2016, it was estimated that we spent 10 hours and 39 minutes per day consuming media in some form—via our smartphones, TVs, computers, tablets, e-readers, and so on.[28] Our attention, as discussed in chapter 7, is a precious commodity. We try hard not to waste it; we spend only 10 to 20 seconds per web page to evaluate whether it's worth our time before moving on to the next one.[29]

If we can barely pay attention for the 10-plus seconds it takes to see whether or not we want to stay on a web page, the notion that we'd take 36 minutes to peruse a boring user agreement—76 times a year, no less—is simply unrealistic.

And our net states in question? There are six we've been looking at, as you know. But they're not just six companies; they're six *parent* companies, and collectively they've acquired 673 other companies, according to transaction data publicly available through Crunchbase. Many of the acquired entities are likely irrelevant to us: Apple bought lots of hardware component manufacturers, for instance; Facebook, lots of messaging startups. But many of them are significant entities that we run into on occasion, though we may not be aware that they're owned by a major net state: Instagram and WhatsApp (owned by Facebook); Skype, LinkedIn, and Nokia (those are Microsoft's); IMDb and Zappos (Amazon bought those); YouTube, Waze, and Zagat (all Google's), to name just a few. And they all have *their* own terms of service as well. Indeed, if we were to map out the universe of the net states and their 673 acquisitions, I'm fairly certain we'd discover that there's no way to both be active in the digital sphere and avoid having to agree to terms of service informed by one of the major net states.

In sum, if we're online, we are bound to our net states, in some way. We've given ourselves over to them, without even knowing what that really means. We can't put this particular genie back in the bottle—the data we've set free has likely passed through so many third parties at this point that it would be virtually impossible to suck it all back in. But we can make a decision to change how we handle our data moving forward, to reclaim power over what we give out, to whom, and for what purposes.

LET'S START PRACTICALLY. HERE ARE THE MAJOR AREAS COVERED IN this book's big-six net state terms of service: (1) your data when you use their product/service; (2) your data once you leave their product/

service; (3) your expectation of privacy while using their product/service; (4) the ground rules about using their product/service; and (5) what access third parties have to your data.

The reason to start with existing areas in the terms of service is that the ultimate goal for crafting a Declaration of Citizen-User Rights is to be a viable challenge to the TOS that net states create. Thus we can't ignore their areas of concern. But instead of focusing on ensuring that the net states don't get sued, *our* document—a set of universal "terms of *rights*"—will ensure that citizen-users don't get taken advantage of.

First, some general observations on terms of service. Based on those that I examined, they tend to be an umbrella set of conditions users must agree to. Yet despite their length, they don't come close to laying out all the details—especially some of the details users are likely most concerned with. For example, all six net state TOS that I reviewed directed readers to a separate "Privacy Policies" document that users were encouraged, but not required, to read. *These* are the policies that really get into detail about what they're doing with our data. By including a blanket statement in the TOS along the lines of "By accepting this agreement, you're also accepting our privacy policies, which you can read here [click here]," companies effectively shuttle their privacy policies even further out of users' reach.

I'll spare you the mind-numbingly boring details. Here's the upshot, which isn't great news but no surprise: if you use any product or service of the six major net states, you don't really have any privacy online. Your data will be kept, analyzed, and shared with or sold to third parties, as they see fit. One notable exception is Apple, sometimes: whatever data is on your phone remains on your phone, not on Apple servers—unless you enable automatic syncing with iCloud. This is one of the reasons iPhones bedevil law enforcement officials: there's no "back door" where Apple can unlock someone's phone, remotely or in person.[30] If you use a strong password—meaning, not the word "password" or "1 2 3 4 5 6," which remain the two most common passwords, amazingly—it's really, really hard for anyone to break into your phone.[31]

The data collected *while* using your phone is all tracked of course. Facebook notes, for instance, "We use the data we have—for example, about the connections you make, the choices and settings you select, and *what you share and do on and off our Products*—to personalize your experience" (italics added).[32] This means that if you have Facebook open on your phone and then open another app, Facebook will collect data on what you're doing on that app, too. *All* the net states (and tons of other companies, too) share or sell—also called "licensing"—your data with unnamed "third parties" as they see fit, for which you have granted them "worldwide" permission, according to their TOS. Amazon goes so far as to say that they'll also share or license your name in addition to your data, if you've ever posted a review or comment.[33]

In short, our data online is pretty exposed. We *say* we care a lot about it—a Pew poll in 2016 reported 75 percent of us as saying privacy is "very important" to us.[34] But we do precious little to actually protect our privacy online. Almost all of us—91 percent—accept TOS without reading them (the figure is closer to 97 percent for millennials).[35] Claiming to care about privacy but not taking action to protect it is such a widespread phenomenon that researchers have dubbed it the "privacy paradox."[36]

The most plausible explanation I've seen for this paradox has to do with our squishy sense of time.[37] Our worries about privacy—a breach, a leak, an abuse of our data—are focused on something possibly happening *in the future*. The future's an abstract concept; it's difficult to visualize in a tangible way. On the flip side, the actions that put our privacy at risk are *in the now*. We're getting something immediately—access to information, a platform or service, the ability to connect, see photos, receive updates. We get visceral, tangible, right-now representations of our friends, our frenemies, our crushes, our curiosities. All you have to do is sign away your privacy with the click of a TOS "accept" button and bam! You're in; you're connected to the object of your desire, whatever that may be. But taking the time to be careful about your privacy takes, well, time—a lot of it—as well as that prized commodity, your attention.

So here's what I propose. We take the time, in these next few pages, to negotiate a better deal for ourselves. I'm not a lawyer, but then I'm not suggesting anything legally binding. Rather, let's set out a few simple, high-level terms that protect users in principle; the lawyers can duke it out and pretty it up some other time. Our focus in the Declaration of Citizen-User Rights is to address the meat of the problem: how to reclaim the right to our privacy—and our sense of power—without having to sacrifice the benefits of technology.

The goal with these rights is not to codify *tactics* for user protection against technology's ill effects—for example, steps to turn off location-tracking on your phone or to prevent your contacts list from being uploaded without your permission. For these sorts of protections, the previously mentioned Center for Humane Technology, founded by former Silicon Valley leaders, provides excellent resources. The goal here is to identify the fundamental rights that users would need to claim to substantively alter the balance of power between us and net states. There are likely dozens of possible rights. Let's start, though, with the three rights poised to make the greatest impact on our lives, expressed here as three principles. Those principles alone make up the Declaration of Citizen-User Rights. They are: Citizen-users have the right to choose how they pay for their own content privacy. Citizen-users have the right to delete their own content from the public record. And citizen-users have the right to know how their data is being used. Now let's look at each of these principles in turn.

#1. Citizen-users have the right to choose how they pay for their own content privacy.

This first principle may seem counterintuitive. Like, why should we pay for something that's already ours? It's *our* information, after all.

While understandable, that argument isn't going to help. We're so far down the road of information-giveaway—since the dawn of the information age, really—that tech companies would have to reinvent their entire business model to simply stop collecting our data if they got nothing in return. Also, to repeat a quote from Google Jigsaw's Scott Carpenter, "Most tech companies are just trying to get off the mat and into the game." The six net states featured in this book are über-rich, but the vast majority of tech companies are barely scraping by. It's simply not viable to operate a business that gives away a free product or service if the provider doesn't get anything of value back from the user.

For a long time, this thing of value was our data. Our privacy. Our sense of power. What I'm suggesting is that we redefine what's valuable. If our data, privacy, and sense of power are precious to us, then we need to offer something else that's valuable. And just about everyone values money.

Here's what we'd get: walls—borders protecting our content. The photos, videos, and text we put online would stay on our chosen platform and that platform only—no sharing, licensing, selling, or in any other way permitting access of that data to unidentified third parties. As it stands, those third parties underwrite the cost for our free access to platforms and services. So in exchange for keeping our data nice and snug on a single net state's system, we as users would have to pay for those walls.

It may not be an immediately attractive notion: demanding the option to pay money for something that's currently free. But remember, it's not *actually* free—we're paying with our data. Therefore, I suggest that Amazon, Apple, Facebook, Google, Microsoft, and Tesla—and any other tech company large enough and wealthy enough to adopt this option without it breaking their ability to do business—provide users with a for-pay option for any service in which our personal data currently serves as payment.

This is not an unusual business model. This is how premium cable TV channels like HBO work: we pay for it, and in exchange we're not forced to watch advertisements. Streaming music services like Pandora work like this as well: there's a free version, with advertisements.

But if you pay for the service, you're not subjected to the interruption of ads.

What I'm proposing goes beyond simply saying that net states should create an option whereby users are invited to pay for a particular service. We're talking about fundamental *rights* here. This first principle suggests that we have the *right* to demand a for-pay service. We have the *right* to declare our data sacred and offer something else of value in return. We have the *right* to demand that our data goes in a vault.

The company itself can still access the data: learn from it, develop products from it, use it to improve their services. But in exchange for our usage, the company would have to agree not to share or sell that data with any unidentified third parties.

This creates a few problems, unfortunately. In a society like ours, where the gulf between rich and poor is already vast and growing vaster, a for-pay option might seem affordable only to the well-heeled. This would leave the struggling lower socioeconomic classes at an even greater disadvantage: they'd still be losing out on personal privacy, while the rich could protect theirs.

Yet, even people without a ton of money do tend to pay for things they really value. As was noted earlier, 77 percent of Americans own smartphones and 73 percent of American households have broadband internet access, for instance.[38] Privacy, like phones and internet, might turn out to be so valuable that it's deemed affordable even by those with few financial resources.

Here's what the financial impact of the internet-privacy-for-pay option might be. The median American household income in 2016 was $57,617. Income varies significantly by state, of course, but that was the nationwide average.[39] Calculating how much of that gets paid in taxes also depends on many variables, like your marital status, whether you have dependents, and what state you live in. But let's take what we know about federal, state, and local tax rates and ballpark the average tax rate for the average American to be 17 percent.[40] That would put the average American's take-home pay at about $47,800, or just under $4K a month.

In 2017 cell phone and broadband bills were $75 to $125 a month.[41] This means that even without accounting for the cost of our actual devices—phone, laptop, television, tablet—we're spending around $150 a month to be digital citizen-users.

Apple and Google Drive (where you can store files in the cloud) already have services that cost money: you can stream unlimited music from Apple for $10 a month, for example, or store gobs of data on Google Drive for $2 a month.[42] In other words, for-pay tech services that already exist don't cost a ton of money. If each of the major net states charged just $5 a month for their privacy-enabled for-pay tier of service, (1) they'd make an absolute fortune, which sounds like an incentive to create such tiers, and (2) we, the citizen-users, would be buying back our data, our privacy, and our sense of power for the cost of a muffin and a Frappuccino at Starbucks.

However. There's another problem with this very transactional approach to our privacy, a more principled and foundational one. What I've suggested is a pragmatic approach that looks forward, forgetting about all the data that has already been taken for free and focusing on the future data that we can protect. But there's another argument that we need to contend with—beyond whether it was ever okay that our data was so freely trafficked without our actual consent (not TOS consent, but conscious, reflective, I-know-what-I'm-getting-myself-into consent).

And that is this: "Information wants to be free." Free and freely shared. It's one of the founding principles of the internet, as we've seen; the central tenet of the tech ethos.

What happens to the European Union's internet experience in light of the General Data Protection Regulation (the GDPR, discussed in chapter 3) will be instructive in seeing how intensive regulation plays out. Users in the EU can invoke their "right to be forgotten," effectively removing their digital footprint from search engines' reach. The question becomes: If we kept our personal data private, would our global internet experience still be full and rich and diverse and accessible, or would the internet become a different place, showing us only the content that users couldn't afford to pay to keep private? Would the internet as we've experienced it effectively shrink and wither, becoming a less vibrant, democratic, and diverse hunting ground for information and connections than we have today?

Unclear. But the status quo—in order to be digitally active, citizen-users have to relinquish their right to privacy—cannot be sustained. Principle 1 is our best option for creating a viable workaround to the privacy issue: one that both net states and citizen-users could benefit from.

#2. Citizen-users have the right to delete their own content from the public record.

Think of this principle as the clean-up clause. It will make more sense with a bit of background. The European Union's General Data Protection Regulation made a bold statement, as we saw earlier, with respect to users' online information. Instead of focusing on the information itself, it focused on the person: the "data subject." In so doing, the entire set of legislation shifted our attention from *content* to *humanity*: "subjects," defined as "people who live in the territory of, enjoy the protection of, and owe allegiance to a sovereign power or state." In the case of the GDPR, the subjects are enjoying the protection of and owe allegiance to the European Union, which, in turn, oversees their protection. And not just generally, but in a very specific area: their data.

In this book, I've used the term "citizen-user." Why not just say "data subject"? Why introduce yet another term into the mix? Two reasons. First, I'm referring not to citizens of a single country, but *all* people who use digital tools and services. Second, I took my cues from Microsoft's Digital Geneva Convention—aka the Cybersecurity Tech Accord—which refers to people as "users," "customers," "civilians," and "citizens." The accord makes it clear that signatories are protecting users *and* customers *and* civilians *and* citizens, regardless of the country in which those individuals reside. I could have called us all "user-customer-civilian-citizens," but that felt a little unwieldy. Hence, citizen-users: "citizen," because we have a responsibility to be active participants in the power structures in which we function; and "users," because we don't physically inhabit the digital ecosystem. Rather, we *use* its tools and services. Those same tools and services—the products of net states—*they're* the things inhabiting the digital ecosystem. We citizen-users are visitors: at times, cruising through; at times, veritable squatters—but only and always transitional actors.

If the digital realm is an ecosystem that we improve or degrade through our presence, principle 2 ensures that it remains a place we'd want to be. Think of a park: it's more fun to have a picnic in a clean park than one strewn with litter; therefore, it's our responsibility to clean up our own litter.

Principle 2 proposes that we claim the right to improve what our presence online does to the overall quality of the place. I am in no way suggesting censorship. If someone criticizes an article I wrote, I shouldn't be able to delete that criticism just because it irks me. However, if I realize that this criticism is valid, that I advanced a line of thinking that wasn't supported by facts and data or was just plain wrongheaded, I should be able to correct my mistake and take that article down. Such action would be a form of informational pollution control, like cleaning up the mess you make after that picnic in a park. The evidence that I wrote something dumb would remain—in the form of whatever criticism was posted about me. But I would at the very least be able to retract the original, offensive material and thus keep from negatively impacting future readers.

The GDPR has a similar provision in its "right to be forgotten" (RTBF) clause. That clause doesn't broadly stipulate that any data subject can demand that any and all content about them be removed from, for example, Google. No, the "right to be forgotten" is balanced out with the "right of freedom of expression and information"—which is, granted, open to interpretation—and with the public interest, such as if certain data is necessary for "legal obligations, for archiving purposes in the public interest, scientific, or historical research purposes or statistical purposes."[43]

It turns out the wiggle room under what's considered to be in the public interest or of research value is fairly broad. In the first year Google processed RTBF requests, it decided to retain 58.2 percent of URLs requested to be taken down[44]—taking down only about half a million links. The indexable internet (the portion that search engines like Google can find) contains roughly 4.44 billion web pages.[45] Even if Google took down half a million web pages each year—roughly 1,369 web pages per day—we're posting several hundred thousand times that amount in new digital data: approximately 1.5 million new websites a day.[46] Even if only 4 percent of those 1.5 million websites can be found on Google, the fact remains that there's a staggering amount of new content that we *can* find, almost all the time. Judiciously removing 1,000 or so links a day—one-third of which link to personally identifiable directory information, according to the registry of removals in the EU—would be like extracting drops of water from an ocean; it would barely register as a ripple.[47] Granted, if the one page you're looking for happens to be one of the pages that's no longer online, it'd be frustrating. But still, it's a worthwhile trade-off to ensure that you regain control over your data, your privacy, and your power.

#3: Citizen-users have the right to know how their data is being used.

Think of this principle as similar to the US 1967 Freedom of Information Act, or FOIA, for net states—with a twist.[48] FOIA is the law that lets citizens request that government agencies share a copy of their information with the public. Of course, you have to first know what information to ask for. A request like "Send me all your emails!" sent to all government agencies will likely be ignored. A request with specific parameters, like "Send me any email traffic pertaining to Benghazi, Libya, from August

2014 to November 2014," sent to a specific agency, such as the United States Mission to the United Nations, is more likely to be processed.

FOIA isn't a perfect system; it can take a few months to get a response. Yet regardless of how clunky and slow it may be, this transparency process exists. And it gives citizens power to review information that their government—whose work they fund through their taxes—produces.

Here's where principle 3 differs from FOIA: it's not just a request for information. Most net states will already let you gain access to the information they have on you. For example, you can download all of your previous Google searches or Facebook posts with a few clicks. But principle 3 states that net states also have to let you know *what they're doing* with the information that they have on you. If they're sharing it with marketers for the purpose of better targeting you with ads, they should say so. If they're licensing it (remember, a friendly way of saying "making money off of it") with data brokers for the purpose of contributing to some third-party profile on you, they should say that. And users should have the right to opt out of certain practices, especially if they're paying for services, as outlined in principle 1.

This issue is also covered in the European Union's GDPR in its "right of access."[49] Not only do tech companies have to let you know what data they have on you, the GDPR requires them to tell you what they're doing with it—"the purpose of processing"—as well as anyone else they're sharing it with: "Any third party recipients of this personal data, both backward or forward looking, especially recipients in third countries."

What this means in practice is that anyone in the EU can ask any company—including the company they work for—to hand over any documents that refer to them, including private personnel records and other potentially sensitive materials.[50] Companies have 30 days to comply or face paying a substantial fine—up to a walloping €20 million (roughly $23.6 million) or 4 percent of the company's total worldwide annual turnover of the preceding financial year, whichever is higher.[51]

While this sounds like a huge victory for citizen-users on the face of it, it creates significant complications for companies. For instance, the General Practitioner Committee of the British Medical Association—representing 160,000 members in the UK medical community—urged members to petition their local officials highlighting "unintended consequences" of the subject access requests specifically.[52] The petition notes that by requiring physicians to review every page of medical records to ensure that no prohibited information was being shared, they were losing on average an hour a day reviewing paperwork—an hour they formerly would have spent "caring for sick patients."[53]

The subject access requests aren't just a time issue; most companies wouldn't know where to look for all the data requested. An astonishing 82 percent of companies surveyed admitted that they didn't know where in their own companies their critical data was located,[54] what with data being stored on individual workstations instead of (or in addition to) a centralized database, in email inboxes, or in multiple databases or documents within the organization that aren't connected or communicating with each other. While "Big Data analytics" are all the rage conceptually, in practice a lot of companies are still playing catch-up to understand even the very basics about their data, like what they have and where it's stored. How to mine it for useful insights is a whole different level of complexity that many companies are still only dreaming to reach.

But the inability of companies to find data—a real and immediate challenge—is no reason for citizen-users not to exert their right to know how their own data is being used.

MOST OF US PROBABLY FIRST ENCOUNTERED THE TERM "ECOSYSTEM" in biology class. It refers, essentially, to any system of parts that are connected to and interact with one another. It's actually a fairly modern notion, first introduced in 1935 by a scientist named A. G. Tansley.[55] He introduced the term as a sort of mash-up of "ecology"—the relationship between organisms and their environment—and "system"—a complex entity made up of interconnected parts. With the introduction of Tansley's term, scientists suddenly had the means to describe something that had always existed but, nameless, hadn't really been considered a topic worthy of study in its own right. Thanks to Tansley, people could look beyond two sets of things—individual creatures and the space in which they lived—and zoom out to a bigger picture, studying the interactions between creatures and their space.

This book is also in its way about creatures and space. There are two sets of creatures: the citizen-users of tech products and services; and net states, the makers and keepers of those products and services. Together, we traverse the same space—the digital. We and our interactions in the digital space generate our own kind of ecosystem: the digital ecosystem.

Ecosystems, in addition to having constituent parts, such as creatures and space, exhibit another important characteristic: they tend toward equilibrium, or a state of relative balance between opposing forces.[56] In nature, this means that enough creatures are born to offset the number who die; that enough nutrients are generated to replenish those that have been consumed.

For our digital ecosystem to be in a state of equilibrium, we—the creatures (both individual and net state) who traverse the digital space—would have to get out as much as we put in. We citizen-users create content; we consume content. We upload photos, videos, and news; we download photos, videos, and news. These activities reflect what *looks* like a relatively steady state, but only if we disregard the net states that are our fellow ecosystem visitors.

With respect to our own content and net states, we're *not* currently in a state of equilibrium. We're not getting out as much as we're putting in. Our content goes to net states, and then disappears behind their firewalls. We can't see it all assembled. It goes into databases that generate insights—insights about our behaviors and habits and preferences—that net states gain benefits from but that we—the creators of the data—don't necessarily know about and certainly don't have access to.

This has to change. And we—the citizen-user content creators, the creatures who populate the digital ecosystem with our thoughts, our memories, our records, and our interactions—already have the power to ensure that it does change.

Without us, the digital ecosystem loses its life force. Websites absent of users interacting with them are virtual ghost towns, with no data to gather and analyze. We, the citizen-users, provide the energy that fuels the ecosystem; the net states provide only the space and pathways within it. So if we want our Declaration of Citizen-User Rights adopted, we have to use our power to withhold information from a website; to boycott usage; switch to a different service or platform that will agree to our terms and respect our rights. We can massively disrupt the equilibrium of the digital ecosystem, should we so desire. We just have to commit to doing it.

In short, it is in the best interests of all parties to the digital ecosystem to ensure that its human population is cared for. We don't need to deny ourselves the benefits that tech provides us. As *New York Times* "Smarter Living" editor Tim Herrera reflected, "When we talk about unplugging, I think what we're really talking about is structuring our lives in ways that allow technology to serve us, rather than the other way around."[57]

Right now, we serve tech companies—we give them our data, they generate insights based on that data, and we never see the results. We can continue to supply tech companies our data, but under different conditions—conditions, laid out in the Declaration of Citizen-User Rights, that give us control over and information about that data. We don't need to unplug and retreat from tech. We just need to adjust the equilibrium in the digital ecosystem.

AS OF 2017, THE WORLD HOSTED 65.6 MILLION PEOPLE WHO HAD BEEN forcibly displaced from their homes.[58] Refugees from war. Asylum seekers in search of protection because of their political or religious beliefs.

People displaced by their own countries, by their own compatriots. Ten million among them are considered officially "stateless" by the United Nations.[59] With these horrifying stats, the year 2017 ushered in a depressing new world record: an all-time high for displaced persons of every kind.

According to the US Department of State, a stateless person is defined as "someone who, under national laws, does not enjoy citizenship—the legal bond between a government and an individual—in any country."[60] Notably, the definition contains the word "enjoy." Citizenship in a country indeed offers something positive: It offers protections. It affords rights. While not all citizenships are created equal—the rights of citizens in the EU, for instance, are now quite different from those of citizens in the US—they do provide individuals with two invaluable benefits.

First, the fact of citizenship establishes a framework of expectations for what it means to be a contributing member of one's society. For instance, most countries have some sort of legal document that details citizens' rights and responsibilities in regard to the state, even if they're as basic as freedom of speech and paying taxes. Second, citizenship, in democracies at least, grants rights that exist independent of who is in power. The president of the United States, for instance, may want to block someone on Twitter, as Donald Trump has expressed in many instances. But the Constitution, which protects freedom of speech, overrules that desire. According to court rulings, the president cannot block you or any other American citizen, because to do so violates the First Amendment, which guarantees freedom of speech. Citizenship in the United States means that you have power that outweighs the wishes of the (otherwise) most powerful person in the land—the president.[61]

In short, rights bring power. They are our protection against potential abuses of power. They are, according to the US Constitution, inviolate—something we're born with as Americans. And we are fortunate to have rights. As Hannah Arendt, a philosopher who was stateless for 17 years, once wrote, "The right to have rights, or the right of every individual to belong to humanity, should be guaranteed by humanity itself."[62] She then added, "It is by no means certain whether this is possible."

Net states, operating internationally, have the opportunity to afford their users rights that may differ from and even exceed rights of any individual's nation-state (or, for the 10 million stateless, stand *in place of* citizenship rights). In this way, the rights afforded by net states rise above those guaranteed by nation-states.

The nation-state is no longer the only game in town. Nation-states can partner with net states, as we saw in the partnerships between Tesla and Vermont, California, and Connecticut. They can establish contracts with net states, as the federal government does in operating almost entirely on net state technology. Conversely, they can fight against net states: with regulations, as in the EU; with taxes, as in Uganda; or with bans, as in Papua New Guinea and Sri Lanka.

Whatever type of relationship nation-states and net states ultimately have, the reality is that there must be some sort of a relationship—and it must be diplomatic in nature, not just economic. As things now stand, net states don't have just our information; they also have power over our rights. If nation-states don't set the terms of how net states must operate with respect to our rights, then net states will write their own rules of engagement, in some cases avoiding regulations, as Facebook did when it moved more than a billion users out of its data center in Ireland—or adapting to them, as Google and Apple have done to comply with the GDPR.

We—as citizens and citizen-users—have the right to have rights: with our nation-states and with our net states. It is up to our home countries to enforce the rights they've granted in our constitutions. And it is up to us as individuals to hold our net states accountable if they fail to grant us protection with respect to those rights. The US Constitution holds that we have the right to "be secure in [our] persons, houses, papers, and effects." Our data may not be who we are, but it is certainly something *of* us. As such, it is something that needs to be secured.

TUCKED AWAY IN THE CORNER OF THE FIRST FLOOR OF THE UNITED Nations General Assembly building—that low, white, asymmetrical building with a dome atop and 192 country flags in front—sits a tiny, inconspicuous, triangular room. It's virtually empty. It has a few rows of benches in the back, an abstract cubist-style mural at the front. In the center stands a six-and-a-half-ton rectangular block of polished iron ore, perennially cool to the touch. It's illuminated by a single beam of light. This is called the meditation room.

The second secretary-general of the United Nations—the everyman's aristocrat, the sky-eyed Swede, Dag Hammarskjöld, who's widely regarded as a founding architect of the organization itself—personally redesigned the space in 1957. The original setup in 1952 was a dreary place, with tightly packed rows of chairs facing what looked like a tree stump holding up a flower arrangement and a tired-looking miniature UN flag.[63] Aesthetically, the room didn't inspire serene thoughts; it looked more like a shrunken, misshapen conference room.

According to a reporter working at the UN at the time, in the middle of one night in 1956, Hammarskjöld, still at work, dragged some of his aides out of their offices. Having been summoned by the secretary-general himself in the wee hours of the morning, they expected to be told bad news from one of the far-off war zones where United Nations personnel were on the ground. Instead, he said, "I want to go down to the meditation room." Because this was the secretary-general himself making this request, they followed—despite it being two in the morning.

Hammarskjöld wanted to talk to them about the utmost importance of redesigning this room. "We want to bring back, in this room, the stillness which we have lost in our streets, and in our conference rooms," he told his aides, "and to bring it back in a setting in which no noise would impinge on our imagination." In the subsequent year, Hammarskjöld took on the meditation room as a personal project, dictating details ranging from the color of the paint on the walls to the fabric of the carpet.[64] In his formal dedication of the room in 1957, four years before he would be killed in a plane crash in the Congo, he called attention to the iron stone altar in particular, asking visitors to ask themselves: What they would make out of iron, such a powerful material—would they use it to craft the means to support war or to sustain life?[65]

He wrote, "Of iron he has constructed tanks, but of iron he has likewise built homes for man. The block of iron ore is part of the wealth we have inherited on this Earth of ours. How are we to use it?"

The meditation room, like most rooms, has a practical function. It's a place to meditate, to reflect. But its real significance is symbolic. This place is something that one of the founding fathers of the United Nations obsessed over getting right. And it's not like he had nothing else to occupy his time: Hammarskjöld, in his tenure as secretary-general, was responsible for onboarding 4,000 new UN staffers and establishing how they'd work and what'd they'd do. And he had a nightmare's worth of thorny diplomatic demands: trying to establish peaceful relations between Israel and the Arab states; negotiating the release of captured US pilots during the Korean War; untangling the Suez Canal crisis; establishing the United Nations Emergency Force, the first-ever officially sanctioned nationless military unit; and attempting to orchestrate the decolonization of the African continent.

Hammarskjöld cared about the meditation room because of what it represented. It would serve as a reminder to every king, queen, president, and prime minister who would someday walk the halls of the United Nations building that they should, at some point, stop to meditate. Stop to think. To wonder: Is this work I do creating a better world? Am I taking the resources at my disposal—the iron ore of my era—and molding them into opportunities for my people? Or am I just making tools that will bring about their destruction?

CONCLUSION

THE NET STATE PATTERN

"Hello? Uh, hello? Hello, Dmitri? Listen, I can't hear too well; do you suppose you could turn the music down just a little? Aha, that's much better," begins the phone conversation between two fictional world leaders, the American president Merkin Muffley and the Soviet premier Dmitri Kissov, in Stanley Kubrick's 1964 classic film *Dr. Strangelove, or: How I Learned to Stop Worrying and Love the Bomb.*

"Now then, Dmitri," says the US president, "you know how we've always talked about the possibility of something going wrong with the bomb." Pause. "The *bomb*, Dmitri. The hydrogen bomb." He continues, "Well now, what happened is, uh, one of our base commanders, he had a sort of—well, he went a little funny in the head. You know. Just a little funny. And uh, he went and did a silly thing." The president sighs. "Well, I'll tell you what he did. He ordered his planes . . . to attack your country."

In the absurd exchange that follows, the US president tries to talk his Soviet counterpart off the ledge, assuring him that this is all just a big misunderstanding and that he is really sorry and they'll figure out some way to deal with it. The scene captured both the frightfully high stakes and the fragility of relations between the United States and the Soviets in the 1960s. The exchange also employed a piece of technology absent in the actual diplomatic communication channels between the two nuclear powers: a dedicated and secure telephone system that linked the two powers.

The satirical film captured the mood of the era in part because of how closely it mirrored reality. Just two years prior, the world had held its breath watching the Cuban missile crisis play out. This 13-day standoff in which the United States and the Soviet Union debated whether to blow each other and the world to smithereens in a nuclear war was, in part, elevated due to faulty communication channels between the two countries.

On October 24, 1962, at 6 p.m., Soviet premier Nikita Khrushchev messaged a fierce 3,000-word missive to the Americans, declaring that "the actions of the United States with regard to Cuba constitute outright banditry or, if you like, the folly of degenerate imperialism." He spelled out the stakes in no uncertain terms, saying, "Unfortunately, such folly can bring grave suffering to the peoples of all countries, and to no lesser degree to the American people themselves." Khrushchev then directly threatened war: "The Soviet government considers that the violation of the freedom to use international waters and international air space is an act of aggression which pushes mankind toward the abyss of a world nuclear-missile war."[1]

Diplomacy is generally a careful affair, in which participants deploy measured language; threats are typically inferred from a combination of painstaking reading-between-the-lines and outright guesswork. Compared to the norm, Khrushchev's letter was about as subtle as a bazooka blast across the bow.

But it was a blast that took an obscene amount of time to register with the Americans. They didn't get the letter decoded and translated until 11 a.m. the next day—almost 12 hours after the Soviets sent it. Given the stakes—mutually assured destruction—and the intensity of Khrushchev's letter, it must have seemed baffling to the Soviets that they didn't receive some sort of response from the US during that initial half day. So while the Americans were scrambling to decode and translate the original letter, the Soviets sent another, even more bellicose letter. The situation quickly elevated from verbal trash talk of war to the actual, imminent risk of it.

While many interceding twists and turns ensued, for the purposes of this discussion I will skip to the end: in a massive diplomatic gamble, the Americans decided to simply claim they had never received the second letter.[2] By essentially pretending it had gotten lost in the mail, the Americans and the Soviets were both able to blame poor communication for this diplomatic nightmare, save a modicum of face, and, lock-eyed with one another at every step, back away from the brink of nuclear war.[3]

With nuclear annihilation averted, both sides vowed to never again let convoluted communication channels stand in the way of avoiding Armageddon. They turned to technology for an answer. The Washington-Moscow hotline was born.

Despite Hollywood depictions of a red telephone, the hotline was actually a clunky teletype machine that routed signals from Washington through a series of relays—via London, Copenhagen, Stockholm, and Helsinki—before they reached Moscow. As we've seen, undersea cables are vulnerable to physical damage, and the cable involved in this linkup was no exception—it was accidentally severed several times, including by a bulldozer in Copenhagen and by a Finnish farmer who inadvertently snagged it while plowing.[4]

Further complicating communication via that particular means, the teletype used wasn't just a machine, but a set of them—one was a backup in case the first failed—each equipped with English and Cyrillic keypads. They were tested on an hourly basis every day since being installed over fifty years ago, on August 30, 1963.[5] The technology's been upgraded since then—we've moved on from teletype to encrypted email by now—but the dedicated communications channel and testing protocols have remained unchanged for over half a century.

The very first test was almost a diplomatic disaster in itself. To ensure that all the keys worked, the inaugural test from the United States consisted of the sentence "The quick brown fox jumped over the lazy dogs"—words that any keyboard user would recognize as a standard test sentence, used because it contains all the letters of the English alphabet. Puzzled at the message, the Soviets on the receiving end inquired as to its meaning. The Americans quickly clarified that this was a standard keyboard test and *not* some cryptic code. The Soviets then replied with their own surprisingly humanizing test message: a lyrical description of a Moscow sunset.

Technology has transformed the *way* we communicate with our global partners. But it cannot replace the guts of *how* we communicate—the need to understand foreign languages, cultures, and the context affecting one's counterparts.

Even among people with similar backgrounds—say, Silicon Valley entrepreneurs—there's only so much communication that technology can effectively convey. As physical creatures, we still assess one another through nonverbal cues—physical appearance, body language, style of dress, posture, tone, volume, and so on. In a seminal study of thousands of interviews and negotiations, researchers determined that body language accounts for upwards of 60 to 80 percent of the impact made around a negotiating table.[6] Furthermore, people establish 60 to 80 percent of their initial opinions about their conversation partners within the first four minutes of their interaction. The words we say matter, but how we say them and who people perceive us to be may matter even more.

As legendary journalist Edward R. Murrow put it, "The really critical link in the international communications chain is the last three feet, which is best bridged by personal contact—one person talking to another."[7] This applies to any communication chain, really, not just diplomatic ones.

And we're beginning to see net states take this concept to heart. As they begin to coordinate efforts around defense and diplomacy on their platforms and services, despite being armed with the most sophisticated communication tools on the planet, Facebook and Google and Amazon and the like are finding ways to bridge the last three feet. They're convening in person.

FACEBOOK, WHILE ITS FOUNDERS MAY NOT HAVE FORESEEN THIS BACK in Zuckerberg's dorm room days, has evolved beyond just social networking. Its people and tools are fighting extremists, as we saw in chapter 4. What's more, in 2018 Facebook increased the size of its internal counterterrorism team by over 30 percent.[8] With approximately 200 staffers solely devoted to counterterrorism, this unit is still only a minuscule portion of Facebook's total workforce of over 25,000 employees. But its growth rate—from unit launch in 2016 to a force 200 strong in just two years—signifies that this is a trend that's unlikely to reverse anytime soon.[9]

Monitoring for offensive or illegal activity, like it or not, occupies a lot of Facebook's energy these days. It's not alone among net states in this, and it's not going it alone either. It's teaming up with its kin. On December 5, 2016, Facebook, Microsoft, Twitter, and YouTube (Google) announced a partnership to stem the spread of online terrorist content.[10] Six months later, the same foursome announced the formation of the Global Internet Forum to Counter Terrorism, a partnership intended to create information-sharing channels between the participants to better combat terrorism using their products.[11]

Whether those initiatives generated any significant results is difficult to discern; the tech companies have released few details publicly since their initial announcement. But what we do know is that there's a noticeably absent player in these partnerships: government.

To be fair, the firms tried reaching out to government, at least once. On May 23, 2018, about a year after their initial convening, Facebook assembled representatives from Amazon, Apple, Google, Microsoft, Oath (a major ad agency for net states), Snap (high-end tech consultants), and

Twitter. It also invited representatives from the US Department of Homeland Security and the FBI. The agenda: to talk about how to prepare for foreign attempts to influence the 2018 midterm elections.[12]

The meeting was a disaster. To put it politely, the event was "tense," according to a participant who did not wish to be named. After sharing privileged information with the US security apparatus, the net states got little in return other than tight-lipped "no comment" responses from their government counterparts. The tech firms left the meeting discouraged. According to one person present, "the encounter led the tech companies to believe they would be on their own to counter election interference."

Just a few months later, on August 23, 2018, a dozen major tech companies held another face-to-face meeting on the same topic.[13]

This time, they didn't bother to invite government.

WHETHER NET STATES ARE INDEED ON THEIR OWN IN FIGHTING FOREIGN infiltration of their platforms or not, they believe that they are. Indeed, they may be better equipped for that fight than is government. In an off-the-record conversation in the spring of 2018, a senior FBI official told me, "The truth is, we need them much more than they need us." When I asked why that was, he said they couldn't hire enough staff with expertise at the intersection of tech and terrorism: federal agencies pay a pittance compared with tech companies. "We're just not competitive."

This isn't a good sign for you, government. I write this as a believer in the power of government; indeed, as a government official myself. If officials don't work with net states, these firms will only expand further into territory that used to be government's primary dominion. If government tells net states that they're on their own in the fight against non-state-sponsored terrorist activities online, and that they'll also have to deal with state-sponsored cyberattacks from Iran, Russia, and other countries alone, they most certainly will do it; and—

unhampered by the need to comply with laws and regulations—they'll likely do it well.

Without government assistance, net states have no choice but to deal with these issues on their own: terrorism on their platforms is an existential threat. Because if net states allow their ecosystem to be polluted by terrorists and by Russian and Iranian (and other) misinformation warfare campaigns, regular users may very well take their profiles and their data and, worse, their attention elsewhere.

My recommendation for government is this: engage. It's not too late to partner with net states. Appoint your own tech ambassadors. Set up diplomatic channels to net state leadership. Because there isn't much time left; the window to net state decision-makers will at some point close. Net states are already taking action in defense and diplomacy without government—no, *above* government. They are already investing in public infrastructure, which they will then possess slices of ownership over. They already manufacture the physical devices stowed in the vast majority of Americans' pockets: our phones. They already supply the platforms and services that ferry the content of a vast majority of Americans' thoughts. In short, net states are already gaining ground into defense, diplomacy, public infrastructure, and citizen services. And they're likely to only gain more ground in these areas as time goes on.

Nation-states—and government generally—will become increasingly irrelevant if they continue to cede power to net states, especially if they do so as a byproduct of inattention. And inattention is what appears to be driving these policies of noncooperation. It's highly doubtful that the Department of Homeland Security and the FBI have any clearly defined policy that stipulates intentional nonpartnership with technology companies. Being a government official, I've seen how policy often gets made: by adopting whatever the default is—by not actually *having* a policy. The default becomes the norm, the norm slides into the expected, and the expected eventually gets cemented as the official. Thoughtful and considered reflection is all too often absent in the process. Especially when it comes to confronting new and not-well-understood tech, government's default response is pretty much always no.

The competence with which net states are filling the leadership void left by the government's default policy could be the wake-up call nation-states need to recognize just how powerful net states have become. Conversely, it could lead to the systemic unraveling of the nation-state system as we know it.

In "The Demise of the Nation State," an article the British Indian novelist Rana Dasgupta wrote for *The Guardian*, he suggests that this wouldn't be such a bad thing.[14] He points to how the nation-state system counters the notion of universal human rights.

"Citizenship," he writes, "is itself the primordial kind of injustice in the world. It functions as an extreme form of inherited property and, like other systems in which inherited privilege is overwhelmingly determinant, it arouses little allegiance in those who inherit nothing." In other words, if you're lucky enough to be born in the United States or Sweden or the UK, you automatically "inherit" a better quality of life than if you happen to be born in Somalia or North Korea, whose governments are either too weak or too dictatorial to protect human rights.

Alex Tabarrok makes a similar case in his 2015 article for *The Atlantic*: "The Case for Getting Rid of Borders—Completely."[15] He contends that "no standard moral framework . . . regards people from foreign lands as less entitled to exercise their rights—or as inherently possessing less moral worth—than people lucky to have been born in the right place at the right time."

There are plenty of others also making similar arguments: advocating for, predicting, or otherwise presaging the beginning of the end of the nation-state system. I don't think we're there yet. But unless the US government takes up the mantle of responsibility set forth in the Constitution—to "promote the general welfare" and "secure the blessings of liberty to ourselves and our posterity"—it will miss its chance to influence how the international order is going to shake out in the coming decades.

"TYRANNY" IS GENERALLY DEFINED AS SOME VERSION OF "OPPRESSIVE power." It's been a loaded word in America since the beginning. The founding fathers were particularly fond of the term. Thomas Jefferson used it liberally in his writings, citing "tyrannical" rule from the United Kingdom twice in the Declaration of Independence. He wrote, in a letter to a colleague in the fight for independence, "I have sworn upon the altar of god eternal hostility against every form of tyranny over the mind of man"—words so emblematic of his philosophy that they're engraved on the Jefferson Memorial in Washington, DC.[16] James Madison, attempting to reassure the American people that the Constitution would protect against tyranny, specifically addressed that matter in the Federalist Papers (no. 46). If *every* country had the separation of powers that's set out in the American Constitution, he said, "the throne of every tyranny in Europe would be speedily overturned."[17] The Second Amendment of the Constitution, dealing with the right to bear arms, was born out of the framers' concern over the possibility of tyrannical rule; it ensured that the citizenry would have the ability to rise up against such forces.[18] Of course, that was then; today, there's virtually no chance that even if you armed every man, woman, and child in the United States we could defeat the steamroller that is the US military apparatus. But the anti-tyranny argument is still a popular one in gun law debates.

This is important background to keep in mind when considering why the government doesn't do more to protect us against potentially damaging net state practices. We as a citizenry grew into our present frame of mind with a deep distrust of government. This is one of the reasons for the establishment of the federal system of separating powers.

"If men were angels, no government would be necessary," as James Madison put it in the Federalist Papers (no. 51), going on to write that "in framing a government which is to be administered by men over men, the great difficulty is this: You must first enable the government to control the governed; and in the next place, oblige it to control itself."[19]

So, controlling itself—through the system of checks and balances—is embedded in the foundation of our government. While the American system didn't come out of thin air—ancient Rome was governed through a separation of powers, and such governance was popular in the writings of a number of influential political philosophers, including Baron de Montesquieu and John Locke—in modern history, the United States was among the first nations to implement the system.[20] In so doing, the US established checks and balances as a hallmark of the modern democratic state that would go on to be adopted by countries the world over.

True to our roots, we persist in our distrust of government. As discussed in chapter 6, Americans seem to accept that net states and marketing firms and data brokers will compile profiles on us in incredible detail. However, we will simply not tolerate the idea that *government*—that entity capable of "tyranny," in the popular imagination—might have anything more than the most basic data about us. When National Security Agency contractor Edward Snowden leaked to the public information about the metadata collection program PRISM in 2013, Americans freaked out, as we saw earlier. The fact that it was only metadata that was collected, and that metadata contains only information *about* communications, not the *content* of communications—not what person X said to person Y—got lost in most of the public discourse surrounding PRISM. The Snowden leaks, as they've come to be known, led Americans to become even more distrustful of how their government handles people's private information. In one poll, 70 percent of respondents reported that they believed the government used the data collected in that program "for purposes beyond anti-terror efforts."[21]

We didn't trust government when we formed as a nation. We don't trust it now. Actually, we don't trust much, when you really get down to it. Owing in part to a toxic partisan political climate following the 2016 presidential election, Americans in early 2018 reported feeling a loss of trust in one another, in the media, in institutions of government, in tech companies—in just about every index measured on a global "Trust Barometer," with trust dropping 23 points from the previous year, placing the US at 18th out of 26 countries surveyed.[22]

But there's an important distinction to be made between our reporting a loss of trust in government and a loss of trust in net states. We *need* our government in order to live safe, orderly lives. As we have seen, government is charged to provide the foundation that permits exactly that; it is to "establish justice," "insure domestic tranquility," "promote the general welfare," and "secure the blessings of liberty."

We don't *need* our net states in the same way. Or do we? Do we need them in order to live safe, orderly lives? Will they work toward justice, tranquility, our general welfare, and our blessings of liberty?

Increasingly, it seems that they may. They're rooting out terrorists on their platforms. They're shutting down accounts promoting hateful rhetoric. They're issuing diplomatic statements condemning practices that violate fundamental human rights. They're giving us tools that shoo us *off* their tools. They are, in their own way, looking out for us—as citizen-users. It may have nothing to do with principles: what's "right" or what's "just." It may have everything to do with protecting their ecosystem, preserving their own health and the health of us as the creatures who coinhabit it. Regardless of the motivations, genuinely protective action on the part of net states appears to be happening. The question remains, though, Can we expect it to *keep* happening? Can we ask for even more of it? And how?

And finally, the question for us to ask ourselves—you and me and everyone we know—is, What to do about government? If we need government—and we would be literally lawless and defenseless without it—how can we rebuild our trust in it; get it fully functioning, to where citizens participate in affairs of the state and feel powerful for their participation?

One last set of quotes from James Madison. He argued, "If Tyranny and Oppression come to this land, it will be in the guise of fighting a foreign enemy."[23] In a 1798 letter to Thomas Jefferson, he wrote, "It is a universal truth that the loss of liberty at home is to be charged to the provisions against danger, real or pretended, from abroad."[24]

We as a citizenry are indeed fighting foreign enemies—from nonstate actors to nation-state actors. We have been fighting foreign enemies in the current war on radical Islamist terror for 18 years, with no end in sight. And we know war well: the United States has been at war for 148 years out of our 243-year existence—and that's excluding the 45-year-long Cold War, our decades-long hostilities with Iran and North Korea, and our relationship with countries that conduct cyberattacks against American infrastructure, media, and our financial industry[25] on a regular basis.

Net states know us to a degree, as individuals, as a people. But they're not *responsible* for us; they are not responsible for our defense and they are not responsible for our well-being. *Government* is. And it's ours, as Roosevelt said: "We are the government—you and I." We face too many foreign enemies to ignore how much we need government to defend us. And we face too many domestic injustices to ignore how much we need government to serve us.

As we search—for information, for advice, for ways to better our days—we must remember to also search deeper: for meaning, for purpose, for something greater than ourselves. We can help ourselves as individuals and as a citizenry by searching for ways to see ourselves more clearly. And we can work with those who wield power over us—our net states and our nation-state—to protect and defend us from enemies foreign and domestic; from our own worst habits; and in pursuit of our better selves.

ON THE OUTSIDE, METEORITES LOOK MUCH LIKE ORDINARY EARTH rocks. Indeed, earth rocks and space rocks appear virtually indistinguishable. But if you split open a meteorite and etch it with weak acid, a pattern emerges. This pattern is called the Widmanstätten pattern. It doesn't occur in any rock on earth.

The pattern looks like a series of geometric cross-hatching, technically described as a "three-dimensional octahedral structure."[26] While the pattern can be reproduced on earth rocks, it doesn't last; its lines distort over time.[27] In the end, the pattern is wholly unique to meteorites. It takes a little work to see it; you have to look past the surface layer, apply a little technique. But once you do the work, the pattern is unmistakable.

Net states are not so different. They look like ordinary companies, just as meteorites look like earth rocks. They are not nation-states, though they increasingly engage in activities that nation-states usually handle. Net states are something unique. And they too have their telltale marks, their Widmanstätten pattern. Net states are tech companies, with a cause. They are tech collectives, with an agenda. They are tech believers, loosely unified. They possess physical resources, political means, and social capital. They are of countries, and beyond them. They are in virtually every aspect of our lives, yet for the most part effectively invisible to us.

We may occasionally wish we could go back to a time when we had control of all our data—when we didn't have to worry that the foolish pictures we took in college would end up on someone's Facebook page, when that professional embarrassment didn't end up being so easy to find on Google. But there's no unringing this bell. Net states are here to stay.

The best we can do is become knowledgeable about how they operate, why they do what they do, and what their underlying agendas are—beyond the obvious one: money. Also, we owe it to ourselves to figure out how to protect our data and our identities while still remaining productive members of the digital ecosystem. We must become the technological version of armchair geologists, able to identify a net state from a regular old tech company or flash-in-the-pan online movement. We must be able to spot their Widmanstätten pattern and, once we do, pay a very particular kind of attention to them.

This book is one of the first attempts I know of to chronicle the existence of net states and analyze how they impact our lives. But the specific examples in this book aren't really the point. Net states adapt, change, evolve, or perish. Whether we're talking about Google or Facebook or a net state that won't launch until tomorrow or a net state that dies before this book goes to print is almost irrelevant. What matters is that we keep the *pattern* of net states in mind: they are of the tech ethos; they advance belief-driven agendas; and they are international in scale and scope. And we empower them each time we use their devices, platforms, and services; we control their fate with every click, keystroke, and swipe.

We have to make those actions matter: they add up, especially with so many of us using their products and services. With net states, our habits are our votes. And the more we understand the pattern of what net states are and how much of our data they have, how deeply into our lives they reach, the more likely we will be able to mobilize like-minded others to push back.

It may feel like too much to expect. Who are we, after all, as individual citizen-users, to demand that multibillion-dollar tech companies change their ways? As individuals, we may not have much clout. But tech itself has given us the ability to mobilize in ways never before possible.

As the novelist Alice Walker once said, “The most common way people give up their power is by thinking they don’t have any.”[28] Every single tech user has power over their net states. We just have to claim it.

ACKNOWLEDGMENTS

There's a story behind this book. I didn't start out thinking I would write one. My original plan was just to get an article published, for which I have Joanna Pearlstein at *WIRED* to thank. It came out in November 2017 and caught the attention of NPR's *Science Friday* program. I thank Ira Flatow, who invited me to appear on air alongside the whip-smart tech journalist Max Reed, for the chance to bring my ideas to the airwaves.

Because it was after that segment that things suddenly started happening: Roger Freet, my esteemed literary agent at Foundry Media, was driving to the package store when my episode of *Science Friday* aired. He said he was listening to the radio, heard my description of net states, and pulled over to take out a pen and write down my name. He found my email address online and reached out with what really is one of the best questions in the world: Would I be interested in turning this idea into a book?

Roger coached me through the grueling process of converting a 1,000-word article into a proper full-length book proposal with incredible patience, generosity of spirit, and plain old blind faith. I am enormously grateful for his encouragement and support, and for guiding the book to its publishing home at HarperOne, with my superstar editor, Miles Doyle.

I happen to know a little bit about editors, having the great good fortune of being married to one of the most successful editors in the business, Jonathan Tepperman (more on him later). One of the best descriptions of Miles's contribution to this book came from my husband. Upon reading my final draft, Jonathan said, "I don't know exactly what Miles did, but you should be extremely happy to have him on your side—this version is light-years better than the first draft I read!" It is indeed, and I owe a massive thanks to Miles for his insightful, thoughtful, and always-encouraging approach to editing this book. I am also deeply indebted to Kathy Reigstad for her surgical-precision copyediting, which transformed this from manuscript to proper book, ready for the world.

I wrote this book while also working a full-time job as press secretary at the New York City Department of Veterans' Services, taking every Sunday from 4 a.m. to noon to write, and squishing in writing during odd hours each week with my boss's encouragement. My boss, Loree Sutton, MD, Commissioner, Brigadier General (ret.), US Army, is a boss people dream of. I couldn't have written this book without her encouragement, support, and flexibility, as well as the professionalism and camaraderie of all my incredible colleagues, especially Jeffrey Roth, Ellen Greeley, Cassandra Alvarez, Jason Parker, Eric Henry, Aquilla Hines, Venkat Motupalli, and, above all, my counterpart, Gabriel Ramos.

In addition to being a government official, I am blessed to be able to teach part-time at Columbia University's School of International and Public Affairs (SIPA). I am eternally grateful for and owe enormous thanks to Anya Schiffrin, Dan McIntyre, and Merit Janow for the chance to be a part of the SIPA family.

It was through SIPA that I met my intrepid research assistant, CJ Dixon, whose discoveries opened my eyes to possibilities I would never have found without him. I also owe a huge debt of gratitude to my "Technology, National Security, and the Citizen" students, whose fascinating insights informed so many topics in this book—with a special thanks to my teaching assistant, Camilia Razavi.

Those who lent me the brainpower, moral support, and wisdom that lifted this book to a whole new level deserve special praise, as they had to wade through my less-than-polished ideas and drafts. Their insights made the book a far better final product, and I am lucky to have them on my side: Alexandra Farkas, Aliya Bhatia, Aquilla Hines, Camilia Razavi, Chet Wichowski, Christine Cearnal, Dahna Black, Danielle Tomson, Dawn Emsellem, Emily Boitel, Gabriel Ramos, Jacqueline Burns Koven, Joel Putnam, Katherine Garcia, Keith Wichowski, Madison Jacox, Marijka Hoczko, Matthew Burton, Michael McColpin, Natasha Cohen, Shalaka Jokshi, and Venkat Motupalli—thanks for being so incredibly generous with your time and mental energy.

As a parent of school-age children, I couldn't do any work at all without a team of responsible, committed caregivers. A huge, heartfelt thank-you to Chanise Martineau for ensuring that my littlest remained happy, safe, and loved while I worked; and my most sincere appreciation for all the dedicated teachers at Congregation Beth Elohim, MS 51, and Williamsburg Prep, who make my children smarter, kinder, and better people by the day.

My family. It gives me tremendous pleasure to recognize them, my inner orbit, the people who most inspire and sustain me: my parents, Chet Wichowski and Marijka Hoczko; my brother and team, Keith Wichowski, Dawn Emsellem, Milo, Andre, and Desi; my Canadian family, the Teppermans—Bill and Rochelle; Noah, Julie, Lily, and Benji; Andrew, Tina, Lia, and Nathan; and my dear and lifelong friends, Christine Cearnal and Alexandra Farkas.

Of course, I would be nothing without my nucleus: my children, Gerome, Novi, and Leo, and my husband, Jonathan. Gerome, Novi, and Leo, watching you grow and guiding your progress into the wonderful human beings you are is the greatest thing I've ever been a part of, and I'm so grateful I get to care for you and just plain delight in you.

Jonathan, being in your presence makes me want to be the best version of myself. I would not have been able to write this book without you—literally, as it was only because you were willing to watch our four-year-old single-handedly every Sunday for a year (a year!) that I had time to write, and in every other possible sense as well. I'm inspired by you; I'm awed by you. I am so very grateful for you. Thank you for being you. Thank you for choosing me.

APPENDIX

CATEGORY	NAME	SUBSIDIARY OF	INVESTED IN BY	ACQUIRED BY
Agriculture & food services	Farmers Business Network		Google	
Agriculture & food services	The Climate Corporation		Google	
Agriculture & food services	Granular		Google	
Agriculture & food services	Bowery Farming, Inc.		Google	
Agriculture & food services	Abundant Robotics		Google	
Agriculture & food services	Solum		Google	
Agriculture & food services	Impossible Foods		Google	
Agriculture & food services	LevelUp		Google	
Agriculture & food services	Soylent		Google	
Agriculture & food services	Juicero		Google	
Agriculture & food services	Ripple Foods		Google	
Agriculture & food services	Clear Labs		Google	
Agriculture & food services	Siddhivinayak Agri Processing		Google	
Agriculture & food services	Whole Foods Market			Amazon
Agriculture & food services	Ocean's Halo		Facebook	
Agriculture & food services	Rise		Google	
Biotech	Evelo Biosciences		Google	
Biotech	BenchSci		Google	
Biotech	ARMO BioSciences		Google	
Biotech	Magenta Therapeutics		Google	
Biotech	23andMe		Google	
Biotech	Arcus Biosciences		Google	

CATEGORY	NAME	SUBSIDIARY OF	INVESTED IN BY	ACQUIRED BY
Biotech	FLX Bio		Google	
Biotech	Forty Seven		Google	
Biotech	IDEAYA Biosciences		Google	
Biotech	SQZ Biotech		Google	
Biotech	BlackThorn Therapeutics		Google	
Biotech	GLO		Google	
Biotech	XtalPi		Google	
Biotech	Gritstone Oncology		Google	
Biotech	Vaccitech		Google	
Biotech	Cala Health		Google	
Biotech	Arsanis		Google	
Biotech	Alector		Google	
Biotech	Obsidian Therapeutics		Google	
Biotech	Fulcrum Therapeutics		Google	
Biotech	Cambridge Epigenetix		Google	
Biotech	iPierian		Google	
Biotech	Adimab		Google	
Biotech	Genomics Medicine Ireland		Google	
Biotech	SynapDx		Google	
Biotech	Transcriptic		Google	
Biotech	SpyBiotech		Google	
Biotech	NeuroVigil			Elon Musk
Biotech	DNAnexus		Microsoft	
Biotech	Rosetta Biosoftware			Microsoft
Drones & delivery	GoJek		Google	
Drones & delivery	EasyPost		Google	
Drones & delivery	Sidecar Technologies		Google	
Drones & delivery	Convoy			Amazon
Drones & delivery	Airware		Google	
Drones & delivery	Skycatch		Google	
Drones & delivery	Cape Productions		Google	
Drones & delivery	Hivemapper		Google	
Drones & delivery	Ascenta (UK)			Facebook
Energy	SolarCity		Google	Tesla
Energy	Transphorm Inc.		Google	
Energy	Cool Planet Energy Systems		Google	

CATEGORY	NAME	SUBSIDIARY OF	INVESTED IN BY	ACQUIRED BY
Energy	EDF Renewable Energy		Google	
Energy	BrightSource Energy		Google	
Energy	Clean Power Finance		Google	
Energy	Alta Wind Energy Center		Google	
Energy	Makani Power		Google	Google
Energy	Center for Resource Solutions		Google	
Energy	Outride			Google
Energy	Zep Solar			SolarCity
Energy	Paramount Solar			SolarCity
Energy	Off Grid Electric		Tesla	
Energy	Valence Technology			Microsoft
Energy	Silevo	Tesla		SolarCity
Energy	Iliosson			SolarCity
Energy	Palo Alto Solar			SolarCity
Energy	Declination Solar			SolarCity
Energy	BuffaloGrid		Microsoft	
Energy	Xively			Google
Environment	Gradient		Google	
Geopolitics	Jigsaw	Google		
Geospatial tech	echoecho		Google	
Geospatial tech	Keyhole Inc.			Google
Geospatial tech	Indoor.io			Apple
Geospatial tech	BroadMap			Apple
Geospatial tech	Locationary			Apple
Geospatial tech	WifiSlam			Apple
Geospatial tech	Flyby Media			Apple
Geospatial tech	Color Labs Inc.			Apple
Geospatial tech	Poly9			Apple
Geospatial tech	Mapsense			Apple
Geospatial tech	Coherent Navigation			Apple
Geospatial tech	Glancee			Facebook
Geospatial tech	MultiMap			Microsoft
Geospatial tech	Spindle			Twitter
Government	Vicinity			Microsoft
Government	Fast Search & Transfer			Microsoft

CATEGORY	NAME	SUBSIDIARY OF	INVESTED IN BY	ACQUIRED BY
Health & wellness	ClassPass		Google	
Health & wellness	ProtoGeo Oy			Facebook
Health & wellness	Beddit			Apple
Health & wellness	Gliimpse			Apple
Health & wellness	Hello Heart		Facebook and Google	
Health & wellness	Fight The Stroke		Microsoft	
Health & wellness	MD.Voice		Microsoft	
Health & wellness	Mgv		Microsoft	
Health & wellness	Medstory			Microsoft
Health & wellness	Global Care Solutions			Microsoft
Health & wellness	Sentillion			Microsoft
Health & wellness	Flatiron Health		Google	
Health & wellness	Clover Health		Google	
Health & wellness	Quartet		Google	
Health & wellness	Collective Health		Google	
Health & wellness	One Medical		Google	
Health & wellness	Foundation Medicine		Google	
Health & wellness	Freenome		Google	
Health & wellness	Vida Health		Google	
Health & wellness	PatientPing		Google	
Health & wellness	Aledade		Google	
Health & wellness	Zephyr Health		Google	
Health & wellness	Walker & Company Brands		Google	
Health & wellness	ZappRx		Google	
Health & wellness	Spruce Health		Google	
Health & wellness	Aspire Health		Google	
Health & wellness	Orange Chef		Google	
Health & wellness	Predilytics		Google	
Health & wellness	Oration		Google	
Health & wellness	TinyRx		Google	
Health & wellness	FitStar		Google	Fitbit
Health & wellness	Navigenics		Google	
Health & wellness	1Life Healthcare		Google	
Health & wellness	FogPharma		Google	
Health & wellness	Wingu		Google	

CATEGORY	NAME	SUBSIDIARY OF	INVESTED IN BY	ACQUIRED BY
Health & wellness	Qliance Medical Management			Amazon
Health & wellness	Zocdoc			Amazon
Health & wellness	Doctor on Demand		Google	
Health & wellness	Senosis Health			Google
Health & wellness	Lift Labs			Google
Health & wellness	Fractyl Laboratories		Google	
Health & wellness	Rare Light			Apple
Health & wellness	Editas Medicine		Google	
Health & wellness	DNAnexus		Google	
Health & wellness	Relay Therapeutics		Google	
Health & wellness	Rani Therapeutics		Google	
Health & wellness	Spero Therapeutics		Google	
Health & wellness	Celsius Therapeutics		Google	
Health & wellness	LifeMine Therapeutics		Google	
Health & wellness	Rodin Therapeutics		Google	
Health & wellness	Carrick Therapeutics		Google	
Home goods	Move Loot		Google	
Infrastructure	Poynt		Google	
Infrastructure	Puppet		Google	
Infrastructure	Stackdriver			Google
Infrastructure	Metaweb Technologies			Google
Infrastructure	Union Bay Networks			Apple
Infrastructure	Equinix		Microsoft	
Infrastructure	SmartPipes		Microsoft	
IOT	Nest Labs	Google	Google	Google
IOT	Sidewalk Labs	Google		
IOT	2lemetry			Amazon
IOT	Styleware			Apple
Manufacturing	Carbon		Google	
Manufacturing	Veo Robotics		Google	
Manufacturing	Plethora		Google	
Manufacturing	CircuitHub		Google	
Manufacturing	HTC-Pixel Phone Division			Google
Manufacturing	Channel Intelligence			Google
Manufacturing	InVisage Technologies			Apple

CATEGORY	NAME	SUBSIDIARY OF	INVESTED IN BY	ACQUIRED BY
Manufacturing	Passif Semiconductor			Apple
Manufacturing	PA Semi			Apple
Manufacturing	Intrinsity			Apple
Manufacturing	Beats Electronics			Apple
Manufacturing	Corning Incorporated		Apple	
Manufacturing	Perbix			Tesla
Manufacturing	Riviera Tool LLC			Tesla
Manufacturing	Grohmann Engineering			Tesla
Manufacturing	Alta Motors		Tesla	
Manufacturing	Pacific Microsonics			Microsoft
Real estate	HomeLight		Google	
Real estate	Cozy		Google	
Real estate	Airbnb			Amazon
Security	eMagin		Apple	
Security	LegbaCore			Apple
Space exploration	SpaceX			Elon Musk
Space exploration	Planetary Resources		Eric Schmidt	
Space exploration	Planetary Resources		Larry Page	
Telecoms	Imagination Technologies		Apple	
Telecoms	Internet.org (Free Basics)	Facebook		
Telecoms	Endaga			Facebook
Telecoms	Orbital Insight, Inc.		Google	
Telecoms	O3b Networks		Google	
Telecoms	Orion Network Systems			Apple
Telecoms	Google Fiber	Google		
Telecoms	Dialpad		Google	
Telecoms	Helium		Google	
Telecoms	Clearwire		Google	
Telecoms	Ubiquisys		Google	
Telecoms	GrandCentral			Google
Telecoms	modu			Google
Telecoms	SayNow			Google
Telecoms	Motorola Mobility			Google
Telecoms	Limes Audio			Google
Telecoms	Finisar		Apple	
Telecoms	2nd Century		Microsoft	

CATEGORY	NAME	SUBSIDIARY OF	INVESTED IN BY	ACQUIRED BY
Telecoms	Nokia			Microsoft
Telecoms	Navic Systems			Microsoft
Telecoms	Skype			Microsoft
Transportation	Urban Engines			Google
Transportation	Waze			Google
Transportation	Tesla		Google	
Transportation	Aptera		Google	
Transportation	Openbay		Google	
Transportation	ZipDash			Google
Transportation	Playment		Google	
Transportation	Scotty Labs		Google	
Transportation	Uber		Microsoft	Amazon
Transportation	Kitty Hawk		Larry Page	
Transportation	Embark			Apple
Transportation	HopStop			Apple
Transportation	Didi Chuxing		Apple	
Transportation	The Boring Company	Tesla		
Wearables	Magic Leap		Google	
Wearables	Bloomlife		Google	
Wearables	WIMM Labs			Google
Wearables	Cronologics Corporation			Google
Wearables	Eyefluence			Google
Wearables	Vocal IQ			Apple

NOTES

INTRODUCTION

1. Peter Cohen, "Macworld Expo Attendance Breaks Records," *Macworld*, March 13, 2007, retrieved June 27, 2018, https://www.macworld.com/article/1056750/macworldexpo.html.
2. "Steve Jobs iPhone 2007 Presentation," YouTube, May 13, 2013, retrieved June 23, 2018, https://www.youtube.com/watch?v=vN4U5FqrOdQ.
3. Judy Bachrach, "WikiHistory: Did the Leaks Inspire the Arab Spring?" *World Affairs Journal*, July/August 2011, retrieved August 5, 2018, http://www.worldaffairsjournal.org/article/wikihistory-did-leaks-inspire-arab-spring.
4. Issie Lapowsky, "Facebook Exposed 87 Million Users to Cambridge Analytica," *WIRED*, April 4, 2018, retrieved December 16, 2018, https://www.wired.com/story/facebook-exposed-87-million-users-to-cambridge-analytica/.
5. "iPhone Sales Said to Hit Half-Million," CNN Money, July 2, 2007, retrieved June 23, 2018, http://money.cnn.com/2007/07/02/technology/iphone_sales/.
6. Dawn Chmielewski, "The Wait Is Over: The Scene from 2007 When the iPhone First Went on Sale," *Los Angeles Times*, June 30, 2007, retrieved June 23, 2018, http://www.latimes.com/business/technology/la-fi-iphone-debut-20070630-story.html.
7. "Number of Smartphone Users Worldwide from 2014 to 2020 (in Billions)," Statista, n.d., retrieved June 27, 2018, https://www.statista.com/statistics/330695/number-of-smartphone-users-worldwide/.
8. "Transcript of Keynote Address at the RSA Conference 2017: The Need for a Digital Geneva Convention," Microsoft News, February 14, 2017, retrieved January 30, 2019, https://news.microsoft.com/uploads/2017/03/Transcript-of-Brad-Smiths-Keynote-Address-at-the-RSA-Conference-2017.pdf.
9. Alexis Wichowski, "Net States Rule the World: Ignore Them at Your Peril," *WIRED*, November 4, 2017, retrieved April 7, 2019, https://www.wired.com/story/net-states-rule-the-world-we-need-to-recognize-their-power/.
10. Gary Davis, "Silk Road's 'Libertas'? 5 Fast Facts You Need to Know," Heavy (website), December 2013, retrieved November 25, 2018, https://heavy.com/news/2013/12/gary-davis-silk-road-2-arrests-libertas/.
11. Aaron Rogan, "Irishman Gary Davis Denies Helping to Run Silk Road Dark Net Website," *Times* (UK), August 24, 2018, retrieved November 25, 2018, https://www.thetimes.co.uk/article/irishman-gary-davis-denies-helping-to-run-silk-road-dark-net-website-3fk93r7bl.

12. Aaron Rogan, "Irishman Gary Davis Denies Helping to Run Silk Road Dark Net Web site," *Times* (UK), August 24, 2018, retrieved February 5, 2019, https://www.thetimes.co.uk/article/irishman-gary-davis-denies-helping-to-run-silk-road-dark-net-website-3fk93r7bl.
13. Steve Kovach, "FBI Says Illegal Drugs Marketplace Silk Road Generated $1.2 Billion in Sales Revenue," Business Insider, October 2, 2013, retrieved February 2, 2019, https://www.businessinsider.com/silk-road-revenue-2013–10.
14. John Biggs, "Alleged Top Moderators of Silk Road 2 Forums Arrested in Ireland, U.S. in International Sweep," TechCrunch, December 20, 2013, retrieved February 5, 2019, https://techcrunch.com/2013/12/20/alleged-top-moderator-of-silk-road-2-forums-arrested-in-ireland/.
15. 47 U.S. Code § 230, "Protection for Private Blocking and Screening of Offensive Material," Communications Decency Act, 1996, Legal Information Institute, retrieved February 2, 2019, https://www.law.cornell.edu/uscode/text/47/230.
16. United States of America v. Microsoft Corporation, 253 F.3d 34 (DC Cir. 2001), Justia Legal Resources, US Federal Law, retrieved February 5, 2019, https://law.justia.com/cases/federal/appellate-courts/F3/253/34/576095.
17. Kate Conger, "Google Plans Not to Renew Its Contract for Project Maven, a Controversial Pentagon Drone AI Imaging Program," Gizmodo Media Group, June 1, 2018, retrieved December 2, 2018, https://gizmodo.com/google-plans-not-to-renew-its-contract-for-project-mave-1826488620?rev=1527878336532.
18. Cheryl Pellerin, "Project Maven to Deploy Computer Algorithms to War Zone by Year's End," US Department of Defense, July 21, 2017, retrieved December 2, 2018, https://dod.defense.gov/News/Article/Article/1254719/project-maven-to-deploy-computer-algorithms-to-war-zone-by-years-end/.
19. Cheryl Pellerin, "Project Maven to Deploy Computer Algorithms to War Zone by Year's End," US Department of Defense, July 21, 2017, retrieved July 3, 2019, https://dod.defense.gov/News/Article/Article/1254719/project-maven-to-deploy-computer-algorithms-to-war-zone-by-years-end/.
20. Kate Conger, "Google Plans Not to Renew Its Contract for Project Maven, a Controversial Pentagon Drone AI Imaging Program," Gizmodo Media Group, June 1, 2018, retrieved December 2, 2018, https://gizmodo.com/google-plans-not-to-renew-its-contract-for-project-mave-1826488620?rev=1527878336532.
21. Kate Conger, "Google Employees Resign in Protest Against Pentagon Contract," Gizmodo Media Group, May 14, 2018, retrieved December 2, 2018, https://gizmodo.com/google-employees-resign-in-protest-against-pentagon-con-1825729300.
22. Daisuke Wakabayashi, "Google Will Not Renew Pentagon Contract That Upset Employees," *New York Times*, June 1, 2018, retrieved December 2, 2018, https://static01.nyt.com/files/2018/technology/googleletter.pdf.
23. "Anonymous Claims to Have Taken Down 20,000 IS Twitter Accounts," BBC News, November 20, 2015, retrieved May 11, 2018, http://www.bbc.co.uk/newsbeat/article/34877968/anonymous-claims-to-have-taken-down-20000-is-twitter-accounts.
24. J. M. Berger, "The Evolution of Terrorist Propaganda: The Paris Attack and Social Media," Brookings Institution, January 27, 2015, retrieved May 11, 2018, https://www.brookings.edu/testimonies/the-evolution-of-terrorist-propaganda-the-paris-attack-and-social-media/.
25. Micah Sifry, "Ambient Awareness," *Civicist*, June 18, 2019, retrieved June 23, 2019, https://civichall.org/civicist/ambient-awareness/.
26. "Facial Recognition System Helps Trace 3,000 Missing Children in 4 Days," NDTV, April 22, 2018, retrieved May 30, 2018, https://www.ndtv.com/india-news/facial-recognition-system-helps-trace-3-000-missing-children-in-4-days-1841192?pfrom=home-topstories.
27. Gabrielle Fonrouge, "Teens Caught in Rough Surf Saved in First Drone Rescue of Its Kind," *New York Post*, January 18, 2018, retrieved May 30, 2018, https://nypost

.com/2018/01/18/drone-makes-first-ever-water-rescue-for-teens-caught-in-rough-surf/.

28. Mitchell Broussard, "Tim Cook Shares Lifesaving Apple Watch Story After Teen Goes into Kidney Failure," MacRumors, May 1, 2018, retrieved May 30, 2018, https://www.macrumors.com/2018/05/01/tim-cook-apple-watch-kidney-failure/.
29. Henry David Thoreau, *Civil Disobedience* (Nashville: American Renaissance Books, 2010), 23.

CHAPTER 1: RISE OF THE CITIZEN-USER

1. Thomas Juster, "Changing Times of American Youth: 1981–2003," Institute for Social Research, University of Michigan, November 2004, retrieved September 9, 2018, http://ns.umich.edu/Releases/2004/Nov04/teen_time_report.pdf; "Prices in the Eighties," In the 80s (website), n.d., retrieved September 9, 2018, http://www.inthe80s.com/prices.shtml.
2. Stewart Wolpin, "The First Cellphone Went on Sale 30 Years Ago for $4,000," Mashable, March 13, 2014, https://mashable.com/2014/03/13/first-cellphone-on-sale/; CPI Inflation Calculator, "U.S. Inflation Rate, $4,000 in 1985 to 2019," retrieved June 23, 2019, http://www.in2013dollars.com/us/inflation/1985?amount=4000.
3. Scott Brown, "*WarGames*: A Look Back at the Film That Turned Geeks and Phreaks into Stars," *WIRED*, July 21, 2008, retrieved September 9, 2018, https://www.wired.com/2008/07/ff-wargames/.
4. "Sir Tim Berners-Lee," National Portrait Gallery, n.d., retrieved April 14, 2019, https://www.npg.org.uk/collections/search/person/mp58761/sir-tim-berners-lee.
5. "Number of Smartphone Users Worldwide from 2014 to 2020 (in Billions)," Statista, n.d., retrieved September 12, 2018, https://www.statista.com/statistics/330695/number-of-smartphone-users-worldwide/.
6. Steven Levy, *Hackers: Heroes of the Computer Revolution* (Garden City, NY: Anchor Press/Doubleday, 1984), 26–36.
7. "Why Half of Developers Don't Have a Computer Science Degree," Next Web, April 23, 2016, retrieved November 25, 2018, https://thenextweb.com/insider/2016/04/23/dont-need-go-college-anymore-programmer/.
8. Christopher Lehmann-Haupt, "Books of the Times; Hackers as Heroes," *New York Times*, December 24, 1984, retrieved September 9, 2018, https://www.nytimes.com/1984/12/24/books/books-of-the-times-hackers-as-heroes.html.
9. Faubrice Florin, *Hackers: Wizards of the Electronic Age* (documentary), 1985, retrieved September 9, 2018, https://www.youtube.com/watch?v=zOP1LNr70aU.
10. Carole Cadwalladr, "Stewart Brand's 'Whole Earth Catalog,' the Book That Changed the World," *Guardian*, May 4, 2013, retrieved May 20, 2018, https://www.theguardian.com/books/2013/may/05/stewart-brand-whole-earth-catalog.
11. "Stewart Brand States Information Wants to Be Free," Getty Images, November 1, 1984, retrieved September 9, 2018, https://www.gettyimages.com/detail/video/at-the-first-hackers-conference-in-1984-steve-wozniak-and-news-footage/146496695.
12. Joshua Gans, "'Information Wants to Be Free': The History of That Quote," *Digitopoly* (blog), 25 October 2015, www.digitopoly.org/2015/10/25/information-wants-to-be-free-the-history-of-that-quote/.
13. This is an idea I first explored in a book chapter on information-sharing and diplomacy: Alexis Wichowski, "'Secrecy Is for Losers': How Diplomats Should Embrace Openness to Protect National Security," in *Digital Diplomacy: Theory and Practice*, ed. Marcus Holmes and Corneliu Bjola (London and New York: Routledge, 2016).
14. "Mission Statement," *Economist*, June 2, 2009, retrieved November 25, 2018, https://www.economist.com/news/2009/06/02/mission-statement.
15. "From the Garage to the Googleplex," *Google: Our Story*, n.d., retrieved November 25, 2018, https://www.google.com/about/our-story/.

16. Christian Zibreg, "Mark Zuckerberg Says You Need to 'Move Fast and Break Things,'" Geek, October 2, 2009, retrieved November 25, 2018, https://www.geek.com/news/mark-zuckerberg-says-you-need-to-move-fast-and-break-things-922432/.
17. Anna Wiener, "The Complicated Legacy of Stewart Brand's 'Whole Earth Catalog,'" *New Yorker*, November 18, 2018, retrieved November 25, 2018, https://www.newyorker.com/news/letter-from-silicon-valley/the-complicated-legacy-of-stewart-brands-whole-earth-catalog.
18. "Home Computers and Internet Use in the United States: August 2000," US Census Bureau, September 2001, retrieved September 9, 2018, https://www.census.gov/prod/2001pubs/p23–207.pdf.
19. Bill Gates, "Content Is King," Microsoft, January 1996, retrieved via Wayback Machine on June 27, 2018, http://web.archive.org/web/20010126005200/http://www.microsoft.com/billgates/columns/1996essay/essay960103.asp.
20. Farhad Manjoo, "Jurassic Web," Slate, February 24, 2009, retrieved June 27, 2018, http://www.slate.com/articles/technology/technology/2009/02/jurassic_web.html.
21. "Historical Daily Viewing Activity Among Households & Persons 2+," Nielsen Company, n.d., retrieved June 27, 2018, http://www.nielsen.com/content/dam/corporate/us/en/newswire/uploads/2009/11/historicalviewing.pdf.
22. "The Best Websites of 1996," *TIME*, June 24, 2001, retrieved June 27, 2018, http://content.time.com/time/magazine/article/0,9171,135245,00.html.
23. Tim O'Reilly, "What Is Web 2.0?" O'Reilly Media, September 30, 2005, retrieved July 1, 2018, https://www.oreilly.com/pub/a/web2/archive/what-is-web-20.html.
24. Gary Marshall, "Happy Birthday, Windows 95: The OS That Changed It All," *TechRadar*, August 24, 2015, retrieved July 1, 2018, https://www.techradar.com/news/software/operating-systems/happy-birthday-windows-95-the-os-that-changed-it-all-711984.
25. Lance Ulanoff, "Remembering the Windows 95 Launch: A Triumph of Marketing," Mashable, August 24, 2015, retrieved June 27, 2018, https://mashable.com/2015/08/24/remembering-windows-95-launch/#X9YZEiwCSiqB.
26. Ian Morris, "Seven Ways Windows 95 Changed the World," *Forbes*, August 24, 2015, retrieved June 27, 2018, https://www.forbes.com/sites/ianmorris/2015/08/24/windows-95-changed-the-world/.
27. Alyssa Newcomb, "Why Windows 95 Was a Game-Changer for Computer Users Everywhere," ABC News, August 24, 2015, retrieved July 1, 2018, https://abcnews.go.com/Technology/windows-95-game-changer-computer-users/story?id=33212036.
28. Brian Feldman, "*U.S.* v. *Microsoft* Proved That Antitrust Can Keep Tech Power in Check," *New York* (magazine), December 12, 2017, retrieved July 1, 2018, http://nymag.com/selectall/2017/12/u-s-v-microsoft-proved-that-antitrust-can-check-tech-power.html.
29. Michael Calore, "April 22, 1993: Mosaic Browser Lights Up Web with Color, Creativity," *WIRED*, April 22, 2010, retrieved July 1, 2018, https://www.wired.com/2010/04/0422mosaic-web-browser/.
30. Sean Cooper, "Whatever Happened to Netscape?" Engadget, May 10, 2014, retrieved July 1, 2018, https://www.engadget.com/2014/05/10/history-of-netscape/.
31. "AOL, Netscape Tie Knot," CNN Money, November 24, 1998, retrieved July 1, 2018, http://money.cnn.com/1998/11/24/technology/aol/.
32. Luisa Kroll, "The World's Billionaires: 25th Anniversary Timeline," *Forbes*, March 7, 2012, retrieved July 1, 2018, https://www.forbes.com/special-report/2012/billionaires-25th-anniversary-timeline.html.
33. Brian Feldman, "*U.S.* v. *Microsoft* Proved That Antitrust Can Keep Tech Power in Check," *New York* (magazine), December 12, 2017, http://nymag.com/intelligencer/2017/12/u-s-v-microsoft-proved-that-antitrust-can-check-tech-power.html.
34. Désiré Athow, "Microsoft's Lawsuit Payouts Amount to Around $9 Billion," *Inquirer*, July 14, 2005, retrieved July 1, 2018, https://www.theinquirer.net/inquirer/news/1048246/microsoft-lawsuit-payouts-usd9-billion.

35. "AT&T Breakup II: Highlights in the History of a Telecommunications Giant," *Los Angeles Times*, September 21, 1995, retrieved July 1, 2018, http://articles.latimes.com/1995-09-21/business/fi-48462_1_system-breakup.
36. "May 15, 1911: Supreme Court Orders Standard Oil to Be Broken Up," *New York Times*, May 25, 2012, retrieved July 1, 2018, https://learning.blogs.nytimes.com/2012/05/15/may-15-1911-supreme-court-orders-standard-oil-to-be-broken-up/.
37. "Final Judgment Civil Action No. 98-1232 (CKK)," US Department of Justice, n.d., retrieved July 1, 2018, https://www.justice.gov/atr/case-document/final-judgment-133; Anna Domanska, "The Famous History of Microsoft's Antitrust Case," *Industry Leaders*, February 12, 2018, retrieved July 1, 2018, https://www.industryleadersmagazine.com/famous-history-microsofts-antitrust-case/; Nancy Gohring, "End of an Era: Microsoft Antitrust Oversight Ends," *Macworld*, May 11, 2011, retrieved July 1, 2018, https://www.macworld.com/article/1159841/microsoft_antitrust.html.
38. David Moschella, "Microsoft Losing in the Court of Public Opinion," *Computerworld*, May 22, 2000, retrieved July 1, 2018, https://www.computerworld.com/article/2594627/app-development/microsoft-losing-in-the-court-of-public-opinion.html.
39. "Pirate Bay Founding Group Disbands," BBC News, June 28, 2010, retrieved May 20, 2018, http://www.bbc.com/news/10433195.
40. "What Is the Pirate Agency?," Piratbyrån, February 9, 2004, retrieved May 22, 2018, https://web.archive.org/web/20040209024510/http://piratbyran.org:80/index.php?view=articles&cat=8.
41. Patrick Burkart, *Pirate Politics: The New Information Policy Contests* (Cambridge, MA: MIT Press, 2014), 17.
42. Tom Lamont, "Napster: The Day the Music Was Set Free," *Guardian*, February 23, 2013, retrieved May 20, 2018, https://www.theguardian.com/music/2013/feb/24/napster-music-free-file-sharing.
43. Paul Resnikoff, "Pressure Intensifies as the Pirate Bay Crosses 300 Million Visitors," Digital Music News, June 11, 2017, retrieved May 20, 2018, https://www.digitalmusicnews.com/2017/06/11/pressure-pirate-bay-300-million/; home page, Pirate Bay, n.d., retrieved May 20, 2018, https://thepiratebay.org/.
44. Jan Libbenga, "Pirate Bay Resurfaces, While Protesters Walk the Street," *Register*, June 5, 2006, retrieved September 12, 2018, https://www.theregister.co.uk/Print/2006/06/05/pirate_bay_reemerges/.
45. Ryan Neal, "The Pirate Bay Moves Again: How Many Domains Are Left?" *International Business Times*, December 18, 2013, retrieved May 20, 2018, http://www.ibtimes.com/pirate-bay-moves-again-how-many-domains-are-left-1514306; Christopher de Looper, "History of the Pirate Bay: Internet Outlaw or Internet File-Sharing Freedom Fighter?," *Tech Times*, December 17, 2014, retrieved May 20, 2018, http://www.techtimes.com/articles/22362/20141217/history-pirate-bay.htm.
46. Julianne Pepitone, "The Pirate Bay Down in Extended Outage," CNN Money, October 3, 2012, retrieved May 20, 2018, http://money.cnn.com/2012/10/02/technology/pirate-bay-down/index.html.
47. "About PPI," Pirate Parties International, n.d., retrieved May 20, 2018, https://pp-international.net/about-ppi/.
48. "Julia Reda, Euro-Pirate: For DE, EU, WWW, and Human Rights," Piraten Partei Deutschland, n.d., retrieved May 20, 2018, https://www.piratenpartei.de/partei/julia-reda/; "European Parliament 2019 Elections Results," Bloomberg, June 23, 2019, retrieved June 23, 2019, https://www.bloomberg.com/graphics/2019-european-parliament-elections/.
49. "Iceland: Individual-Taxes on Personal Income," Pricewaterhouse Coopers, n.d., retrieved July 20, 2018, http://taxsummaries.pwc.com/ID/Iceland-Individual-Taxes-on-personal-income; Jon Henley, "Iceland PM Steps Aside After Protests over Panama Papers Revelations," *Guardian*, April 5, 2016, retrieved June 3, 2018, https://www.theguardian.com/world/2016/apr/05/iceland-prime-minister-resigns-over-panama-papers-revelations.

50. "Iceland's Pirate Party Invited to Form Government," *Guardian*, December 2, 2016, retrieved May 20, 2018, https://www.theguardian.com/world/2016/dec/02/iceland-pirate-party-invited-form-government-coalition.
51. Derek Thompson, "Google's CEO: 'The Laws Are Written by Lobbyists'" *Atlantic*, October 1, 2010, retrieved November 23, 2018, https://www.theatlantic.com/technology/archive/2010/10/googles-ceo-the-laws-are-written-by-lobbyists/63908/.
52. "Facebook: Facts & Figures for 2010," *Digital Buzz* (blog), March 22, 2010, retrieved November 25, 2018, http://www.digitalbuzzblog.com/facebook-statistics-facts-figures-for-2010/.
53. Rebecca Rosen, "So, Was Facebook Responsible for the Arab Spring After All?," *Atlantic*, September 23, 2011, retrieved November 25, 2018, https://www.theatlantic.com/technology/archive/2011/09/so-was-facebook-responsible-for-the-arab-spring-after-all/244314/.
54. Aaron Smith, "Smartphone Adoption and Usage," Pew Research Center: Internet and Technology, July 11, 2011, retrieved November 25, 2018, http://www.pewinternet.org/2011/07/11/smartphone-adoption-and-usage/.
55. K. S. Young, "Internet Addiction: A New Clinical Phenomenon and Its Consequences," *American Behavioral Scientist* 48, no. 4 (2004): 402–15.
56. C. Shirky, "The Political Power of Social Media: Technology, the Public Sphere, and Political Change," *Foreign Affairs* (2011): 28–41.
57. Technically, it's Habermas's "habilitation thesis," the European equivalent of a doctoral dissertation; it's the final requirement one must pass in order to be able to teach as a professor in a European university.
58. Jürgen Habermas, *The Structural Transformation of the Public Sphere: An Inquiry into a Category of Bourgeois Society* (Cambridge, MA: MIT Press, 1962/1989).
59. Thomas File, "Voting in America: A Look at the 2016 Presidential Election," US Census Bureau, May 10, 2017, retrieved May 20, 2018, https://www.census.gov/newsroom/blogs/random-samplings/2017/05/voting_in_america.html.
60. Gregory Wallace, "Voter Turnout at 20-Year Low in 2016," CNN, November 30, 2016, retrieved May 20, 2018, https://www.cnn.com/2016/11/11/politics/popular-vote-turnout-2016/index.html.
61. Charlotte Alter, "Voter Turnout in Midterm Elections Hits 72-Year Low," *TIME*, November 10, 2014, retrieved May 20, 2018, http://time.com/3576090/midterm-elections-turnout-world-war-two/.
62. Philip Bump, "Donald Trump Will Be President Thanks to 80,000 People in Three States," *Washington Post*, December 1, 2016, retrieved May 20, 2018, https://www.washingtonpost.com/news/the-fix/wp/2016/12/01/donald-trump-will-be-president-thanks-to-80000-people-in-three-states/?noredirect=on&utm_term=.8d271fef3166; "Average Annual Population of NYC Neighborhoods, 2011–2015," New York State Department of Health, n.d., retrieved May 20, 2018, https://www.health.ny.gov/statistics/cancer/registry/appendix/neighborhoodpop.htm.
63. "Mobile Fact Sheet," Pew Research Center: Internet and Technology, June 12, 2019, retrieved July 3, 2019, http://www.pewinternet.org/fact-sheet/mobile/.
64. Tina Lu, "Smartphone Users Replace Their Device Every Twenty-One Months," Counterpoint Research, October 13, 2017, retrieved June 3, 2018, https://www.counterpointresearch.com/smartphone-users-replace-their-device-every-twenty-one-months/.
65. Linda Poon, "Why Won't You Be My Neighbor?," CityLab, August 19, 2015, retrieved June 3, 2018, https://www.citylab.com/equity/2015/08/why-wont-you-be-my-neighbor/401762/.
66. Aaron Smith, "U.S. Smartphone Use in 2015," Pew Research Center: Internet and Technology, April 1, 2015, retrieved on May 20, 2018, http://www.pewinternet.org/2015/04/01/us-smartphone-use-in-2015/.
67. Katherine Stewart, "Online Voting: The Solution to Declining Political Engagement?," *RAND*, March 2018, retrieved May 20, 2018, https://www.rand.org/blog/2018/03/online-voting-the-solution-to-declining-political-engagement.html.

68. "Voting and Registration in the Election of November 2008," US Census Bureau, July 2012, retrieved May 20, 2018, https://www.census.gov/prod/2010pubs/p20–562.pdf.
69. Paul Hiebert, "Half of Young People Hold onto Their Phone Throughout the Day," YouGov, July 6, 2016, retrieved May 20, 2018, https://today.yougov.com/topics/lifestyle/articles-reports/2016/07/06/young-people-carry-phone-in-hand.
70. Shannon Liao, "Google Admits It Tracked User Location Data Even When the Setting Was Turned Off," Verge, November 21, 2017, retrieved May 20, 2018, https://www.theverge.com/2017/11/21/16684818/google-location-tracking-cell-tower-data-android-os-firebase-privacy.
71. Gopal Sathe, "It's Not Just Facebook, All Apps Are Scooping Up Your Data," Gadgets NDTV, March 28, 2018, retrieved June 3, 2018, https://gadgets.ndtv.com/apps/features/its-not-just-facebook-all-apps-are-scooping-up-your-data-1829861.
72. Narseo Vallina-Rodriguez, "7 in 10 Smartphone Apps Share Your Data with Third-Party Services," *Scientific American*, May 30, 2017, retrieved June 3, 2018, https://www.scientificamerican.com/article/7-in-10-smartphone-apps-share-your-data-with-third-party-services/.
73. Rick Noack, "How Long Would It Take to Read the Terms of Your Smartphone Apps? These Norwegians Tried It Out," *Washington Post*, May 28, 2016, retrieved June 3, 2018, https://www.washingtonpost.com/news/worldviews/wp/2016/05/28/how-long-would-it-take-to-read-the-terms-of-your-smartphone-apps-these-norwegians-tried-it-out/?utm_term=.e440a179ba77.
74. Andy Greenberg, "Who Reads the Fine Print Online? Less Than One Person in 1000," *Forbes*, April 8, 2010, retrieved June 2, 2018, https://www.forbes.com/sites/firewall/2010/04/08/who-reads-the-fine-print-online-less-than-one-person-in-1000/#1c2a32967017.
75. Buckminster Fuller, "2025, If . . ." *Whole Earth Catalog*, Spring 1975, retrieved June 2, 2018, http://www.wholeearth.com/issue/2005/article/94/2025.if.

CHAPTER 2: NET STATES IRL

1. Robinson Meyer, "What's Happening with the Relief Effort in Puerto Rico?" *Atlantic*, October 4, 2017, retrieved November 15, 2017, www.theatlantic.com/science/archive/2017/10/what-happened-in-puerto-rico-a-timeline-of-hurricane-maria/541956/.
2. Mindy Weisberger, "Harvey vs. Katrina: How Do These Monster Storms Compare?" Live Science, August 29, 2017, retrieved November 15, 2017, www.livescience.com/60257-harvey-vs-katrina-storm-comparison.html.
3. Brian Resnick, "Why Hurricane Maria Is Such a Nightmare for Puerto Rico," Vox, September 21, 2017, retrieved November 15, 2017, www.vox.com/science-and-health/2017/9/21/16345176/hurricane-maria-2017-puerto-rico-san-juan-meteorology-wind-rain-power.
4. "Historic Hurricane Harvey's Recap," Weather Channel, August 29, 2017, retrieved November 15, 2017, weather.com/storms/hurricane/news/tropical-storm-harvey-forecast-texas-louisiana-arkansas.
5. Eva Moravec, "Texas Officials: Hurricane Harvey Death Toll at 82, 'Mass Casualties Have Absolutely Not Happened,'" *Washington Post*, September 14, 2017, retrieved August 5, 2018, https://www.washingtonpost.com/national/texas-officials-hurricane-harvey-death-toll-at-82-mass-casualties-have-absolutely-not-happened/2017/09/14/bff3ffea-9975-11e7-87fc-c3f7ee4035c9_story.html?utm_term=.168ee7353810.
6. "Historic Disaster Response to Hurricane Harvey in Texas," Federal Emergency Management Agency, September 22, 2017, retrieved May 30, 2018, https://www.fema.gov/news-release/2017/09/22/historic-disaster-response-hurricane-harvey-texas.
7. Nishant Kishore et al., "Mortality in Puerto Rico After Hurricane Maria," *New England Journal of Medicine*, May 29, 2018, retrieved May 30, 2018, https://www.nejm.org/doi/full/10.1056/NEJMsa1803972.

8. Luis Ferré-Sadurní, "In Puerto Rico, the Storm 'Destroyed Us,'" *New York Times*, September 21, 2017, retrieved May 11, 2018, https://www.nytimes.com/2017/09/21/us/hurricane-maria-puerto-rico.html.
9. Nishant Kishore et al., "Mortality in Puerto Rico After Hurricane Maria," *New England Journal of Medicine*, May 29, 2018, retrieved August 5, 2018, https://www.nejm.org/doi/full/10.1056/NEJMsa1803972.
10. "DoD Accelerates Hurricane Relief, Response Efforts in Puerto Rico," US Department of Defense, September 30, 2017, retrieved November 15, 2017, www.defense.gov/News/Article/Article/1330501/dod-accelerates-hurricane-relief-response-efforts-in-puerto-rico.
11. Julia Jacabo, "Hurricane Maria Death Toll Jumps to 34 in Puerto Rico," ABC News, October 3, 2017, retrieved December 11, 2018, https://abcnews.go.com/US/hurricane-maria-death-toll-jumps-34-puerto-rico/story?id=50267118; Alexis Campbell, "FEMA Has Yet to Authorize Full Disaster Help for Puerto Rico," Vox, October 16, 2017, retrieved December 11, 2018, https://www.vox.com/policy-and-politics/2017/10/3/16400510/fema-puerto-rico-hurricane; Leanna Garfield, "Hurricane Maria Decimated Puerto Rico's Food Supply: Here's What the Island's Farms Look Like Now," Business Insider, September 29, 2017, retrieved December 11, 2018, https://www.businessinsider.com/photos-hurricane-maria-puerto-rico-farms-damage-caribbean-2017-9.
12. Brian Kahn, "Puerto Rico Has a Once in a Lifetime Opportunity to Rethink How It Gets Electricity," Earther (website), October 4, 2017, retrieved December 11, 2018, https://earther.gizmodo.com/puerto-rico-has-a-once-in-a-lifetime-opportunity-to-ret-1819143446.
13. Christina Mercer, "History of PayPal: 1998 to Now," Techworld, November 25, 2015, retrieved November 15, 2017, www.techworld.com/picture-gallery/business/history-of-paypal-1998-now-3630386/.
14. Ian Kar, "The Design That Got MIT Engineers the Top Spot at Elon Musk's Hyperloop Competition," Quartz, February 8, 2016, retrieved November 15, 2017, www.qz.com/611000/the-design-that-got-mit-engineers-the-top-spot-at-elon-musks-hyperloop-competition/.
15. "SpaceX Dragon Docked to Space Station," NASA, May 5, 2015, retrieved November 15, 2017, www.nasa.gov/mission_pages/station/multimedia/gallery/iss034e060718.html.
16. Brian Fung, "Here's How Far Every Tesla Model S Can Go on a Single Charge," *Washington Post*, December 30, 2014, retrieved November 15, 2017, www.washingtonpost.com/news/the-switch/wp/2014/12/30/heres-how-far-every-tesla-model-s-can-go-on-a-single-charge/?utm_term=.5574ef4d7c7c.
17. Elon Musk, "Making Life Multi-Planetary," *New Space* 6, no. 1 (March 1, 2018), retrieved May 11, 2018, https://doi.org/10.1089/space.2018.29013.emu.
18. "Cost and Payment Options," Tesla Support, n.d., retrieved December 11, 2018, https://www.tesla.com/support/energy/learn/solar-panels/cost-and-payment-options.
19. Carmine Gallo, "Tesla's Elon Musk Lights Up Social Media with a TED Style Keynote," *Forbes*, May 4, 2015, retrieved May 11, 2018, https://www.forbes.com/sites/carminegallo/2015/05/04/teslas-elon-musk-lights-up-social-media-with-a-ted-style-keynote/.
20. Tony Moore, "Queensland's Solar Energy Uptake Dims," *Brisbane Times*, August 12, 2015, retrieved December 11, 2018, https://www.brisbanetimes.com.au/national/queensland/queenslands-solar-uptake-dims-20150812-gixuy1.html; Alison Brown, "Queensland City Tops National Solar Panel List," *Sydney Morning Herald*, June 16, 2018, retrieved December 11, 2018, https://www.smh.com.au/environment/sustainability/queensland-city-tops-national-solar-panel-list-20180615-p4zlni.html; Joshua Robertson, "Queensland Installs Australia's First Powerwall Battery for Solar Trial," *Guardian*, January 18, 2016, retrieved December 11, 2018, www.theguardian.com/environment/2016/jan/18/queensland-installs-australias-first-powerwall-battery-for-solar-trial.
21. Peter Dockrill, "This Island in American Samoa Is Almost 100% Powered by Tesla Solar

Panels," Science Alert, November 23, 2016, retrieved December 11, 2018, https://www.sciencealert.com/this-island-in-american-samoa-is-almost-100-powered-by-tesla-solar-panels.

22. Austin Carr, "The Real Story Behind Elon Musk's $2.6 Billion Acquisition of SolarCity and What It Means for Tesla's Future—Not to Mention the Planet's," *Fast Company*, June 7, 2017, retrieved December 11, 2018, https://www.fastcompany.com/40422076/the-real-story-behind-elon-musks-2-6-billion-acquisition-of-solarcity-and-what-it-means-for-teslas-future-not-to-mention-the-planets.
23. Danielle Muoio, "Tesla Is Ramping Up Battery Production to Aid Puerto Rico—But Installations Costs Are Reportedly Skyrocketing," Business Insider, October 9, 2017, retrieved December 11, 2018, https://www.insider.com/tesla-ramps-battery-production-puerto-rico-2017-10.
24. "How Much Electricity on Average Do Homes in Your State Use? (Ranked by State)," *Electric Choice* (blog), n.d., retrieved December 11, 2018, https://www.electricchoice.com/blog/electricity-on-average-do-homes/.
25. Fred Lambert, "Tesla Powerwalls and Powerpacks Keep the Lights on at 662 Locations in Puerto Rico During Island-Wide Blackout," Electrek, April 18, 2018, retrieved June 3, 2018, https://electrek.co/2018/04/18/tesla-powerwall-powerpack-puerto-rico-blackout-elon-musk/.
26. Richard Wolf, "Timeline: Same-Sex Marriage Through the Years," *USA Today*, June 24, 2015, retrieved June 3, 2018, https://www.usatoday.com/story/news/politics/2015/06/24/same-sex-marriage-timeline/29173703/.
27. Anastasia Pantsios, "Burlington, Vermont, Becomes First U.S. City to Run On 100% Renewable Electricity," EcoWatch, February 10, 2015, retrieved November 15, 2018, https://www.ecowatch.com/burlington-vermont-becomes-first-u-s-city-to-run-on-100-renewable-elec-1882012101.html.
28. "Vermont: State Profile and Energy Estimates," US Energy Information Administration, n.d., retrieved June 3, 2018, https://www.eia.gov/state/?sid=VT.
29. Lisa Rathke, "Vermont Reduces Incentives for Renewable Energy Program," Associated Press, May 4, 2018, retrieved November 15, 2018, https://www.apnews.com/020a179d40b54c81b47b30cb14070870.
30. "Totals and Averages: Vermont," US Climate Data, n.d., retrieved June 3, 2018, https://www.usclimatedata.com/climate/vermont/united-states/3215.
31. "State of Vermont: Energy Sector Risk Profile," US Department of Energy, 2015, retrieved June 3, 2018, https://www.energy.gov/sites/prod/files/2015/06/f22/VT_Energy%20Sector%20Risk%20Profile.pdf.
32. "Historical Monthly Snowfall, Burlington, VT," National Weather Service, n.d., retrieved June 3, 2018, https://www.weather.gov/btv/historicalSnow.
33. Robert Walton, "Tesla, Green Mountain Power Roll Out $15/Month BTM Battery Program," Utility Dive, May 16, 2017, retrieved June 3, 2018, https://www.utilitydive.com/news/tesla-green-mountain-power-roll-out-15month-btm-battery-program/442783/.
34. Fred Lambert, "Tesla Deployed a New Powerpack System in the UK," Electrek, April 26, 2018, retrieved December 11, 2018, https://electrek.co/2018/04/26/tesla-powerpack-system-uk/.
35. Fred Lambert, "Tesla Is Installing Powerwalls and Solar Power on 50,000 Homes to Create Biggest Virtual Power Plant in the World," Electrek, February 4, 2018, retrieved June 3, 2018, https://electrek.co/2018/02/04/tesla-powerwall-solar-virtual-power-plant/; Fred Lambert, "Tesla Deploys Another Big 50 MWh Powerpack Project in Australia," Electrek, November 16, 2018, retrieved December 11, 2018, https://electrek.co/2018/11/16/tesla-powerpack-project-australia-megapack/.
36. Timothy Seppala, "Tesla Powerwall Systems Help Some Hawaii Schools Beat the Heat," Engadget, February 26, 2018, retrieved June 3, 2018, https://www.engadget.com/2018/02/26/hawaii-powerwall-school-air-conditioning/; Fred Lambert, "Tesla Deployed over 300 Powerwalls in Hawaiian Schools to Cool Down Hot Classrooms,"

Electrek, February 23, 2018, retrieved December 11, 2018, https://electrek.co/2018/02/23/tesla-powerwall-hawaii-school/.

37. Fred Lambert, "Tesla Deploys New Powerpack Project That Could Save Colorado Ratepayers $1 Million Per Year," Electrek, October 14, 2018, retrieved December 11, 2018, https://electrek.co/2018/10/14/tesla-powerpack-project-save-colorado-ratepayers/.
38. Simon Alvarez, "Tesla to Partner with NY Utility Company on Battery Storage System," Teslarati, February 15, 2018, retrieved December 11, 2018, https://www.teslarati.com/tesla-con-edison-utility-new-york-powerpack-battery-system/.
39. "Utilities," Tesla, n.d., retrieved June 3, 2018, https://www.tesla.com/utilities.
40. "Google X: Project Loon," n.d., retrieved May 11, 2018, https://x.company/loon/.
41. "FCC Grants Experimental License for Project Loon in Puerto Rico," Federal Communications Commission, October 7, 2017, retrieved May 11, 2018, https://www.fcc.gov/document/fcc-grants-experimental-license-project-loon-puerto-rico.
42. Katie Roof, "Google Parent Alphabet Looks to Restore Cell Service in Puerto Rico with Project Loon Balloons," TechCrunch, October 7, 2017, retrieved December 11, 2018, https://techcrunch.com/2017/10/07/google-parent-alphabet-looks-to-restore-cell-service-in-puerto-rico-with-project-loon-balloons/.
43. Richard Lawler, "Project Loon's LTE Balloons Are Floating over Puerto Rico," Engadget, October 20, 2017, retrieved November 15, 2017, www.engadget.com/2017/10/20/project-loon-att-puerto-rico/.
44. Alastair Westgarth, "Turning on Project Loon in Puerto Rico," *Loon* (blog), October 20, 2017, retrieved December 11, 2018, https://medium.com/loon-for-all/turning-on-project-loon-in-puerto-rico-f3aa41ad2d7f.
45. Louis Matsakis, "What Would Really Happen If Russia Attacked Undersea Internet Cables?" *WIRED*, January 5, 2018, retrieved May 30, 2018, https://www.wired.com/story/russia-undersea-internet-cables/; Prachi Bhardwaj, "Fiber Optic Wires, Servers, and More Than 550,000 Miles of Underwater Cables," Business Insider, June 23, 2018, retrieved June 23, 2019, https://www.businessinsider.com/how-internet-works-infrastructure-photos-2018-5; Tyler Cooper, "Google and Other Tech Giants Are Quietly Buying Up the Most Important Part of the Internet," April 6, 2019, retrieved June 23, 2019, https://venturebeat.com/2019/04/06/google-and-other-tech-giants-are-quietly-buying-up-the-most-important-part-of-the-internet/.
46. Alan Mauldin, "Frequently Asked Questions: Submarine Cables 101," *Telegeography* (blog), February 14, 2017, retrieved July 8, 2018, https://blog.telegeography.com/frequently-asked-questions-about-undersea-submarine-cables.
47. Bobbie Johnson, "How One Clumsy Ship Cut Off the Web for 75 Million People," *Guardian*, February 1, 2008, retrieved May 30, 2018, https://www.theguardian.com/business/2008/feb/01/internationalpersonalfinancebusiness.internet.
48. Alexandra Chang, "Why Undersea Internet Cables Are More Vulnerable Than You Think," *WIRED*, April 2, 2014, retrieved May 30, 2018, https://www.wired.com/2013/04/how-vulnerable-are-undersea-internet-cables/.
49. Will Oremus, "The Global Internet Is Being Attacked by Sharks, Google Confirms," Slate, August 15, 2014, retrieved May 30, 2018, http://www.slate.com/blogs/future_tense/2014/08/15/shark_attacks_threaten_google_s_undersea_internet_cables_video.html.
50. Lydia Saad, "The '40-Hour' Workweek Is Actually Longer—by Seven Hours," Gallup, August 29, 2014, retrieved July 8, 2018, https://news.gallup.com/poll/175286/hour-workweek-actually-longer-seven-hours.aspx.
51. "Average Hours Per Day Parents Spent Caring for and Helping Household Children as Their Main Activity," American Time Use Survey, Bureau of Labor Statistics, June 28, 2018, retrieved July 8, 2018, https://www.bls.gov/charts/american-time-use/activity-by-parent.htm.
52. Jeffrey Jones, "In U.S., 40% Get Less Than Recommended Amount of Sleep," Gallup, December 19, 2013, retrieved July 8, 2018, https://news.gallup.com/poll/166553/less-recommended-amount-sleep.aspx.

53. "Time Spent in Leisure and Sports Activities for the Civilian Population by Selected Characteristics, Averages Per Day, 2017 Annual Averages," American Time Use Survey, Bureau of Labor Statistics, June 28, 2018, retrieved July 8, 2018, https://www.bls.gov/news.release/atus.t11a.htm.
54. Joan Engebretson, "iGR: Average Monthly Broadband Usage Is 190 Gigabytes Monthly Per Household," Telecompetitor, September 26, 2016, retrieved July 8, 2018, https://www.telecompetitor.com/igr-average-monthly-broadband-usage-is-190-gigabytes-monthly-per-household/.
55. Alex Choros, "How Much Data Do You Really Need?" WhistleOut, March 26, 2018, retrieved July 8, 2018, https://www.whistleout.com.au/Broadband/Guides/Broadband-Usage-Guide#per-activity-usage.
56. "Tech Giants Are Building Their Own Undersea Fibre-Optic Networks," *Economist*, October 7, 2017, retrieved July 8, 2018, https://www-economist-com.ezproxy.cul.columbia.edu/business/2017/10/07/tech-giants-are-building-their-own-undersea-fibre-optic-networks.
57. Alan Mauldin, "A Complete List of Content Providers' Submarine Cable Holdings," *Telegeography* (blog), November 9, 2017, retrieved July 8, 2018, https://blog.telegeography.com/telegeographys-content-providers-submarine-cable-holdings-list.
58. Jameson Zimmer, "Google Owns 63,605 Miles and 8.5% of Submarine Cables Worldwide," Broadband Now, September 12, 2018, retrieved December 11, 2018, https://broadbandnow.com/report/google-content-providers-submarine-cable-ownership/.
59. "Locations," Google Cloud, n.d., retrieved July 8, 2018, https://cloud.google.com/cdn/docs/locations.
60. Christina Farr, "Google Has Hired Geisinger's David Feinberg to Lead Its Health Strategy," CNBC, November 8, 2018, retrieved December 11, 2018, https://www.cnbc.com/2018/11/08/google-hires-geisinger-ceo-david-feinberg-to-oversee-health.html.
61. Dieter Bohn, "Google Fit Is Getting Redesigned with New Health-Tracking Rings," Verge, August 21, 2018, retrieved December 11, 2018, https://www.theverge.com/2018/8/21/17761768/google-fit-redesign-heart-points-aha-whowear-os-android.
62. Leo Kelion, "Google-Nest Merger Raises Privacy Issues," BBC News, February 8, 2018, retrieved December 11, 2018, https://www.bbc.com/news/technology-42989073.
63. Anne-Marie Slaughter, "Three Responsibilities Every Government Has Towards Its Citizens," World Economic Forum, February 13, 2017, retrieved May 12, 2018, https://www.weforum.org/agenda/2017/02/government-responsibility-to-citizens-anne-marie-slaughter/.
64. Central Intelligence Agency, "Roadways," *The World Factbook*, n.d., retrieved May 12, 2018, https://www.cia.gov/library/publications/the-world-factbook/fields/2085.html.
65. David Brooks, "The Stem and the Flower," *New York Times*, December 2, 2013, retrieved May 12, 2018, https://www.nytimes.com/2013/12/03/opinion/brooks-the-stem-and-the-flower.html.
66. World Bank, "Power Outages in Firms in a Typical Month (Number)," World Bank Enterprise Surveys 2006–2017, n.d., retrieved May 12, 2018, https://data.worldbank.org/indicator/IC.ELC.OUTG?end=2017&start=2017&view=map&year_high_desc=false.
67. "Global Terrorism Index 2017," Institute for Economics & Peace, n.d., retrieved May 12, 2018, http://visionofhumanity.org/app/uploads/2017/11/Global-Terrorism-Index-2017.pdf.
68. This included the investing and research arms of their organizations as well as their founders. For instance, for Google, I also included Alphabet, X (formerly known as Google X), GV (formerly known as Google Ventures), as well as investments by cofounders Sergey Brin and Larry Page. For Microsoft, I also included Microsoft Research. For Amazon, I also included Jeff Bezos and Bezos Expeditions. For Tesla, I also included Elon Musk, SolarCity, SpaceX, and the Boring Company.
69. Multiple pages at Crunchbase, retrieved May 16, 2018, https://www.crunchbase.com/.
70. Jeff Desjardins, "The Jeff Bezos Empire in One Giant Chart," Visual Capitalist, Janu-

ary 11, 2019, retrieved June 23, 2019, https://www.visualcapitalist.com/jeff-bezos-empire-chart/.

71. Brian Caulfield, "Planetary Resources Co-Founder Aims to Create Space 'Gold Rush,'" *Forbes*, April 25, 2012, retrieved April 17, 2018, www.forbes.com/sites/briancaulfield/2012/04/20/planetary-resources-co-founder-aims-to-create-a-gold-rush-in-space/#1b0b3cc56b9b.
72. "What Are Rare Earths?," Rare Earth Technology Alliance, n.d., retrieved April 25, 2018, www.rareearthtechalliance.com/What-are-Rare-Earths.
73. H. Ali Saleem et al., "Mineral Supply for Sustainable Development Requires Resource Governance," *Nature* 543 (March 15, 2017): 367–72, retrieved May 30, 2018, https://www.nature.com/articles/nature21359.
74. "Russian Deputy PM Sees No Reason for Competing with Musk on Launch Vehicles Market," TASS Russian News Agency, April 17, 2018, retrieved May 12, 2018, http://tass.com/science/1000229.
75. "Origins of the Commercial Space Industry," Federal Aviation Administration, n.d., retrieved May 12, 2018, https://www.faa.gov/about/history/milestones/media/Commercial_Space_Industry.pdf.
76. "UCS Satellite Database," Union of Concerned Scientists, n.d., retrieved May 12, 2018, https://www.ucsusa.org/nuclear-weapons/space-weapons/satellite-database#.WvbC-JM-dE5.
77. Steve Dent, "SpaceX Price Hikes Will Make ISS Cargo Missions More Costly," Engadget, April 27, 2018, retrieved May 12, 2018, https://www.engadget.com/2018/04/27/spacex-price-hikes-iss-resupply-costs/.
78. Tim Fernholz, "Jeff Bezos Says He's Putting Billions into Space; He's Not Alone," Quartz, April 11, 2017, retrieved May 12, 2018, https://qz.com/955427/jeff-bezos-says-hes-putting-billions-into-his-space-company-blue-origin-hes-not-alone/.
79. "Technology Grows Exponentially," Big Think, n.d., retrieved May 12, 2018, http://bigthink.com/think-tank/big-idea-technology-grows-exponentially.
80. "Major Countries in Rare Earth Mine Production Worldwide from 2012 to 2017," Statista, n.d., retrieved May 12, 2018, www.statista.com/statistics/268011/top-countries-in-rare-earth-mine-production/.
81. Mayuko Yatsu, "Revisiting Rare Earths: The Ongoing Efforts to Challenge China's Monopoly," *Diplomat*, August 29, 2017, retrieved May 30, 2018, https://thediplomat.com/2017/08/revisiting-rare-earths-the-ongoing-efforts-to-challenge-chinas-monopoly/.
82. Charlotte McLeod, "Top Rare Earth Reserves by Country," *Rare Earth Investing News*, May 22, 2019, retrieved June 23, 2019, https://investingnews.com/daily/resource-investing/critical-metals-investing/rare-earth-investing/rare-earth-reserves-country/.
83. Amber Pariona, "The Poles of Inaccessibility of Our Planet," WorldAtlas, August 5, 2016, retrieved May 12, 2018, www.worldatlas.com/articles/the-poles-of-inaccessibility-of-our-planet.html; Andrew Higgins, "China's Ambitious New 'Port': Landlocked Kazakhstan," *New York Times*, January 1, 2018, retrieved May 12, 2018, www.nytimes.com/2018/01/01/world/asia/china-kazakhstan-silk-road.html.
84. Lazaro Gamio, "The Staggering Scale of China's Belt and Road Initiative," Axios, January 19, 2018, retrieved May 30, 2018, https://www.axios.com/staggering-scale-china-infrastructure-142f3b1d-82b5–47b8–8ca9–57beb306f7df.html.
85. "Rare Earths: Battling China's Monopoly," MINING, September 16, 2011, retrieved May 12, 2018, www.mining.com/rare-earths-battling-chinas-monopoly-after-molycorps-debacle/.
86. Charles Clover, "China's Xi Hails Belt and Road as 'Project of the Century,'" *Financial Times*, May 14, 2017, retrieved May 12, 2018, www.ft.com/content/88d584a2–385e-11e7–821a-6027b8a20f23.
87. Max Roser, "The World Is Much Better; The World Is Awful; The World Can Be Much Better," *Our World in Data* (blog), October 31, 2018, retrieved December 11, 2018, https://ourworldindata.org/much-better-awful-can-be-better.

88. Fred Lambert, "Tesla Unveils 'First of Many Solar+Storage Projects' at Hospital in Puerto Rico," Electrek, October 24, 2017, retrieved December 11, 2018, https://electrek.co/2017/10/24/tesla-unveils-first-solar-storage-projects-hospital-puerto-rico/.
89. J. J. Gallagher, "Struggling Puerto Rico Children's Hospital Gets Solar Power from Tesla," ABC News, October 25, 2017, retrieved December 11, 2018, https://abcnews.go.com/US/struggling-puerto-rico-childrens-hospital-solar-power-tesla/story?id=50721869.

CHAPTER 3: PRIVACY ALLIES AND ADVERSARIES

1. Jonathan Stempel, "Lawsuit Says Google Tracks Phone Users Regardless of Privacy Settings," Reuters, August 20, 2018, retrieved September 12, 2018, https://www.reuters.com/article/us-alphabet-google-privacy-lawsuit/lawsuit-says-google-tracks-phone-users-regardless-of-privacy-settings-idUSKCN1L51M3.
2. "Patacsil v. Google, Inc.," Case No. 5:2018cv05062, Justia Legal Resources, Dockets & Filings, August 17, 2018, retrieved September 12, 2018, https://dockets.justia.com/docket/california/candce/5:2018cv05062/330787.
3. Ryan Nakashima, "'Location History' Off? Google's Still Tracking You," Associated Press, August 13, 2018, retrieved September 12, 2018, https://apnews.com/828aefab64d4411bac257a07c1af0ecb; "About CSE Security Services," CSE Security Services, n.d., retrieved September 12, 2018, https://www.cseservices.net/about.html.
4. "Class Action Complaint," case no. 18-5062, US District Court, Northern District of California, DocumentCloud, August 17, 2018, retrieved September 12, 2018, https://www.documentcloud.org/documents/4777351-Gov-Uscourts-Cand-330787-1-0.html.
5. "AB-375 Privacy: Personal Information: Businesses," California Legislative Informa tion, June 29, 2018, retrieved September 12, 2018, https://leginfo.legislature.ca.gov/faces/billTextClient.xhtml?bill_id=201720180AB375.
6. Greg Ferenstein, "Google's Vint Cerf Says 'Privacy May Be an Anomaly.' Historically, He's Right," TechCrunch, November 20, 2013, retrieved August 12, 2018, https://techcrunch.com/2013/11/20/googles-cerf-says-privacy-may-be-an-anomaly-historically-hes-right/.
7. John Methven, "Why We Sleep Together," *Atlantic*, June 11, 2014, retrieved August 12, 2018, https://www.theatlantic.com/health/archive/2014/06/why-we-sleep-together/371477/.
8. Greg Ferenstein, "The Birth and Death of Privacy: 3,000 Years of History," Medium, November 25, 2015, retrieved August 12, 2018, https://medium.com/the-ferenstein-wire/the-birth-and-death-of-privacy-3–000-years-of-history-in-50-images-614c26059e.
9. R. E. Behrman and M. J. Field, eds., *When Children Die: Improving Palliative and End-of-Life Care for Children and Their Families* (Washington, DC: National Academies Press, 2003).
10. John Whiting, *Culture and Human Development: The Selected Papers of John Whiting*, edited by Eleanor Hollenberg Chasdi (Cambridge, UK: Cambridge University Press, 2006).
11. "The Coercive Acts," Our American Revolution, n.d., retrieved August 12, 2018, http://www.ouramericanrevolution.org/index.cfm/page/view/m0197.
12. Mark Kinsler, "The Luxury of a Private Telephone Line: 1939," *Lancaster Eagle Gazette*, August 3, 2014, retrieved August 12, 2018, https://www.lancastereaglegazette.com/story/news/history/remember-when/2014/08/03/luxury-private-telephone-line/13547025/.
13. "The Rise of Mobile Phones: 20 Years of Global Adoption," *Cartesian* (blog), n.d., retrieved August 12, 2018, https://blog.cartesian.com/the-rise-of-mobile-phones-20-years-of-global-adoption.
14. Jacob Kastrenakes, "Google's Chief Internet Evangelist Says 'Privacy May Actually Be an Anomaly,'" Verge, November 20, 2013, retrieved August 12, 2018, https://www.theverge.com/2013/11/20/5125922/vint-cerf-google-internet-evangelist-says-privacy-may-be-anomaly.

15. "Employment by Industry 1910 and 2015," Bureau of Labor Statistics, n.d., retrieved August 12, 2018, https://www.bls.gov/opub/ted/2016/employment-by-industry-1910-and-2015.htm.
16. Hilda Bastian, "The Hawthorne Effect: An Old Scientists' Tale Lingering 'in the Gunsmoke of Academic Snipers,'" *Absolutely Maybe* (blog), *Scientific American*, July 26, 2013, retrieved August 12, 2018, https://blogs.scientificamerican.com/absolutely-maybe/the-hawthorne-effect-an-old-scientistse28099-tale-lingering-e2809cin-the-gunsmoke-of-academic-sniperse2809d/.
17. David Evans, "The Hawthorne Effect: What Do We Really Learn from Watching Teachers (and Others)?," *Development Impact* (blog), World Bank, February 17, 2014, retrieved August 12, 2018, http://blogs.worldbank.org/impactevaluations/hawthorne-effect-what-do-we-really-learn-watching-teachers-and-others.
18. Sander van der Linden, "How the Illusion of Being Observed Can Make You a Better Person," *Scientific American*, November 3, 2011, retrieved August 12, 2018, https://www.scientificamerican.com/article/how-the-illusion-of-being-observed-can-make-you-better-person/; R. Oda, Y. Kato, and K. Hiraishi, "The Watching-Eye Effect on Prosocial Lying," *Evolutionary Psychology* 13, no. 3 (2015), https://doi.org/10.1177/1474704915594959.
19. Padraig Flanagan, "Crime . . . the Answer's Staring You in the Face: The Striking New Police Poster That Has Slashed Thefts by Up to 40 Per Cent," *Daily Mail*, June 1, 2013, retrieved August 17, 2018, http://www.dailymail.co.uk/news/article-2334577/Crime—answers-staring-face-The-striking-new-police-poster-slashed-thefts-40-cent.html.
20. Tom Stafford, "What Causes That Feeling of Being Watched?," BBC News, May 15, 2017, retrieved August 12, 2018, http://www.bbc.com/future/story/20170512-what-causes-that-feeling-of-being-watched.
21. Daniel Levitin, "Why It's So Hard to Pay Attention, Explained by Science," *Fast Company*, September 23, 2015, retrieved August 12, 2018, https://www.fastcompany.com/3051417/why-its-so-hard-to-pay-attention-explained-by-science.
22. Kimron Shapiro, ed., *The Limits of Attention: Temporal Constraints in Human Information Processing* (New York: Oxford Univ. Press, 2001).
23. Tim Hornyak, "Fujitsu Supercomputer Simulates 1 Second of Brain Activity," CNET, August 5, 2013, retrieved August 17, 2018, https://www.cnet.com/news/fujitsu-supercomputer-simulates-1-second-of-brain-activity/.
24. "Apple iMac 'Core i5' 2.8 21.5-Inch (Late 2015) Specs," EveryMac, n.d., retrieved August 17, 2018, https://everymac.com/systems/apple/imac/specs/imac-core-i5–2.8–21-inch-aluminum-late-2015-specs.html.
25. Chris Smith, "Our Brains Might Hold 1 Petabyte of Data, Which Is Almost the Entire Internet," *BGR*, January 21, 2016, retrieved August 17, 2018, https://bgr.com/2016/01/21/brain-memory-capacity-petabyte/; Stephanie Pappas, "How Big Is the Internet, Really?," Live Science, March 18, 2016, retrieved August 17, 2018, https://www.livescience.com/54094-how-big-is-the-internet.html.
26. "Vehicle Size Classes Used in the Fuel Economy Guide," US Department of Energy, n.d., retrieved August 17, 2018, https://www.fueleconomy.gov/feg/info.shtml#sizeclasses.
27. Tripp Mickle, "Apple Unveils Ways to Help Limit iPhone Usage," *Wall Street Journal*, June 8, 2018, retrieved August 17, 2018, https://www.wsj.com/articles/apple-unveils-ways-to-help-limit-iphone-usage-1528138570.
28. Douglas MacMillan, "Google Wants You to Get Off Your Phone Every Once in a While," *Wall Street Journal*, May 8, 2018, retrieved August 17, 2018, https://www.wsj.com/articles/new-android-version-will-tell-users-how-long-they-have-been-on-various-apps-1525813268?mod=searchresults&page=1&pos=3&mod=article_inline.
29. Jacob Ward, "Why Data, Not Privacy, Is the Real Danger," NBC, February 4, 2019, retrieved February 5, 2019, https://www.nbcnews.com/business/business-news/why-data-not-privacy-real-danger-n966621.
30. Issie Lapowsky, "Facebook Exposed 87 Million Users to Cambridge Analytica," *WIRED*,

April 4, 2018, retrieved December 16, 2018, https://www.wired.com/story/facebook-exposed-87-million-users-to-cambridge-analytica/.

31. Will Oremus, "The Real Scandal Isn't What Cambridge Analytica Did," Slate, March 20, 2018, retrieved December 16, 2018, https://slate.com/technology/2018/03/the-real-scandal-isnt-cambridge-analytica-its-facebooks-whole-business-model.html.
32. Ben Popken, "Congress Ratchets Up Pressure on Facebook, Twitter, Google," NBC News, November 1, 2017, retrieved December 16, 2018, https://www.nbcnews.com/tech/tech-news/congress-ratchets-pressure-facebook-twitter-google-n816521.
33. Zach Wichter, "2 Days, 10 Hours, 600 Questions: What Happened When Mark Zuckerberg Went to Washington," *New York Times*, April 12, 2018, retrieved December 16, 2018, https://www.nytimes.com/2018/04/12/technology/mark-zuckerberg-testimony.html.
34. Sean Burch, "'Senator, We Run Ads': Hatch Mocked for Basic Facebook Question to Zuckerberg," SFGATE, April 10, 2018, retrieved December 16, 2018, https://www.sfgate.com/entertainment/the-wrap/article/Senator-We-Run-Ads-Hatch-Mocked-for-Basic-12822523.php.
35. Issie Lapowsky, "The Sundar Pichai Hearing Was a Major Missed Opportunity," *WIRED*, December 11, 2018, retrieved December 16, 2018, https://www.wired.com/story/congress-sundar-pichai-google-ceo-hearing/.
36. Derek Hawkins, "The Cybersecurity 202: Why a Privacy Law Like GDPR Would Be a Tough Sell in the U.S.," *Washington Post*, May 25, 2018, retrieved September 12, 2018, https://www.washingtonpost.com/news/powerpost/paloma/the-cybersecurity-202/2018/05/25/the-cybersecurity-202-why-a-privacy-law-like-gdpr-would-be-a-tough-sell-in-the-u-s/5b07038b1b326b492dd07e83/?utm_term=.772dcf3c7ae2.
37. Ian Bogost, "Welcome to the Age of Privacy Nihilism," *Atlantic*, August 23, 2018, retrieved September 12, 2018, https://www.theatlantic.com/technology/archive/2018/08/the-age-of-privacy-nihilism-is-here/568198/.
38. Lee Rainie, "Theme 3: Trust Will Not Grow, but Technology Usage Will Continue to Rise as a 'New Normal' Sets In," Pew Research Center: Internet and Technology, August 10, 2017, retrieved September 12, 2018, http://www.pewinternet.org/2017/08/10/theme-3-trust-will-not-grow-but-technology-usage-will-continue-to-rise-as-a-new-normal-sets-in/.
39. J. Lau, B. Zimmerman, and F. Schaub, "Alexa, Are You Listening? Privacy Perceptions, Concerns and Privacy-Seeking Behaviors with Smart Speakers," *Proceedings of the ACM on Human-Computer Interaction* 2, CSCW issue (2018): 102.
40. Kaleigh Rogers, "People Who Buy Smart Speakers Have Given Up on Privacy, Researchers Find," Vice Motherboard, November 26, 2018, retrieved December 16, 2018, https://motherboard.vice.com/en_us/article/vba7xj/people-who-buy-smart-speakers-have-given-up-on-privacy-researchers-find.
41. Bernard Marr, "Where Can You Buy Big Data? Here Are the Biggest Consumer Data Brokers," *Forbes*, September 7, 2017, retrieved December 16, 2018, https://www.forbes.com/sites/bernardmarr/2017/09/07/where-can-you-buy-big-data-here-are-the-biggest-consumer-data-brokers/#3ea1f8556c27.
42. Yael Grauer, "What Are 'Data Brokers,' and Why Are They Scooping Up Information About You?," Vice Motherboard, March 27, 2018, retrieved June 6, 2018, https://motherboard.vice.com/en_us/article/bjpx3w/what-are-data-brokers-and-how-to-stop-my-private-data-collection.
43. Nick Routley, "The Multi-Billion-Dollar Industry That Makes Its Living from Your Data," Visual Capitalist, April 14, 2018, retrieved June 6, 2018, http://www.visualcapitalist.com/personal-data-ecosystem/.
44. "Data Brokers: A Call for Transparency and Accountability," Federal Trade Commission, May 2014, retrieved June 8, 2018, https://www.ftc.gov/system/files/documents/reports/data-brokers-call-transparency-accountability-report-federal-trade-commission-may-2014/140527databrokerreport.pdf, p. 8.

45. "Media Advertising Spending in the United States from 2015 to 2021 (in Billion U.S. Dollars)," Statista, n.d., retrieved December 16, 2018, https://www.statista.com/statistics/272314/advertising-spending-in-the-us/.
46. "Terms of Use," Expedia, n.d., retrieved June 10, 2018, https://www.expedia.com/p/info-other/legal.htm.
47. "Terms and Conditions," Kayak, n.d., retrieved June 10, 2018, https://www.kayak.com/terms-of-use.
48. "Terms of Use Agreement," Match.com, n.d., retrieved June 10, 2018, https://www.match.com/registration/membagr.aspx?lid=4.
49. "Terms and Conditions," OkCupid, n.d., retrieved June 10, 2018, https://www.okcupid.com/legal/terms.
50. Julie Belluz, "Grindr Is Revealing Its Users' HIV Status to Third-Party Companies," Vox, April 3, 2018, retrieved June 6, 2018, https://www.vox.com/2018/4/2/17189078/grindr-hiv-status-data-sharing-privacy.
51. "FTC Recommends Congress Require the Data Broker Industry to Be More Transparent and Give Consumers Greater Control over Their Personal Information," Federal Trade Commission, May 27, 2014, retrieved December 16, 2018, https://www.ftc.gov/news-events/press-releases/2014/05/ftc-recommends-congress-require-data-broker-industry-be-more.
52. Craig Timberg, "Brokers Use 'Billions' of Data Points to Profile Americans," *Washington Post*, May 27, 2014, retrieved June 6, 2018, https://www.washingtonpost.com/business/technology/brokers-use-billions-of-data-points-to-profile-americans/2014/05/27/b4207b96-e5b2–11e3-a86b-362fd5443d19_story.html?utm_term=.c4b1dca54f3d.
53. Julia Angwin, "Why Online Tracking Is Getting Creepier," ProPublica, June 12, 2014, retrieved June 6, 2018, https://www.propublica.org/article/why-online-tracking-is-getting-creepier.
54. Jeff Pollard, "Equifax Does More Than Credit Scores," *Forbes*, September 8, 2017, retrieved June 6, 2018, https://www.forbes.com/sites/forrester/2017/09/08/equifax-does-more-than-credit-scores/#175a6ba119d8.
55. Selena Larson, "The Hacks That Left Us Exposed in 2017," CNN, December 20, 2017, retrieved June 6, 2018, http://money.cnn.com/2017/12/18/technology/biggest-cyberattacks-of-the-year/index.html?iid=EL.
56. Sarah Perez, "47.3 Million U.S. Adults Have Access to a Smart Speaker, Report Says," TechCrunch, March 7, 2018, retrieved June 6, 2018, https://techcrunch.com/2018/03/07/47–3-million-u-s-adults-have-access-to-a-smart-speaker-report-says/.
57. Kevin Murnane, "Amazon's Alexa Hacked to Surreptitiously Record Everything It Hears," *Forbes*, April 25, 2018, retrieved June 6, 2018, https://www.forbes.com/sites/kevinmurnane/2018/04/25/amazons-alexa-hacked-to-surreptitiously-record-everything-it-hears/#549b96dc4fe2.
58. Rob LeFebvre, "Amazon May Give Developers Your Private Alexa Transcripts," Engadget, July 12, 2017, retrieved June 6, 2018, https://www.engadget.com/2017/07/12/amazon-developers-private-alexa-transcripts/.
59. "Bill of Rights—Full Text," Bill of Rights Institute, n.d., retrieved June 6, 2018, https://www.billofrightsinstitute.org/founding-documents/bill-of-rights/.
60. "GDPR Key Changes," EUGDPR, n.d., retrieved June 6, 2018, https://www.eugdpr.org/key-changes.html.
61. Sarah Perez, "Report: Smartphone Owners Are Using 9 Apps Per Day, 30 Per Month," TechCrunch, May 4, 2017, retrieved June 10, 2018, https://techcrunch.com/2017/05/04/report-smartphone-owners-are-using-9-apps-per-day-30-per-month/.
62. "Article 51: Supervisory Authority," General Data Protection Regulation (website). n.d., retrieved June 6, 2018, https://gdpr-info.eu/art-51-gdpr/.
63. For a handy plain-language explanation of the GDPR, check out "GDPR Requirements in Plain English," *Varonis* (blog), retrieved June 6, 2018, https://blog.varonis.com/gdpr

-requirements-list-in-plain-english/; "Controversial Topics," EUGDPR, n.d., retrieved June 6, 2018, https://www.eugdpr.org/controversial-topics.html.

64. Adam Satariono, "U.S. News Outlets Block European Readers over New Privacy Rules," *New York Times*, May 25, 2018, retrieved June 6, 2018, https://www.nytimes.com/2018/05/25/business/media/europe-privacy-gdpr-us.html.
65. Emma Kidwell, "Ragnorak Online Shutting Down European Servers After 14 Years," Gamasutra, April 25, 2018, retrieved June 23, 2019, https://www.gamasutra.com/view/news/317050/Ragnarok_Online_shutting_down_European_servers_after_14_years.php; Natasha Lomas, "Unroll.me to Close to EU Users Saying It Can't Comply with GDPR," TechCrunch, May 5, 2018, retrieved June 10, 2018, https://techcrunch.com/2018/05/05/unroll-me-to-close-to-eu-users-saying-it-cant-comply-with-gdpr/.
66. Alex Hern, "Facebook Moves 1.5bn Users out of Reach of New European Privacy Law," *Guardian*, April 19, 2018, retrieved June 6, 2018, https://www.theguardian.com/technology/2018/apr/19/facebook-moves-15bn-users-out-of-reach-of-new-european-privacy-law.
67. "Privacy Act of 1974," US Department of Justice, n.d., retrieved September 13, 2018, https://www.justice.gov/opcl/privacy-act-1974.
68. "5 U.S. Code §552a: Records Maintained on Individuals," Legal Information Institute, Cornell Law Library, n.d., retrieved June 6, 2018, https://www.law.cornell.edu/uscode/text/5/552a.
69. Julie Brill, "Millions Use Microsoft's GDPR Privacy Tools to Control Their Data—Including 2 Million Americans," *Microsoft on the Issues* (blog), September 17, 2018, retrieved September 30, 2018, https://blogs.microsoft.com/on-the-issues/2018/09/17/millions-use-microsofts-gdpr-privacy-tools-to-control-their-data-including-2-million-americans/.
70. Elias Biryabarema, "Uganda Imposes Tax on Social Media Use," Reuters, May 31, 2018, retrieved June 6, 2018, https://www.reuters.com/article/us-uganda-internet/uganda-imposes-tax-on-social-media-use-idUSKCN1IW2IK.
71. Tara Francis Chan, "A South Pacific Nation Is Banning Facebook for a Month as the Region Grapples with Fake News and Censorship," Business Insider, May 29, 2018, retrieved June 6, 2018, http://www.businessinsider.com/papua-new-guinea-will-temporarily-ban-facebook-2018–5.
72. Rosie Perper, "After Declaring a State of Emergency, Sri Lanka Has Banned Facebook, Instagram, and WhatsApp," Business Insider, March 7, 2018, retrieved June 6, 2018, http://www.businessinsider.com/sri-lanka-bans-facebook-to-stop-violence-2018–3.
73. Rich Miller, "Google Gets Patent for Data Center Barges," Data Center Knowledge, April 29, 2009, retrieved June 6, 2018, www.datacenterknowledge.com/archives/2009/04/29/google-gets-patent-for-data-center-barges.
74. Arman Shehabi et al., "United States Data Center Energy Usage Report," Ernest Orlando Lawrence Berkeley National Laboratory, June 2016, retrieved June 6, 2018, http://eta-publications.lbl.gov/sites/default/files/lbnl-1005775_v2.pdf.
75. Christopher Helman, "Berkeley Lab: It Takes 70 Billion Kilowatt Hours a Year to Run the Internet," *Forbes*, June 28, 2016, retrieved June 6, 2018, https://www.forbes.com/sites/christopherhelman/2016/06/28/how-much-electricity-does-it-take-to-run-the-internet/#202ebfe91fff.
76. "Nordic Data Center Construction Market: Industry Outlook and Forecast 2018–2023," Research and Markets, Business Wire, May 23, 2018, retrieved June 6, 2018, https://www.businesswire.com/news/home/20180523006288/en/Nordic-Data-Center-Construction-Market–Industry.
77. "Microsoft Plumbs Ocean's Depths to Test Underwater Data Center," *New York Times*, January 31, 2016, retrieved June 6, 2018, https://www.nytimes.com/2016/02/01/technology/microsoft-plumbs-oceans-depths-to-test-underwater-data-center.html?_r=0.
78. "How Much Water Is in the Ocean?" National Oceanic and Atmospheric Administration, n.d., retrieved June 6, 2018, https://oceanservice.noaa.gov/facts/oceanwater.html.

79. "Project Natick," Microsoft, retrieved June 6, 2018, http://natick.research.microsoft.com/.

CHAPTER 4: INFORMATION-AGE WARFIGHTERS

1. "Captured ISIS Fighter on How He Was Betrayed," YouTube, October 23, 2014, retrieved January 11, 2019, https://www.youtube.com/watch?v=mtj3M_Bb0z8&list=PL0I4bTGBHIMeAgZoMlCEKD5TzKWOEYAdJ&index=2&internalcountrycode=AE.
2. "The Redirect Method: A Blueprint for Bypassing Extremism," Google Jigsaw, n.d., retrieved May 11, 2018, https://redirectmethod.org/downloads/RedirectMethod-FullMethod-PDF.pdf, p. 4.
3. J. Uchill, "Exclusive: 'Redirecting' Extremists Away from Radical Content," Axios, March 28, 2018, retrieved May 11, 2018, https://www.axios.com/redirecting-extremists-away-from-radical-content-352b74e7-a3e0–4e21–9253–40a95089aa9c.html.
4. J. Uchill, "Exclusive: 'Redirecting' Extremists Away from Radical Content," Axios, March 28, 2018, retrieved May 11, 2018, https://www.axios.com/redirecting-extremists-away-from-radical-content-352b74e7-a3e0–4e21–9253–40a95089aa9c.html.
5. "The Redirect Method: A Blueprint for Bypassing Extremism," Google Jigsaw, n.d., retrieved May 11, 2018, https://redirectmethod.org/downloads/RedirectMethod-FullMethod-PDF.pdf, p.13; note that the total of videos played breaks down to 347,195 minutes in Arabic and 152,875 minutes in English.
6. J. Uchill, "Exclusive: 'Redirecting' Extremists Away from Radical Content," Axios, March 28, 2018, retrieved May 11, 2018, https://www.axios.com/redirecting-extremists-away-from-radical-content-352b74e7-a3e0–4e21–9253–40a95089aa9c.html.
7. Charles Bagli, "$2.4 Billion Deal for Chelsea Market Enlarges Google's New York Footprint," *New York Times*, February 7, 2018, retrieved January 11, 2019, https://www.nytimes.com/2018/02/07/nyregion/google-chelsea-market-new-york.html.
8. Eric Schmidt, "Google Ideas Becomes Jigsaw," Medium, February 16, 2016, retrieved May 11, 2018, https://medium.com/jigsaw/google-ideas-becomes-jigsaw-bcb5bd08c423#.oi1ydhk9m.
9. "Total Number of Websites," Internet Live Stats, n.d., retrieved July 20, 2018, http://www.internetlivestats.com/total-number-of-websites/.
10. Kimberly Hutcherson, "Six Days After a Ransomware Cyberattack, Atlanta Officials Are Filling Out Forms by Hand," CNN, March 28, 2018, retrieved November 20, 2018, https://www.cnn.com/2018/03/27/us/atlanta-ransomware-computers/index.html.
11. Jon Fingas, "Atlanta Ransomware Attack May Cost Another $9.5 Million to Fix," Engadget, June 6, 2018, retrieved November 20, 2018, https://www.engadget.com/2018/06/06/atlanta-ransomware-attack-struck-mission-critical-services/.
12. David Allison, "City of Atlanta Paid 8 Firms $2.7M to Combat Ransomware Attack," *Atlanta Business Chronicle*, April 24, 2018, retrieved November 20, 2018, https://www.bizjournals.com/atlanta/news/2018/04/24/city-of-atlanta-paid-8-firms-2–7m-to-combat.html.
13. Larry Greenemeier, "Urban Bungle: Atlanta Cyber Attack Puts Other Cities on Notice," *Scientific American*, April 4, 2018, retrieved November 20, 2018, https://www.scientificamerican.com/article/urban-bungle-atlanta-cyber-attack-puts-other-cities-on-notice/.
14. "The Tactics & Tropes of the Internet Research Agency," New Knowledge, December 17, 2018, retrieved January 13, 2019, https://www.newknowledge.com/disinforeport.
15. Alexis Wichowski, "Net States Rule the World: Ignore Them at Your Peril," *WIRED*, November 4, 2017, retrieved July 20, 2018, https://www.wired.com/story/net-states-rule-the-world-we-need-to-recognize-their-power/.
16. Eric Schmitt, "In Battle to Defang ISIS, U.S. Targets Its Psychology," *New York Times*, December 28, 2014, www.nytimes.com/2014/12/29/us/politics/in-battle-to-defang-isis-us-targets-its-psychology-.html?_r=0.

17. Renard Sexton, "Analysis: Did U.S. Aid Win Hearts and Minds in Afghanistan? Yes and No," *Washington Post*, January 6, 2017, www.washingtonpost.com/news/monkey-cage/wp/2017/01/06/did-u-s-nonmilitary-aid-win-hearts-and-minds-in-afghanistan-yes-and-no/.
18. "How Extremists and Terror Groups Hijacked Social Media," BBC News, December 13, 2017, retrieved June 10, 2018, https://www.bbc.co.uk/bbcthree/article/16b6c718–17d4–426d-add4–625af822e8d2.
19. Martin Evans, "Facebook Accused of Introducing Extremists to One Another Through 'Suggested Friends' Feature," *Telegraph*, May 5, 2018, retrieved June 10, 2018, https://www.telegraph.co.uk/news/2018/05/05/facebook-accused-introducing-extremists-one-another-suggested/.
20. Natasha Bernal, "WhatsApp Admits Encryption Has Helped Criminals as It Plans Crackdown on Illegal Behaviour," *Telegraph*, July 4, 2018, retrieved July 20, 2018, https://www.telegraph.co.uk/technology/2018/07/04/whatsapp-admits-encryption-has-helped-criminals-plans-crackdown/.
21. Andrew Russell, "'Rough Consensus and Running Code' and the Internet-OSI Standards War," *IEEE Annals of the History of Computing* 28, no. 3 (2006): 48–61.
22. Philip Tetlock, *Expert Political Judgment: How Good Is It? How Can We Know?* (Princeton, NJ: Princeton Univ. Press, 2006).
23. Stewart Brand, "Philip Tetlock," Long Now Foundation, n.d., retrieved June 10, 2018, www.longnow.org/seminars/02007/jan/26/why-foxes-are-better-forecasters-than-hedgehogs/.
24. "Philip E. Tetlock," School of Arts and Sciences, University of Pennsylvania, n.d., retrieved June 10, 2018, https://www.sas.upenn.edu/tetlock/bio.
25. Stewart Brand, "Philip Tetlock," Long Now Foundation, n.d., retrieved June 20, 2018, www.longnow.org/seminars/02007/jan/26/why-foxes-are-better-forecasters-than-hedgehogs/.
26. "The Good Judgment Project," IARPA "Be the Future," n.d., retrieved June 20, 2018, www.iarpa.gov/index.php/newsroom/iarpa-in-the-news/2015/439-the-good-judgment-project.
27. Tara Isabella Burton, "Future: Could You Be a 'Super-Forecaster'?," BBC, January 20, 2015, retrieved June 20, 2018, www.bbc.com/future/story/20150120-are-you-a-super-forecaster.
28. "From Theory to Practice," Good Judgment, n.d., retrieved June 20, 2018, https://goodjudgment.com/ourStory.html.
29. Matthew Burton, "So You Want to Be an Intelligence Analyst," MatthewBurton.org, October 10, 2013, retrieved June 10, 2018, http://matthewburton.org/become-an-intelligence-analyst/.
30. Leo Shane, "Up-or-Out Rules Get New Scrutiny from Congress," *Military Times*, January 25, 2018, retrieved June 10, 2018, https://www.militarytimes.com/news/pentagon-congress/2018/01/25/congress-again-weighing-changes-to-militarys-officer-promotion-rules/.
31. Stanley McChrystal, *Team of Teams: New Rules of Engagement for a Complex World* (New York: Penguin Books, 2015).
32. Stanley McChrystal, "How the Military Can Teach Us to Adapt." *TIME*, June 9, 2015, retrieved July 2, 2019, https://time.com/3904177/mcchrystal-team-of-teams/.
33. Robert Cross and Andrew Parker, *The Hidden Power of Social Networks: Understanding How Work Really Gets Done* (Brighton, MA: Harvard Business Review Press, 2004).
34. Malcolm Gladwell, "On Connectors, Mavens, and Salesmen: How New Ideas Spread Like Seeds," *Fast Company*, July 30, 2012, www.fastcompany.com/1799410/connectors-mavens-and-salesmen-how-new-ideas-spread-seeds.
35. Chris Fussell, *One Mission: How Leaders Build a Team of Teams* (New York: Penguin Books, 2017).
36. Ilan Mochari, "How General McChrystal's Share-and-Care Meetings Changed a Siloed

Culture," *Inc.*, May 15, 2015, retrieved January 13, 2019, https://www.inc.com/ilan-mochari/genearl-mcchrystal-meetings.html.

37. "Trident Juncture 2018," NATO, n.d., retrieved January 13, 2019, https://www.nato.int/cps/en/natohq/157833.htm#glance.
38. Anders Fjellestad, "Cold Knowledge Is Hot in NATO," Forsvaret, April 18, 2018, retrieved January 13, 2019, https://forsvaret.no/en/newsroom/news-stories/cold-knowledge-is-hot-in-nato.
39. Gerard O'Dwyer, "Finland, Norway Press Russia on Suspected GPS Jamming During NATO Drill," Defense News, November 16, 2018, retrieved January 13, 2019, https://www.defensenews.com/global/europe/2018/11/16/finland-norway-press-russia-on-suspected-gps-jamming-during-nato-drill/.
40. "Traffic Statistics," Avinor, n.d., retrieved January 13, 2019, https://avinor.no/en/corporate/airport/oslo/about-us/traffic_statistics/trafikkstatistikk.
41. Ryan Browne, "Russia Jammed GPS During Major NATO Military Exercise with US Troops," CNN, November 14, 2018, retrieved January 13, 2019, https://edition.cnn.com/2018/11/14/politics/russia-nato-jamming/index.html.
42. Courtney Kube, "Russia Has Figured Out How to Jam U.S. Drones in Syria, Officials Say," NBC News, April 10, 2018, retrieved January 13, 2019, https://www.nbcnews.com/news/military/russia-has-figured-out-how-jam-u-s-drones-syria-n863931.
43. Walker D. Mills, "Dear Google, Please Help Your Country Defend Itself," Defense News, January 27, 2019, retrieved February 4, 2019, https://www.defensenews.com/smr/cultural-clash/2019/01/28/dear-google-please-help-your-country-defend-itself/.
44. "Army Futures Command," n.d., retrieved January 13, 2019, https://www.army.mil/futures#org-about.
45. David Vergun, "Army Network Modernization Efforts Spearheaded by New Cross-Functional Teams," Army News Service, March 21, 2018, retrieved January 13, 2019, https://www.army.mil/article/202500/army_network_modernization_efforts_spearheaded_by_new_cross_functional_teams.
46. Alyssa Goard, "Department of Defense, Army Futures Invest in Austin Hackathon Ideas," KXAN News, September 30, 2018, retrieved January 13, 2019, https://www.kxan.com/news/local/austin/department-of-defense-army-futures-invest-in-austin-hackathon-ideas/1488776718.
47. "U.S. Navy Brings Back Navigation by the Stars for Officers," NPR, February 22, 2016, retrieved January 13, 2019, https://www.npr.org/2016/02/22/467210492/u-s-navy-brings-back-navigation-by-the-stars-for-officers.
48. Julian Borger, "Trump Urges World to Reject Globalism in UN Speech That Draws Mocking Laughter," *Guardian*, September 26, 2018, retrieved September 27, 2018, https://www.theguardian.com/us-news/2018/sep/25/trump-united-nations-general-assembly-speech-globalism-america.
49. "Protecting Users and Customers Everywhere," Cybersecurity Tech Accord, n.d., retrieved June 23, 2018, https://cybertechaccord.org/accord/.
50. Brad Smith, "The Need for a Digital Geneva Convention," *Microsoft on the Issues* (blog), February 14, 2017, retrieved June 23, 2019, https://blogs.microsoft.com/on-the-issues/2017/02/14/need-digital-geneva-convention/.
51. Press releases from the Office of Press Relations, US Department of State, n.d., retrieved July 15, 2018, https://www.state.gov/r/pa/prs/index.htm.
52. Ian Sherr, "Facebook, Apple, Microsoft and More Tech Companies Are Condemning Trump," CNET, June 26, 2018, retrieved July 15, 2018, https://www.cnet.com/news/facebook-apple-microsoft-and-more-tech-companies-are-condemning-trump/.
53. Ian Sherr, "Apple, Google, Facebook, and More Speak Out on DACA," CNET, September 6, 2017, retrieved July 15, 2018, https://www.cnet.com/news/tech-has-its-say-about-president-trump-and-dreamers-daca-microsoft-zuckerberg/.
54. Patrick Cockburn, "Preview 2018: After a String of Defeats in Iraq and Syria What 2018 Means for ISIS," *Independent*, January 1, 2018, retrieved July 20, 2018, https://www

.independent.co.uk/news/world/middle-east/isis-defeat-syria-latest-iraq-2018-survive-a8107891.html.

55. Jason Burke, "Rise and Fall of ISIS: Its Dream of a Caliphate Is Over, So What Now?," *Guardian*, October 21, 2017, retrieved July 20, 2018, https://www.theguardian.com/world/2017/oct/21/isis-caliphate-islamic-state-raqqa-iraq-islamist.
56. Jim Sciutto, "ISIS Can 'Muster' Between 20,000 and 31,500 Fighters, CIA Says," CNN, September 12, 2014, retrieved July 20, 2018, https://www.cnn.com/2014/09/11/world/meast/isis-syria-iraq/index.html.
57. Tim Lister et al., "ISIS Goes Global: 143 Attacks in 29 Countries Have Killed 2,043," CNN, February 12, 2018, retrieved July 20, 2018, https://www.cnn.com/2015/12/17/world/mapping-isis-attacks-around-the-world/index.html.
58. "Foreign Terrorist Organizations," US Department of State, n.d., retrieved July 20, 2018, https://www.state.gov/j/ct/rls/other/des/123085.htm.
59. Katie Worth, "Lone Wolf Attacks Are Becoming More Common—And More Deadly," PBS, July 14, 2016, retrieved July 20, 2018, https://www.pbs.org/wgbh/frontline/article/lone-wolf-attacks-are-becoming-more-common-and-more-deadly/.
60. "Mass shooting" is defined by the Gun Violence Archive as an incident in which four or more people are shot, not including the shooter. "Mass Shootings in 2018," Gun Violence Archive, n.d., retrieved August 5, 2018, http://www.gunviolencearchive.org/reports/mass-shooting.
61. Amy Cohen et al., "Rate of Mass Shootings Has Tripled Since 2011, Harvard Research Shows," *Mother Jones*, October 11, 2014, retrieved July 20, 2018, https://www.motherjones.com/politics/2014/10/mass-shootings-increasing-harvard-research/.
62. Sherry Towers et al., "Contagion in Mass Killings and School Shootings," *PLoS One* 10, no. 7 (2015): e0117259.
63. "'Media Contagion' Is Factor in Mass Shootings, Study Says," press release from the American Psychological Association, August 4, 2016, retrieved July 20, 2018, http://www.apa.org/news/press/releases/2016/08/media-contagion.aspx.
64. "Mass Shootings in 2019," Gun Violence Archive, n.d., retrieved June 23, 2019, https://www.gunviolencearchive.org/reports/mass-shooting.
65. "'Media Contagion' Is Factor in Mass Shootings, Study Says," press release from the American Psychological Association, August 4, 2016, retrieved June 23, 2019, http://www.apa.org/news/press/releases/2016/08/media-contagion.aspx.
66. Christal Hayes, "After Florida Shooting, More Than 600 Copycat Threats Have Targeted Schools," *USA Today*, March 7, 2018, retrieved July 20, 2018, https://www.usatoday.com/story/news/2018/03/07/within-nine-days-after-florida-shooting-there-were-more-than-100-threats-schools-across-u-s-its-not/359986002/.
67. Kevin Drum, "How Many Threats Can the FBI Evaluate on a Daily Basis?," *Mother Jones*, February 18, 2018, retrieved July 20, 2018, https://www.motherjones.com/kevin-drum/2018/02/how-many-threats-can-the-fbi-evaluate-on-a-daily-basis/.

CHAPTER 5: A GREAT WALL OF WATCHERS

1. Shannon Liao, "Google Admits It Tracked User Location Data Even When the Setting Was Turned Off," Verge, November 21, 2017, www.theverge.com/2017/11/21/16684818/google-location-tracking-cell-tower-data-android-os-firebase-privacy.
2. "State Council Notice Concerning Issuance of the Planning Outline for the Construction of a Social Credit System (2014–2020)," China Copyright & Media, June 24, 2014, retrieved June 20, 2018, https://chinacopyrightandmedia.wordpress.com/2014/06/14/planning-outline-for-the-construction-of-a-social-credit-system-2014–2020/.
3. "Millions in China with Bad 'Social Credit' Barred from Buying Plane, Train Tickets," Channel News Asia, March 16, 2018, retrieved June 20, 2018, https://www.channelnewsasia.com/news/asia/china-bad-social-credit-barred-from-buying-train-plane-tickets-10050390.

4. Maya Wang, "China's Chilling 'Social Credit' Blacklist," Human Rights Watch, December 12, 2017, retrieved June 20, 2018, https://www.hrw.org/news/2017/12/12/chinas-chilling-social-credit-blacklist.
5. Rachel Botsman, "Big Data Meets Big Brother as China Moves to Rate Its Citizens," *WIRED*, October 21, 2017, retrieved June 20, 2018, https://www.wired.co.uk/article/chinese-government-social-credit-score-privacy-invasion.
6. "Opinions Concerning Accelerating the Construction of Credit Supervision, Warning and Punishment Mechanisms for Persons Subject to Enforcement for Trust-Breaking," China Copyright & Media, September 25, 2016, retrieved June 20, 2018, https://chinacopyrightandmedia.wordpress.com/2016/09/25/opinions-concerning-accelerating-the-construction-of-credit-supervision-warning-and-punishment-mechanisms-for-persons-subject-to-enforcement-for-trust-breaking/.
7. Yuan Yang, "China Penalises 6.7m Debtors with Travel Ban," *Financial Times*, February 15, 2017, retrieved July 5, 2018, https://www.ft.com/content/ceb2a7f0-f350–11e6–8758–6876151821a6.
8. Celia Hatton, "China 'Social Credit': Beijing Sets Up Huge System," BBC News, October 26, 2015, retrieved June 20, 2018, https://www.bbc.com/news/world-asia-china-34592186.
9. Simina Mistreanu, "Life Inside China's Social Credit Laboratory," *Foreign Policy*, April 3, 2018, retrieved June 20, 2018, http://foreignpolicy.com/2018/04/03/life-inside-chinas-social-credit-laboratory/.
10. "What's Your Citizen 'Trust Score'? China Moves to Rate Its 1.3 Billion Citizens," WUNC Public Radio, November 10, 2017, retrieved June 20, 2018, http://wunc.org/post/whats-your-citizen-trust-score-china-moves-rate-its-13-billion-citizens#stream/0.
11. Leslie Holmes, *Communism: A Very Short Introduction* (New York: Oxford University Press, 2009), 9.
12. "China Pulls Further Ahead of US in Mobile Payments with Record US$12.8 Trillion in Transactions," *South China Morning Post*, February 20, 2018, retrieved July 5, 2018, https://www.scmp.com/tech/apps-gaming/article/2134011/china-pulls-further-ahead-us-mobile-payments-record-us128-trillion.
13. Cheang Ming, "FICO with Chinese Characteristics: Nice Rewards, but Punishing Penalties," CNBC, March 16, 2017, retrieved July 5, 2018, https://www.cnbc.com/2017/03/16/china-social-credit-system-ant-financials-sesame-credit-and-others-give-scores-that-go-beyond-fico.html.
14. Ann Carrns, "For Those in Their 20s, a Finding That They Don't Manage Debt Well," *New York Times*, November 20, 2013, retrieved July 5, 2018, https://www.nytimes.com/2013/11/20/business/for-those-in-their-20s-a-finding-that-they-dont-manage-debt-well.html.
15. "Planning Outline for the Construction of a Social Credit System (2014–2020)," China Copyright & Media, June 14, 2014, retrieved July 5, 2018, https://chinacopyrightandmedia.wordpress.com/2014/06/14/planning-outline-for-the-construction-of-a-social-credit-system-2014–2020/.
16. Mara Hvistendhal, "Inside China's Vast New Experiment in Social Ranking," *WIRED*, December 14, 2017, retrieved July 5, 2018, https://www.wired.com/story/age-of-social-credit.
17. Jonathan Margolis, "A Big Brother Approach Has Qualities That Would Benefit Society," *Financial Times*, October 31, 2017, retrieved July 5, 2018, https://www.ft.com/content/ffe78e52-bd54–11e7–823b-ed31693349d3.
18. M. J. Franklin, "'People You May Know' Is the Perfect Demonstration of Everything That's Wrong with Facebook," Yahoo News, May 15, 2018, retrieved February 4, 2019, https://finance.yahoo.com/news/apos-people-may-know-apos-142239841.html.
19. Alex Langone, "Facebook Admits It May Collect Data About Your Calls and Text Messages. Here's How to Turn It Off," *TIME*, March 26, 2018, retrieved September 19, 2019, https://time.com/5215274/facebook-messenger-android-call-text-message-data.
20. Russell Brandom, "Shadow Profiles Are the Biggest Flaw in Facebook's Privacy De-

fense," Verge, April 11, 2018, retrieved June 27, 2019, https://www.theverge.com/2018/4/11/17225482/facebook-shadow-profiles-zuckerberg-congress-data-privacy.

21. Violet Blue, "Firm: Facebook's Shadow Profiles Are 'Frightening' Dossiers on Everyone," ZDNet, June 24, 2013, retrieved February 4, 2019, https://www.zdnet.com/article/firm-facebooks-shadow-profiles-are-frightening-dossiers-on-everyone/.
22. Jessica Guynn, "Bye Facebook, Hello Instagram: Users Make Beeline for Facebook-Owned Social Network," *USA Today*, March 22, 2018, retrieved February 4, 2019, https://www.usatoday.com/story/tech/news/2018/03/22/bye-facebook-hello-instagram-users-make-beeline-facebook-owned-social-network/433361002/.
23. Steven Musil, "Facebook 'Unintentionally Uploaded' Email Contacts from 1.5M Users," CNET, April 17, 2019, retrieved June 27, 2019, https://www.cnet.com/news/facebook-unintentionally-uploaded-email-contacts-from-1-5m-users/.
24. Leo Mirani, "Millions of Facebook Users Have No Idea They're Using the Internet," Quartz, February 9, 2015, retrieved June 27, 2019, https://qz.com/333313/milliions-of-facebook-users-have-no-idea-theyre-using-the-internet/.
25. Sam Weber, "How Online Social Movements Translate to Offline Results," PBS NewsHour, June 10, 2017, retrieved June 27, 2019, https://www.pbs.org/newshour/show/online-social-movements-translate-offline-results.
26. "More than 12M 'Me Too' Facebook Posts, Comments, Reactions in 24 Hours," CBS News, October 17, 2017, retrieved June 27, 2019, https://www.cbsnews.com/news/metoo-more-than-12-million-facebook-posts-comments-reactions-24-hours/.
27. Issie Lapowsky, "For Philando Castile, Social Media Was the Only 911," *WIRED*, July 7, 2016, retrieved June 27, 2019, https://www.wired.com/2016/07/philando-castile-social-media-911/.
28. Jessi Hempel, "Social Media Made the Arab Spring, But It Couldn't Save It," *WIRED*, January 26, 2016, retrieved June 27, 2019, https://www.wired.com/2016/01/social-media-made-the-arab-spring-but-couldnt-save-it/.
29. "Nearly Half of Americans Get News Through Facebook," Pew Research Center: Fact Tank, April 9, 2018, retrieved June 27, 2019, https://www.pewresearch.org/fact-tank/2019/05/16/facts-about-americans-and-facebook/ft_18-04-06_facebooknews/.
30. Li Tao, "Jaywalkers Under Surveillance in Shenzhen Soon to Be Punished via Text Messages," *South China Morning Post*, March 27, 2018, retrieved June 20, 2018, http://www.scmp.com/tech/china-tech/article/2138960/jaywalkers-under-surveillance-shenzhen-soon-be-punished-text.
31. Tara Chan, "22 Eerie Photos Show How China Uses Facial Recognition to Track Its Citizens as They Travel, Shop—and Even Use Toilet Paper," Business Insider, February 12, 2018, retrieved June 20, 2018, http://www.businessinsider.com/how-china-uses-facial-recognition-technology-surveillance-2018–2#customers-of-china-merchants-bank-scan-their-faces-instead-of-their-bank-cards-at-some-1000-atms-users-still-need-to-enter-a-pin-though-11.
32. "Facial Recognition Surveillance Test Extended at Berlin Train Station," Deutsche Welle, December 15, 2017, retrieved June 20, 2018, http://www.dw.com/en/facial-recognition-surveillance-test-extended-at-berlin-train-station/a-41813861.
33. Lynsey Chutel, "China Is Exporting Facial Recognition Software to Africa, Expanding Its Vast Database," Quartz, May 25, 2018, retrieved June 20, 2018, https://qz.com/1287675/china-is-exporting-facial-recognition-to-africa-ensuring-ai-dominance-through-diversity/.
34. Paul Mozur, "Inside China's Big Tech Conference, New Ways to Track Citizens," *New York Times*, December 5, 2017, retrieved June 20, 2018, https://www.nytimes.com/2017/12/05/business/china-internet-conference-wuzhen.html.
35. "Backing Big Brother: Chinese Facial Recognition Firms Appeal to Funds," Reuters, November 12, 2017, retrieved June 20, 2018, https://www.reuters.com/article/us-china-facialrecognition-analysis/backing-big-brother-chinese-facial-recognition-firms-appeal-to-funds-idUSKBN1DD00A.

36. "SenseTime," Crunchbase, n.d., retrieved September 19, 2019, https://www.crunchbase.com/organization/sensetime#section-investors.
37. Will Knight, "Paying with Your Face," *MIT Technology Review*, February 22, 2017, retrieved June 20, 2018, https://www.technologyreview.com/s/603494/10-breakthrough-technologies-2017-paying-with-your-face/; "Comet Labs," Crunchbase, n.d., retrieved September 19, 2019, https://www.crunchbase.com/search/funding_rounds/field/organizations/num_investments/comet-labs.
38. James Millward, "What It's Like to Live in a Surveillance State," *New York Times*, February 3, 2018, retrieved June 20, 2018, https://www.nytimes.com/2018/02/03/opinion/sunday/china-surveillance-state-uighurs.html.
39. Paul Mozur, "One Month, 500,000 Face Scans: How China Is Using A.I. to Profile a Minority," *New York Times*, April 14, 2019, retrieved July 2, 2019, https://www.nytimes.com/2019/04/14/technology/china-surveillance-artificial-intelligence-racial-profiling.html.
40. Kurban Niyaz, "Authorities Require Uyghurs in Xinjiang's Aksu to Get Barcodes on Their Knives," Radio Free Asia, October 11, 2017, retrieved July 2, 2019, https://www.rfa.org/english/news/uyghur/authorities-require-uyghurs-in-xinjiangs-aksu-to-get-barcodes-on-their-knives-10112017143950.html.
41. Mallory Locklear, "Microsoft Calls on Congress to Regulate Facial Recognition," *Engadget*, July 13, 2018, retrieved August 5, 2018, https://www.engadget.com/2018/07/13/microsoft-congress-regulate-facial-recognition/.
42. "Study: Your Facebook Profile Picture Can Boost Your Chances of Being Interviewed by as Much as 40%," Workopolis, January 29, 2016, retrieved July 5, 2018, https://careers.workopolis.com/advice/study-facebook-profile-picture-directly-affects-your-chances-of-being-interviewed-for-a-job/.
43. "Data Brokers: A Call for Transparency and Accountability," Federal Trade Commission, May 2014, retrieved June 8, 2018, https://www.ftc.gov/system/files/documents/reports/data-brokers-call-transparency-accountability-report-federal-trade-commission-may-2014/140527databrokerreport.pdf, p. 8.
44. George Lowery, "Gut Instinct: We Can Identify Criminals on Sight, Study Finds," *Cornell Chronicle*, April 7, 2011, retrieved July 5, 2018, http://news.cornell.edu/stories/2011/04/study-we-can-spot-criminals-photos.
45. Dick Ahlstrom, "'Gut Instinct' Makes for Happy Relationship, Researchers Say," *Irish Times*, November 28, 2013, retrieved July 5, 2018, https://www.irishtimes.com/news/science/gut-instinct-makes-for-happy-relationship-researchers-say-1.1610919.
46. Sarah Young, "Gut Feelings Really Do Stop You from Making Mistakes, Study Finds," *Independent*, March 23, 2018, retrieved July 5, 2018, https://www.independent.co.uk/life-style/gut-feelings-instinct-stop-mistakes-brain-signal-influence-emotions-study-findings-a8270611.html.
47. Vivian Lam, "'We Know Very Little About the Brain': Experts Outline Challenges in Neuroscience," *Scope* (Stanford Medicine blog), November 8, 2016, retrieved July 5, 2018, https://scopeblog.stanford.edu/2016/11/08/challenges-in-neuroscience-in-the-21st-century/.
48. "Google Search Statistics," Internet Live Stats, n.d., retrieved August 12, 2018, http://www.internetlivestats.com/google-search-statistics/#ref-7.
49. "How Search Works," Google Search, n.d., retrieved August 12, 2018, https://www.google.com/search/howsearchworks/responses/#?modal_active=none.
50. "Number of Mobile Phone Search Users in the United States from 2014 to 2020 (in Millions)," Statista, n.d., retrieved July 2, 2019, https://www.statista.com/statistics/368746/us-mobile-search-users/; "Distribution of Total and Mobile Organic Search Visits in the United States as of 1st Quarter 2019, by Engine," Statista, April 24, 2019, retrieved July 2, 2019, https://www.statista.com/statistics/625554/mobile-share-of-us-organic-search-engine-visits/.
51. J. R. Hebert et al., "Social Desirability Bias in Dietary Self-Report May Compromise the

Validity of Dietary Intake Measures," *International Journal of Epidemiology* 24, no. 2 (1995): 389–98.

52. J. A. Karp and D. Brockington, "Social Desirability and Response Validity: A Comparative Analysis of Overreporting Voter Turnout in Five Countries," *Journal of Politics* 67, no. 3 (2005): 825–40.
53. C. G. Davis, J. Thake, and N. Vilhena, "Social Desirability Biases in Self-Reported Alcohol Consumption and Harms," *Addictive Behaviors* 35, no. 4 (2010): 302–11.
54. J. A. Catania et al., "Methodological Problems in AIDS Behavioral Research: Influences on Measurement Error and Participation Bias in Studies of Sexual Behavior," *Psychological Bulletin* 108, no. 3 (1990): 339.
55. Z. Lee and L. Woodliffe, "Donor Misreporting: Conceptualizing Social Desirability Bias in Giving Surveys," *International Journal of Voluntary and Nonprofit Organizations* 21, no. 4 (2010): 569–87.
56. Mark Murphy, "The Dunning-Kruger Effect Shows Why Some People Think They're Great Even When Their Work Is Terrible," *Forbes*, January 24, 2017, retrieved August 12, 2018, https://www.forbes.com/sites/markmurphy/2017/01/24/the-dunning-kruger-effect-shows-why-some-people-think-theyre-great-even-when-their-work-is-terrible/#440e90e85d7c.
57. Sharon Epperson, "What It Takes to Become a Millionaire," CNBC, September 19, 2011, retrieved August 12, 2018, https://www.cnbc.com/id/44559645.
58. D. D. Johnson and J. H. Fowler, "The Evolution of Overconfidence," *Nature* 477, no. 7364 (2011): 317–20.
59. "The Evolution of Overconfidence," *MIT Technology Review*, September 24, 2009, retrieved August 17, 2018, https://www.technologyreview.com/s/415454/the-evolution-of-overconfidence/.

CHAPTER 6: THE ALL-KNOWING INTERNET OF THINGS

1. "The Internet of Things: The Future of Consumer Adoption," Accenture, 2014, retrieved July 8, 2018, www.accenture.com/t20150624T211456__w__/us-en/_acnmedia/Accenture/Conversion-Assets/DotCom/Documents/Global/PDF/Technology_9/Accenture-Internet-Things.pdf.
2. D. S. Wie et al., "Wafer-Recyclable, Environment-Friendly Transfer Printing for Large-Scale Thin-Film Nanoelectronics," *Proceedings of the National Academy of Sciences* 115 (2018): 201806640, retrieved August 5, 2018, http://www.pnas.org/content/115/31/E7236.
3. Purdue University, "Electronic Stickers to Streamline Large-Scale 'Internet of Things,'" Science Daily, July 16, 2018, retrieved July 25, 2018, www.sciencedaily.com/releases/2018/07/180716164508.htm.
4. "Ericsson Mobility Report," Ericsson, June 2016, retrieved July 8, 2018, www.ericsson.com/assets/local/mobility-report/documents/2016/Ericsson-mobility-report-june-2016.pdf.
5. Lee Rainie, "The Internet of Things Connectivity Binge: What Are the Implications?," Pew Research Center: Internet and Technology, June 6, 2017, retrieved July 8, 2018, www.pewinternet.org/2017/06/06/the-internet-of-things-connectivity-binge-what-are-the-implications/.
6. Kashmir Hill and Surya Mattu, "The House That Spied on Me," Gizmodo Media Group, February 7, 2018, retrieved July 8, 2018, https://gizmodo.com/the-house-that-spied-on-me-1822429852.
7. Sapna Maheshwari, "How Smart TVs in Millions of U.S. Homes Track More Than What's on Tonight," *New York Times*, July 5, 2018, retrieved July 8, 2018, https://www.nytimes.com/2018/07/05/business/media/tv-viewer-tracking.html.
8. Kashmir Hill and Surya Mattu, "The House That Spied on Me," Gizmodo Media Group, February 7, 2018, retrieved July 8, 2018, https://gizmodo.com/the-house-that-spied-on-me-1822429852.

9. 115th Congress (2017–2018), "S.J.Res.34: A Joint Resolution . . . Relating to 'Protecting the Privacy of Customers of Broadband and Other Telecommunications Services,'" Congress.gov, retrieved July 8, 2018, https://www.congress.gov/bill/115th-congress/senate-joint-resolution/34.
10. Andy Greenberg, "Marketing Firm Exactis Leaked a Personal Info Database with 340 Million Records," *WIRED*, June 27, 2018, retrieved July 8, 2018, https://www.wired.com/story/exactis-database-leak-340-million-records/.
11. "Public Trust in Government: 1958–2017," Pew Research Center: US Politics and Policy, December 14, 2017, retrieved August 5, 2018, http://www.people-press.org/2017/12/14/public-trust-in-government-1958–2017/.
12. This does not include a brief bump in public trust in government following the attacks of 9/11.
13. Niall McCarthy, "The Institutions Americans Trust Most and Least in 2018," *Forbes*, June 29, 2018, retrieved August 5, 2018, https://www.forbes.com/sites/niallmccarthy/2018/06/29/the-institutions-americans-trust-most-and-least-in-2018-infographic/#2475b3452fc8.
14. Home page, Sidewalk Labs, n.d., retrieved February 4, 2019, https://www.sidewalklabs.com/.
15. Home page, Sidewalk Toronto, n.d., retrieved July 8, 2018, https://sidewalktoronto.ca/; Elizabeth Woyke, "A Smarter Smart City," *MIT Technology Review*, February 21, 2018, retrieved July 8, 2018, https://www.technologyreview.com/s/610249/a-smarter-smart-city/.
16. "New District in Toronto Will Tackle the Challenges of Urban Growth," Sidewalk Toronto, October 17, 2017, retrieved July 8, 2018, https://sidewalktoronto.ca/wp-content/uploads/2018/04/Sidewalk-Toronto-Press-Release.pdf.
17. "Project Vision," Sidewalk Labs, October 17, 2017, retrieved July 8, 2018, https://sidewalktoronto.ca/wp-content/uploads/2017/10/Sidewalk-Labs-Vision-Sections-of-RFP-Submission.pdf, p. 37.
18. Nancy Scola, "Google Is Building a City of the Future in Toronto. Would Anyone Want to Live There?" *Politico*, July/August 2018, retrieved August 5, 2018, https://www.politico.com/magazine/story/2018/06/29/google-city-technology-toronto-canada-218841.
19. Elon Musk, "Elon Musk First Principle Reasoning TED," YouTube, August 25, 2016, retrieved July 13, 2018, https://www.youtube.com/watch?v=0JQXoSmC1rs.
20. Martine Powers, "The Real Reason Why Boston's Geography Is the Most Confusing Thing Ever," Boston.com, April 2, 2014, retrieved July 13, 2018, https://www.boston.com/news/local-news/2014/04/02/the-real-reason-why-bostons-geography-is-the-most-confusing-thing-ever.
21. Charlie Sorrel, "How the iPhone Made Accessibility Accessible to Everyone," Cult of Mac, June 28, 2017, retrieved February 4, 2019, https://www.cultofmac.com/488348/iphone-accessibility/.
22. Vindu Goel, "At $1,000, Apple's iPhone X Crosses a Pricing Threshold," *New York Times*, September 10, 2017, retrieved February 4, 2019, https://www.nytimes.com/2017/09/10/technology/apple-iphone-price.html.
23. "Your New iPhone XS," Apple, n.d., retrieved February 4, 2019, https://www.apple.com/shop/buy-iphone/iphone-xs/6.5-inch-display-512gb-gold-unlocked?afid=p238|sND54E8jO-dc_mtid_1870765e38482_pcrid_324289715177_&cid=aos-us-kwgo-pla-iphone—slid--product-MT5J2LL/A.
24. Justin Pritchard, "What Is the Average Monthly Mortgage Payment?," Balance, November 5, 2018, retrieved February 4, 2019, https://www.thebalance.com/average-monthly-mortgage-payment-4154282.
25. "Map of Adaptive Traffic Control Systems in the US," Google Maps, n.d., retrieved July 8, 2018, https://www.google.com/maps/d/u/0/viewer?ll=37.300274999999985%2C-99.40429699999999&spn=27.845221%2C57.041016&hl=en&t=h&gl=us&oe=UTF8&msa=0&z=5&source=embed&ie=UTF8&mid=17gN_VQSZfNA-oIj7UB_QstT_z6E.

26. "Adaptive Traffic Control Systems: Domestic and Foreign State of Practice," National Academies of Science, Medicine, and Engineering, 2010, retrieved July 8, 2018, https://www.nap.edu/download/14364.
27. Trevor Reed, "Global Traffic Scorecard," INRIX, February 2019, retrieved June 27, 2019, http://inrix.com/scorecard.
28. Ashley Halsey, "There Are About 1.7 Million Rear-End Collisions on U.S. Roads Each Year. Here's How to Stop Them," *Washington Post*, June 8, 2015, retrieved July 8, 2018, https://www.washingtonpost.com/news/dr-gridlock/wp/2015/06/08/there-are-about-1–7-million-rear-end-collisions-on-u-s-roads-each-year-heres-how-to-stop-them/?utm_term=.298fc176f77e.
29. Michael Fleeman, "Google's Waze App Fires Back Against Police Criticism," HuffPost, March 29, 2015, retrieved July 13, 2018, https://www.huffingtonpost.com/2015/01/27/waze-police-tracking-criticism_n_6558408.html.
30. "Does Waze Put Police Officers' Lives at Risk?," CBS News, January 29, 2015, retrieved July 13, 2018, https://www.cbsnews.com/news/does-waze-put-police-officers-lives-at-risk/.
31. "Mayor de Blasio and NYCEDC Launch Sixth Annual NYC Bigapps Competition: Challenging NYC's Tech Innovators to Find Lasting Solutions to Four of the City's Most Pressing Issues," *NYC Office of the Mayor*, July 17, 2015, retrieved July 13, 2018, https://www1.nyc.gov/office-of-the-mayor/news/494–15/mayor-de-blasio-nycedc-launch-sixth-annual-nyc-bigapps-competition–challenging-nyc-s-tech.
32. "Critical Infrastructure Sectors," US Department of Homeland Security, n.d., retrieved February 4, 2019, https://www.dhs.gov/cisa/critical-infrastructure-sectors.
33. Robert Atkinson et al., "A Policymaker's Guide to Digital Infrastructure," Information Technology and Innovation Foundation, May 2016, retrieved July 8, 2018, http://www2.itif.org/2016-policymakers-guide-digital-infrastructure.pdf?_ga=2.125865421.1763289522.1531062137–2090438079.1531062137.
34. Andrew Hawkins, "Elon Musk's Boring Company Approved to Build High-Speed Transit Between Downtown Chicago and O'Hare Airport," Verge, June 13, 2018, retrieved July 8, 2018, https://www.theverge.com/2018/6/13/17462496/elon-musk-boring-company-approved-tunnel-chicago.
35. Elizabeth Shockman, "US Governments Have Made Some Big Flops in the World of Tech. Here's Why," Science Friday, November 9, 2015, retrieved July 8, 2018, https://www.pri.org/stories/2015–11–09/us-governments-have-made-some-big-flops-world-tech-heres-why.
36. Niam Yaraghi, "Doomed: Challenges and Solutions to Government IT Projects," Brookings Institution, August 25, 2015, retrieved July 8, 2018, https://www.brookings.edu/blog/techtank/2015/08/25/doomed-challenges-and-solutions-to-government-it-projects/.
37. Venkat Motupalli, "How Big Data Is Changing Democracy," *Journal of International Affairs*, June 22, 2018, retrieved July 8, 2018, https://jia.sipa.columbia.edu/how-big-data-changing-democracy#4.
38. Derek Powazek, "Nobody Uses Their Real Name Online, and Other Outdated Notions," Gizmodo Media Group, August 25, 2011, retrieved July 5, 2018, https://gizmodo.com/5834112/nobody-uses-their-real-name-online-and-other-outdated-notions.
39. Brent Smith, "Two Decades of Recommender Systems at Amazon.com," IEEE Computer Society, May/June 2017, retrieved July 5, 2018, https://www.computer.org/csdl/mags/ic/2017/03/mic2017030012.html.
40. "E-Stats," US Census Bureau, March 7, 2001, retrieved July 5, 2018, https://www.census.gov/content/dam/Census/library/publications/2001/econ/1999estatstext.pdf.
41. Michael Martinez, "Amazon: Jeff Bezos' Juggernaut Began with a Recommender System That Launched a Thousand Algorithms," IEEE Computer Society, June 20, 2017, retrieved July 5, 2018, https://publications.computer.org/internet-computing/2017/06/20/amazon-jeff-bezos-juggernaut-began-with-a-recommender-system-that-launched-a-thousand-algorithms/.

42. "And the Emmy Goes to . . . Amazon Instant Video!" Business Wire, August 2, 2013, retrieved July 5, 2018, https://www.businesswire.com/news/home/20130802005131/en/Emmy-To%E2%80%A6Amazon-Instant-Video.
43. B. Lika, K. Kolomvatsos, and S. Hadjiefthymiades, "Facing the Cold Start Problem in Recommender Systems," *Expert Systems with Applications* 41, no. 4 (2014): 2065–73.
44. Mehrdad Fatourechi, "The Evolving Landscape of Recommendation Systems," TechCrunch, September 28, 2015, retrieved August 5, 2018, https://techcrunch.com/2015/09/28/the-evolving-landscape-of-recommendation-systems/.
45. "Amazon's Privacy Notice," Amazon, n.d., retrieved July 5, 2018, https://www.amazon.com/gp/help/customer/display.html?nodeId=468496.
46. "Percentage of Paid Units Sold by Third-Party Sellers on Amazon Platform as of 1st Quarter 2018," Statista, n.d., retrieved July 5, 2018, https://www.statista.com/statistics/259782/third-party-seller-share-of-amazon-platform/.
47. Ranju Das, "Amazon Rekognition Announces Real-Time Face Recognition, Support for Recognition of Text in Image, and Improved Face Detection," *AWS Machine Learning* (blog), November 21, 2017, retrieved June 20, 2018, https://aws.amazon.com/blogs/machine-learning/amazon-rekognition-announces-real-time-face-recognition-support-for-recognition-of-text-in-image-and-improved-face-detection/.
48. "Amazon Rekognition Customer Use Cases," Amazon Web Services, n.d., retrieved June 20, 2018, https://aws.amazon.com/rekognition/customers/.
49. Chris Greenwood, "One CCTV Camera for Every 10 People: Report Says There Are Now Six Million Across the UK," *Daily Mail*, October 25, 2016, retrieved February 4, 2019, https://www.dailymail.co.uk/news/article-3872818/One-CCTV-camera-10-people-Report-says-six-million-UK-useless.html.
50. Jon Russell, "China's CCTV Surveillance Network Took Just 7 Minutes to Capture BBC Reporter," TechCrunch, February 4, 2018, retrieved February 4, 2019, https://techcrunch.com/2017/12/13/china-cctv-bbc-reporter/.
51. "Amazon Rekognition FAQs: Data Privacy," Amazon Web Services, n.d., retrieved February 5, 2019, https://aws.amazon.com/rekognition/faqs/.
52. American Civil Liberties Union, records request to the Orlando Police Department, January 18, 2018, retrieved July 2, 2019, https://www.aclunc.org/docs/20180522_ARD.pdf#page=53.
53. Nick Wingfield, "Amazon Pushes Facial Recognition to Police. Critics See Surveillance Risk," *New York Times*, May 22, 2018, retrieved June 20, 2018, https://www.nytimes.com/2018/05/22/technology/amazon-facial-recognition.html.
54. Brian Heater, "Congress Members Demand Answers from Amazon About Facial Recognition Software," TechCrunch, July 26, 2018, retrieved February 5, 2019, https://techcrunch.com/2018/07/26/congress-members-demand-answers-from-amazon-about-facial-recognition-software/.
55. Larry Hardesty, "Study Finds Gender and Skin-Type Bias in Commercial Artificial-Intelligence Systems," *MIT News*, February 11, 2018, retrieved June 20, 2018, http://news.mit.edu/2018/study-finds-gender-skin-type-bias-artificial-intelligence-systems-0212; Molly Wood, "Why Facial Recognition Software Has Trouble Recognizing People of Color," Marketplace, February 13, 2018, retrieved June 27, 2019, https://www.marketplace.org/2018/02/13/why-algorithms-may-have-trouble-recognizing-your-face/.
56. Derek Hawkins, "Lawmakers Worry Amazon's Facial Recognition Tech Could Reinforce Racial Profiling," *Washington Post*, May 29, 2018, retrieved February 5, 2019, https://www.washingtonpost.com/news/powerpost/paloma/the-cybersecurity-202/2018/05/29/the-cybersecurity-202-lawmakers-worry-amazon-s-facial-recognition-tech-could-reinforce-racial-profiling/5b0c10741b326b492dd07eb8/?utm_term=.1f866dbed0a5.
57. Evgeny Morozov, "Will Tech Giants Move on from the Internet, Now We've All Been Harvested?" *Guardian*, January 27, 2018, retrieved July 14, 2018, https://www.theguardian.com/technology/2018/jan/28/morozov-artificial-intelligence-data-technology-online.

58. "The Fate of Online Trust in the Next Decade," Pew Research Center: Internet and Technology, August 10, 2017, retrieved July 14, 2018, http://www.pewinternet.org/2017/08/10/theme-3-trust-will-not-grow-but-technology-usage-will-continue-to-rise-as-a-new-normal-sets-in/.

CHAPTER 7: THE MIND, IMMERSED

1. D. Glassbrenner, "Estimating the Lives Saved by Safety Belts and Air Bags," *Age* 5 (2016): 12; "Air Bags," National Highway Traffic Safety Administration, US Department of Transportation, n.d., retrieved April 21, 2019, https://www.nhtsa.gov/equipment/air-bags.
2. "National Household Travel Survey Daily Travel Quick Facts," *Bureau of Transportation Statistics*, n.d., retrieved August 5, 2018, www.rita.dot.gov/bts/sites/rita.dot.gov.bts/files/subject_areas/national_household_travel_survey/daily_travel.html.
3. B. R. Burdett et al., "Not All Minds Wander Equally: The Influence of Traits, States and Road Environment Factors on Self-Reported Mind Wandering During Everyday Driving," US National Library of Medicine, National Institutes of Health, October 2016, retrieved August 5, 2018, www.ncbi.nlm.nih.gov/pubmed/27372440.
4. "Distracted Driving 2015," National Highway Traffic Safety Administration, US Department of Transportation, March 2017, retrieved August 5, 2018, https://www.nhtsa.gov/sites/nhtsa.dot.gov/files/documents/812_381_distracteddriving2015.pdf.
5. Ioannis Pavlidis, "Dissecting Driver Behaviors Under Cognitive, Emotional, Sensorimotor, and Mixed Stressors," *Nature*, May 12, 2016, retrieved August 5, 2018, www.nature.com/articles/srep25651.pdf.
6. University of Houston, "'Sixth Sense' Protects Drivers Except When Texting," Science Daily, May 12, 2016, retrieved August 5, 2018, www.sciencedaily.com/releases/2016/05/160512084547.htm.
7. J. L. Nasar and D. Troyer, "Pedestrian Injuries Due to Mobile Phone Use in Public Places," *Accident Analysis and Prevention* 57 (2013): 91–95, https://doi.org/10.1016/j.aap.2013.03.021.
8. "Head Up, Phone Down When Headed Back to School," National Safety Council, n.d., retrieved April 21, 2019, https://www.nsc.org/home-safety/safety-topics/distracted-walking/teens.
9. "Apple iMac G3/333 (Fruit Colors) Specs," EveryMac, n.d., retrieved July 13, 2018, https://everymac.com/systems/apple/imac/specs/imac_333.html.
10. "iMac (Rev. C)," Apple History, n.d., retrieved August 5, 2018, https://apple-history.com/imacrevc.
11. Lea Winerman, "Smarter Than Ever?," *American Psychological Association Monitor* 44, no. 3 (March 2013): 30, retrieved July 14, 2018, http://www.apa.org/monitor/2013/03/smarter.aspx.
12. Kenneth Simon, *Digest of Educational Statistics*, Office of Education Bulletin, no. 4 (Washington, DC: US Government Printing Office, 1965).
13. "Educational Attainment in the United States: 2015," US Census Bureau, n.d., retrieved July 13, 2018, https://www.census.gov/content/dam/Census/library/publications/2016/demo/p20–578.pdf.
14. "Percentage of Americans with College Degrees Rises, Paying for Degrees Tops Financial Challenges," PBS NewsHour, April 22, 2014, retrieved July 13, 2018, https://www.pbs.org/newshour/education/percentage-americans-college-degrees-rises-paying-degrees-tops-financial-challenges; J. Shaw, "Statistics of College Graduates," *Quarterly Publications of the American Statistical Association* 17, no. 131 (1920): 335–41, retrieved July 13, 2018, http://www.jstor.org/stable/2965351.
15. M. Tommasi et al., "Increased Educational Level Is Related with Higher IQ Scores but Lower G-Variance: Evidence from the Standardization of the WAIS-R for Italy," *Intelligence* 50 (2015): 68–74; Stuart J. Ritchie and Elliot M. Tucker-Drob, "How Much Does

Education Improve Intelligence? A Meta-Analysis," *Psychological Science* 29, no. 8 (2018): 1358–69.

16. John Horrigan, "Lifelong Learning and Technology," Pew Research Center: Internet and Technology, March 22, 2016, retrieved August 5, 2018, http://www.pewinternet.org/2016/03/22/lifelong-learning-and-technology/.
17. Kristen Purcell, "Americans Feel Better Informed Thanks to the Internet," Pew Research Center: Internet and Technology, December 8, 2014, retrieved August 5, 2018, http://www.pewinternet.org/2014/12/08/better-informed/.
18. Lea Winerman, "Smarter Than Ever?," *American Psychological Association Monitor* 44, no. 3 (March 2013): 30, retrieved July 14, 2018, http://www.apa.org/monitor/2013/03/smarter.aspx.
19. Karen Turner, "Lots of Coders Are Self-Taught, According to Developer Survey," *Washington Post*, March 30, 2016, retrieved July 13, 2018, https://www.washingtonpost.com/news/the-switch/wp/2016/03/30/lots-of-coders-are-self-taught-according-to-developer-survey/?utm_term=.9e26cc363750.
20. Kristen Purcell, "Search and Email Still Top the List of Most Popular Online Activities," Pew Research Center: Internet and Technology, August 9, 2011, retrieved July 15, 2018, http://www.pewinternet.org/2011/08/09/search-and-email-still-top-the-list-of-most-popular-online-activities/; "Number of Explicit Core Search Queries Powered by Search Engines in the United States as of January 2019 (in Billions)," Statista, n.d., retrieved June 27, 2019, https://www.statista.com/statistics/265796/us-search-engines-ranked-by-number-of-core-searches/.
21. "Year in Search: 2017," Google Trends, n.d., retrieved July 13, 2018, https://trends.google.com/trends/yis/2017/US/.
22. Doug Criss, "The Las Vegas Attack Is the Deadliest Mass Shooting in Modern US History," CNN, October 2, 2017, retrieved July 13, 2018, https://www.cnn.com/2017/10/02/us/las-vegas-attack-deadliest-us-mass-shooting-trnd/index.html.
23. Nick Statt, "Google Will Stop Showing Search Results as You Type Because It Makes No Sense on Mobile," Verge, July 26, 2017, retrieved July 13, 2018, https://www.theverge.com/2017/7/26/16034844/google-kills-off-instant-search-for-mobile-consistency.
24. Tomas Chamorro-Premuzic, "Curiosity Is as Important as Intelligence," *Harvard Business Review*, August 27, 2014, retrieved July 13, 2018, https://hbr.org/2014/08/curiosity-is-as-important-as-intelligence.
25. J. Holmes, *Nonsense: The Power of Not Knowing* (New York: Broadway Books, 2016).
26. P. Merrotsy, "Tolerance of Ambiguity: A Trait of the Creative Personality?," *Creativity Research Journal* 25, no. 2 (2013): 232–37.
27. Amy Thompson, "How Learning a New Language Improves Tolerance," Conversation, December 11, 2016, retrieved July 13, 2018, https://theconversation.com/how-learning-a-new-language-improves-tolerance-68472; S. M. Andersen and A. H. Schwartz, "Intolerance of Ambiguity and Depression: A Cognitive Vulnerability Factor Linked to Hopelessness," *Social Cognition* 10, no. 3 (1992): 271–98.
28. "Americans Check Their Phones 80 Times a Day: Study," NBC News, November 11, 2017, retrieved July 14, 2018, https://www.nbcnews.com/nightly-news/video/americans-check-their-smartphones-80-times-a-day-study-1093857859881.
29. "Internet/Broadband Fact Sheet," Pew Research Center: Internet and Technology, February 5, 2018, retrieved June 20, 2018, http://www.pewinternet.org/fact-sheet/internet-broadband/.
30. Monica Anderson, "10% of Americans Don't Use the Internet. Who Are They?" Pew Research Center: Fact Tank, April 22, 2019, retrieved July 2, 2019, https://www.pewresearch.org/fact-tank/2019/04/22/some-americans-dont-use-the-internet-who-are-they/.
31. Nick Bilton, "Steve Jobs Was a Low-Tech Parent," *New York Times*, September 10, 2014, retrieved June 20, 2018, https://www.nytimes.com/2014/09/11/fashion/steve-jobs-apple-was-a-low-tech-parent.html.

32. Leanna Garfield, "These Tech Execs Have Regrets About the World-Changing Sites They Helped Create," Business Insider, December 23, 2017, retrieved June 20, 2018, http://www.businessinsider.com/social-media-affects-society-tech-execs-2017–12#former-facebook-president-sean-parker-has-said-the-network-changes-your-relationship-with-society-2.
33. Steve Kovach, "Former Facebook Exec Feels 'Tremendous Guilt' for What He Helped Make," Business Insider, December 11, 2017, retrieved June 20, 2018, http://www.businessinsider.com/former-facebook-exec-chamath-palihapitiya-social-media-damaging-society-2017-12.
34. David Ginsberg, "Hard Questions: Is Spending Time on Social Media Bad for Us?" Facebook Newsroom, December 14, 2017, retrieved September 19, 2019, https://newsroom.fb.com/news/2017/12/hard-questions-is-spending-time-on-social-media-bad-for-us/.
35. "Why a Dodgy Social Media Profile Is Better Than None," Undercover Recruiter, retrieved June 20, 2018, https://theundercoverrecruiter.com/why-no-social-media-presence-is-bigger-red-flag/.
36. M. Pielot et al., "An In-Situ Study of Mobile Phone Notifications," *Proceedings of the 16th International Conference on Human-Computer Interaction with Mobile Devices & Services* (September 2014): 233–242.
37. S. K. Kim et al., "An Analysis of the Effects of Smartphone Push Notifications on Task Performance with Regard to Smartphone Overuse Using ERP," *Computational Intelligence and Neuroscience* 3 (2016): 1–8.
38. M. Drouin, D. H. Kaiser, and D. A. Miller, "Phantom Vibrations Among Undergraduates: Prevalence and Associated Psychological Characteristics," *Computers in Human Behavior* 28, no. 4 (2012): 1490–96.
39. "Ledger of Harms," Center for Humane Technology, December 14, 2018, retrieved February 5, 2019, https://ledger.humanetech.com/.
40. "Americans Check Their Phones 80 Times a Day: Study," NBC News, November 11, 2017, retrieved July 14, 2018, https://www.nbcnews.com/nightly-news/video/americans-check-their-smartphones-80-times-a-day-study-1093857859881.
41. Natalie Fratto, "Screw Emotional Intelligence: Here's the Key to the Future of Work," *Fast Company*, January 29, 2018, retrieved July 15, 2018, https://www.fastcompany.com/40522394/screw-emotional-intelligence-heres-the-real-key-to-the-future-of-work.
42. Vivian Giang, "These Are the Long-Term Effects of Multitasking," *Fast Company*, March 1, 2016, retrieved July 13, 2018, https://www.fastcompany.com/3057192/these-are-the-long-term-effects-of-multitasking.
43. S. Leroy, "Why Is It So Hard to Do My Work? The Challenge of Attention Residue When Switching Between Work Tasks," *Organizational Behavior and Human Decision Processes* 109, no. 2 (2009): 168–81.
44. Vivian Giang, "These Are the Long-Term Effects of Multitasking," *Fast Company*, March 1, 2016, retrieved July 13, 2018, https://www.fastcompany.com/3057192/these-are-the-long-term-effects-of-multitasking.
45. Verne Kopytoff, "Where Pagers Haven't Gone Extinct Yet," *Fortune*, July 16, 2013, retrieved July 15, 2018, http://fortune.com/2013/07/16/where-pagers-havent-gone-extinct-yet/.
46. R. Kegan and L. L. Lahey, *Immunity to Change: How to Overcome It and Unlock Potential in Yourself and Your Organization* (Brighton, MA: Harvard Business Review Press, 2009).
47. J. Mittell, "Narrative Complexity in Contemporary American Television," *Velvet Light Trap* 58, no. 1 (2006): 29–40.
48. A. J. Westphal et al., "Episodic Memory Retrieval Benefits from a Less Modular Brain Network Organization," *Journal of Neuroscience* (2017): 2509–16, retrieved July 13, 2018, https://www.ncbi.nlm.nih.gov/pubmed/28242796.
49. Paul Zak, "Why Your Brain Loves Good Storytelling," *Harvard Business Review*, October 28, 2014, retrieved July 13, 2018, https://hbr.org/2014/10/why-your-brain-loves-good-storytelling.
50. Annie Murphy Paul, "Your Brain on Fiction," *New York Times*, March 17, 2012, retrieved

July 13, 2018, https://www.nytimes.com/2012/03/18/opinion/sunday/the-neuroscience-of-your-brain-on-fiction.html?pagewanted=all.

51. B. Deen and G. McCarthy, "Reading About the Actions of Others: Biological Motion Imagery and Action Congruency Influence Brain Activity," *Neuropsychologia* 48, no. 6 (2010): 1607–15.
52. Caroline Cakebread, "Millennials Are Making Advertisers' Lives Harder by Watching Less TV Than Their Predecessors," Business Insider, August 25, 2017, retrieved July 13, 2018, http://www.businessinsider.com/millennials-watching-less-tv-than-previous-generations-chart-2017–8.
53. "Reading on the Rise: A New Chapter in American Literacy," National Endowment for the Arts, 2009, retrieved July 15, 2018, https://www.arts.gov/sites/default/files/ReadingonRise.pdf.
54. Andrew Perrin, "Who Doesn't Read Books in America?" Pew Research Center: Internet and Technology, March 23, 2018, retrieved July 15, 2018, http://www.pewinternet.org/2016/09/01/book-reading-2016/.
55. I first explored this idea in a coauthored piece with John S. Johnson: "Immerse Me: How to Get Lost in Story (Story Not Required)," TechCrunch, November 26, 2015, retrieved July 15, 2018, https://techcrunch.com/2015/11/26/immerse-me-how-to-get-lost-in-story-actual-story-not-required/.
56. Oscar Raymundo, "Tim Cook: Augmented Reality Will Be an Essential Part of Your Daily Life, Like the iPhone," *Macworld*, October 3, 2016, retrieved June 27, 2019, https://www.macworld.com/article/3126607/tim-cook-augmented-reality-will-be-an-essential-part-of-your-daily-life-like-the-iphone.html.
57. "16 Cool Things You Can Do with Google Glass," *PC Magazine*, April 15, 2014, retrieved July 15, 2018, https://www.pcmag.com/feature/308711/16-cool-things-you-can-do-with-google-glass/15.
58. Victoria Turk, "One Year On, Who Still Plays Pokémon Go?," *WIRED*, July 6, 2017, retrieved July 15, 2018, http://www.wired.co.uk/article/pokemon-go-first-anniversary-who-still-plays.
59. "Why Pokémon Go Has Been a Viral Success," Business Insider, July 13, 2016, retrieved July 15, 2018, http://www.businessinsider.com/why-pokemon-go-has-been-a-viral-success-2016-7.
60. "Apple: Augmented Reality," Apple, n.d., retrieved July 15, 2018, https://www.apple.com/ios/augmented-reality/.
61. Lee Rainie et al., "The Rise of E-Reading," Pew Research Center: Internet and Technology, April 4, 2012, retrieved July 15, 2018, http://libraries.pewinternet.org/2012/04/04/the-rise-of-e-reading/.
62. Larry Magid, "How (and Why) to Turn Off Google's Personalized Search Results," *Forbes*, January 13, 2012, retrieved July 15, 2018, https://www.forbes.com/sites/larrymagid/2012/01/13/how-and-why-to-turn-off-googles-personalized-search-results/#2e4b284a38f2.
63. Lucas Matney, "Apple Delivers Big Updates to Its Augmented Reality Platform," TechCrunch, June 4, 2018, retrieved July 15, 2018, https://techcrunch.com/2018/06/04/apple-delivers-big-updates-to-its-augmented-reality-platform/.
64. Art Lindsley, "C. S. Lewis on Absolutes," *Knowing & Doing*, Fall 2002, retrieved June 27, 2019, http://www.cslewisinstitute.org/CS_Lewis_on_Absolutes, p. 6.
65. Michelle Fitzsimmons, "Apple's Tim Cook: 'AR Has the Ability to Amplify Human Performance,'" TechRadar, February 2, 2018, retrieved July 15, 2018, https://www.techradar.com/news/apples-tim-cook-ar-has-the-ability-to-amplify-human-performance.
66. J. A. Bonus, A. Peebles, M. L. Mares, and I. G. Sarmiento, "Look on the Bright Side (of Media Effects): Pokémon Go as a Catalyst for Positive Life Experiences," *Media Psychology* 21, no. 2 (2018): 263–87.
67. Stefan Hall, "Augmented and Virtual Reality: The Promise and Peril of Immersive Technologies," McKinsey, October 2017, retrieved August 8, 2018, https://www.mc

kinsey.com/industries/media-and-entertainment/our-insights/augmented-and-virtual-reality-the-promise-and-peril-of-immersive-technologies.

68. David Phelan, "Apple CEO Tim Cook: As Brexit Hangs over UK, 'Times Are Not Really Awful, There's Some Great Things Happening,'" *Independent*, February 10, 2017, retrieved August 8, 2018, https://www.independent.co.uk/life-style/gadgets-and-tech/features/apple-tim-cook-boss-brexit-uk-theresa-may-number-10-interview-ustwo-a7574086.html.
69. Shara Tibken, "Apple's Working on a Powerful, Wireless Headset for Both AR, VR," CNET, April 27, 2018, retrieved August 8, 2018, https://www.cnet.com/news/apple-is-working-on-an-ar-augmented-reality-vr-virtual-reality-headset-powered-by-a-wireless-wigig-hub/?ftag=COS-05–10aaa0b&linkId=51043835.

CHAPTER 8: A DECLARATION OF CITIZEN-USER RIGHTS

1. J. E. Cohen, "The Historical Memory of American Presidents in the Mass Public," *Social Sciences* 7, no. 3 (2018): 36.
2. Rob Neufeld, "Portrait of the Past: Theodore Roosevelt in Asheville," *Citizen Times*, September 25, 2014, retrieved August 17, 2018, https://www.citizen-times.com/story/life/2014/09/25/portrait-past-theodore-roosevelt-asheville/16239399/.
3. R. C. V. Meyers, *Theodore Roosevelt, Patriot and Statesman: The True Story of an Ideal American* (Philadelphia: Ziegler, 1902).
4. "Trust in Judicial Branch Up, Executive Branch Down," Gallup, September 20, 2017, retrieved August 12, 2018, https://news.gallup.com/poll/219674/trust-judicial-branch-executive-branch-down.aspx; "Confidence in Institutions," Gallup, n.d., retrieved August 12, 2018, https://news.gallup.com/poll/1597/confidence-institutions.aspx.
5. "73% Say Freedom of Speech Worth Dying For," *Rasmussen Reports*, August 23, 2017, retrieved August 12, 2018, http://www.rasmussenreports.com/public_content/lifestyle/general_lifestyle/august_2017/73_say_freedom_of_speech_worth_dying_for.
6. "Freedom Rings in Places You Might Not Expect," Gallup, June 27, 2018, retrieved August 12, 2018, https://news.gallup.com/opinion/gallup/235973/freedom-rings-places-not-expect.aspx?g_source=link_newsv9&g_campaign=item_236411&g_medium=copy.
7. "Mass Shootings," Gun Violence Archive, n.d., retrieved August 20, 2018, https://www.gunviolencearchive.org/reports/mass-shooting?year=2015; "mass shooting" is defined as more than four people killed in an incident, not including the shooter and not including those injured, "General Methodology," Gun Violence Archive, n.d., retrieved August 20, 2018, http://www.gunviolencearchive.org/methodology.
8. "Drug Overdose Death Data," Centers for Disease Control and Prevention, n.d., retrieved August 20, 2018, https://www.cdc.gov/drugoverdose/data/statedeaths.html.
9. "Opioid Overdose Crisis," National Institute on Drug Abuse, January 2019, retrieved July 2, 2019, https://www.drugabuse.gov/drugs-abuse/opioids/opioid-overdose-crisis#one.
10. "Deaths: Final Data for 2015," *National Vital Statistics Reports* (a publication of the Centers for Disease Control and Prevention) 66, no. 6 (November 27, 2017), retrieved August 20, 2018, https://www.cdc.gov/nchs/data/nvsr/nvsr66/nvsr66_06.pdf.
11. "Deaths: Final Data for 2015," *National Vital Statistics Reports* (a publication of the Centers for Disease Control and Prevention) 66, no. 6 (November 27, 2017), retrieved August 20, 2018, https://www.cdc.gov/nchs/data/nvsr/nvsr66/nvsr66_06.pdf.
12. It should be noted that prior to the 1950s, youth suicides were generally categorized by medical professionals as "accidents" and rarely discussed outside academic circles. The earliest reporting on the topic in academia emerged in the 1940s, for instance: K. Friedlander, "On the 'Longing to Die,'" *International Journal of Psychoanalysis* 21, (1940): 416; Jason Hanna, "Suicides Under Age 13: One Every 5 Days," CNN, August 14, 2017, retrieved August 20, 2018, https://www.cnn.com/2017/08/14/health/child-suicides/index.html.
13. "A Global Report on the Decline of Democracy," Council on Foreign Relations, April 17,

2018, retrieved August 20, 2018, https://www.cfr.org/news-releases/global-report-decline-democracy.

14. "Attacks on the Record: The State of Global Press Freedom, 2017–2018," Freedom House, n.d., retrieved August 20, 2018, https://freedomhouse.org/report/special-reports/attacks-record-state-global-press-freedom-2017–2018.
15. "Overview Fact Sheet," Freedom House, 2017, retrieved August 20, 2018, https://freedomhouse.org/report/overview-fact-sheet.
16. Niraj Chokshi, "It's Not Just You: 2017 Was Rough for Humanity, Study Finds," *New York Times*, September 12, 2018, retrieved September 16, 2018, https://www.nytimes.com/2018/09/12/world/humanity-stress-sadness-pain.html?rref=collection%2Fsectioncollection%2Fscience&login=smartlock&auth=login-smartlock.
17. "'Depression: Let's Talk' Says WHO, as Depression Tops List of Causes of Ill Health," World Health Organization, June 30, 2017, retrieved August 20, 2018, http://www.who.int/mediacentre/news/releases/2017/world-health-day/en/.
18. Megan A. Moreno et al., "Feeling Bad on Facebook: Depression Disclosures by College Students on a Social Networking Site," *Depression and Anxiety* 28, no. 6 (2011): 447–455.
19. "Ransomware Prevention and Response for CISOs," Federal Bureau of Investigation, n.d., retrieved August 20, 2018, https://www.fbi.gov/file-repository/ransomware-prevention-and-response-for-cisos.pdf/view.
20. Natasha Lomas and Romain Dillet, "Terms of Service: The Biggest Lie of Our Industry," TechCrunch, August 21, 2015, retrieved August 19, 2018, https://techcrunch.com/2015/08/21/agree-to-disagree/.
21. "Apple Media Services Terms and Conditions," Apple, n.d., retrieved August 19, 2018, https://www.apple.com/legal/internet-services/itunes/us/terms.html.
22. "Google Terms of Service," Google, n.d., retrieved August 19, 2018, https://policies.google.com/terms.
23. "Terms of Service," Facebook, n.d., retrieved August 19, 2018, https://www.facebook.com/terms.php.
24. "Conditions of Use," Amazon, n.d., retrieved August 17, 2018, https://www.amazon.com/gp/help/customer/display.html?nodeId=508088.
25. "Terms of Service," Microsoft, n.d., retrieved August 19, 2018, https://www.microsoft.com/en-us/servicesagreement/.
26. A. M. McDonald and L. F. Cranor, "The Cost of Reading Privacy Policies," *I/S: A Journal of Law and Policy for the Information Society* 4 (2008): 543; Alexis Madrigal, "Reading the Privacy Policies You Encounter in a Year Would Take 76 Work Days," *Atlantic*, March 1, 2012, retrieved August 19, 2018, https://www.theatlantic.com/technology/archive/2012/03/reading-the-privacy-policies-you-encounter-in-a-year-would-take-76-work-days/253851/.
27. "The Rumors Are True: We Spend More and More Time Online," TechCrunch, December 23, 2009, retrieved August 19, 2018, https://techcrunch.com/2009/12/23/harris-interactive-poll/.
28. Jacqueline Howard, "Americans Devote More Than 10 Hours a Day to Screen Time, and Growing," CNN, July 29, 2016, retrieved August 19, 2018, https://www.cnn.com/2016/06/30/health/americans-screen-time-nielsen/index.html.
29. "How Long Do Users Stay on Web Pages?," Nielsen & Norman Group, September 12, 2011, retrieved August 19, 2018, https://www.nngroup.com/articles/how-long-do-users-stay-on-web-pages/.
30. Malcolm Owen, "Apple, Other Tech Companies Continue to Resist Encryption Backdoor Proposals by FBI, U.S. DOJ," AppleInsider, May 2, 2018, retrieved August 20, 2018, https://appleinsider.com/articles/18/05/02/apple-other-tech-companies-continue-to-resist-encryption-backdoor-proposals-by-fbi-us-doj.
31. Kirsten Korosec, "The 25 Most Common Passwords of 2017 Include 'Star Wars,'" *Fortune*, December 19, 2017, retrieved August 20, 2018, http://fortune.com/2017/12/19/the-25-most-used-hackable-passwords-2017-star-wars-freedom/.

32. "Terms of Service," Facebook, n.d., retrieved August 19, 2018, https://www.facebook.com/terms.php.
33. "Conditions of Use," Amazon, n.d., retrieved August 17, 2018, https://www.amazon.com/gp/help/customer/display.html?nodeId=508088.
34. Lee Rainie, "The State of Privacy in Post-Snowden America," Pew Research Center: Fact Tank, September 21, 2016, retrieved August 20, 2018, http://www.pewresearch.org/fact-tank/2016/09/21/the-state-of-privacy-in-america/.
35. Caroline Cakebread, "You're Not Alone, No One Reads Terms of Service Agreements," Business Insider, November 11, 2017, retrieved August 20, 2018, https://www.businessinsider.com/deloitte-study-91-percent-agree-terms-of-service-without-reading-2017–11.
36. S. Kokolakis, "Privacy Attitudes and Privacy Behaviour: A Review of Current Research on the Privacy Paradox Phenomenon," *Computers and Security* 64 (2017): 122–34.
37. C. Hallam and G. Zanella, "Online Self-Disclosure: The Privacy Paradox Explained as a Temporally Discounted Balance Between Concerns and Rewards," *Computers in Human Behavior* 68 (2017): 217–27.
38. Aaron Smith, "Record Shares of Americans Now Own Smartphones, Have Home Broadband," Pew Research Center: Fact Tank, January 12, 2017, retrieved August 20, 2018, http://www.pewresearch.org/fact-tank/2017/01/12/evolution-of-technology/.
39. Gloria Guzman, "Household Income: 2016," US Census Bureau, September 2017, retrieved August 20, 2018, https://www.census.gov/content/dam/Census/library/publications/2017/acs/acsbr16–02.pdf.
40. Selena Maranjian, "'What Tax Bracket Am I In?'—'It's Complicated,'" *Motley Fool*, April 26, 2013, retrieved August 20, 2018, https://www.fool.com/investing/general/2013/04/26/what-tax-bracket-am-i-in-its-complicated.aspx.
41. Bree Fowler, "Best Low-Cost Cell-Phone Plans," *Consumer Reports*, August 12, 2018, retrieved August 20, 2018, https://www.consumerreports.org/cell-phone-plans/best-low-cost-cell-phone-plans/; Eli Blumenthal, "Internet Bill Too High? Here's How to Save," *USA Today*, January 8, 2018, retrieved August 20, 2018, https://www.usatoday.com/story/money/personalfinance/budget-and-spending/2018/01/08/internet-bill-too-high-heres-how-save/429890001/.
42. "Apple Music Plans," Apple, n.d., retrieved August 20, 2018, https://www.apple.com/apple-music/plans/; "Google One," Google, n.d., retrieved August 20, 2018, https://one.google.com/getupdates.
43. "Right to Be Forgotten," General Data Protection Regulation (GDPR), n.d., retrieved August 20, 2018, https://gdpr-info.eu/issues/right-to-be-forgotten/.
44. Lucas Laursen, "How Google Handled a Year of 'Right to Be Forgotten' Requests," *IEEE Computer Society*, April 23, 2015, retrieved August 20, 2018, https://spectrum.ieee.org/telecom/internet/how-google-handled-a-year-of-right-to-be-forgotten-requests.
45. "The Size of the World Wide Web (the Internet)," World Wide Web Size, n.d., retrieved August 20, 2018, http://www.worldwidewebsize.com/.
46. "Verisign Domain Name Industry Brief; Internet Grows to 330.6 Million Domain Names in Q1 2017," Verisign, July 18, 2017, retrieved August 20, 2018, https://blog.verisign.com/domain-names/verisign-domain-name-industry-brief-internet-grows-to-330–6-million-domain-names-in-q1–2017/.
47. Rachel Kraus, "Europeans Asked Google for Their 'Right to Be Forgotten' 2.4 million times," Mashable, February 28, 2018, retrieved August 20, 2018, https://mashable.com/2018/02/27/right-to-be-forgotten-google-transparency-report/#r1C3LUVYSOqN.
48. "Freedom of Information Act," FOIA, n.d., retrieved August 22, 2018, https://www.foia.gov/faq.html.
49. "Right of Access," General Data Protection Regulation, EU, n.d., retrieved August 22, 2018, https://www.gdpreu.org/the-regulation/list-of-data-rights/right-of-access/.
50. Olivia Solon, "New Europe Law Makes It Easy to Find Out What Your Boss Has Said About You," *Guardian*, April 24, 2018, retrieved August 23, 2018, https://www.theguardian.com/technology/2018/apr/23/europe-gdpr-data-law-employer-employee.

51. "Article 83: General Conditions for Imposing Administrative Fines," General Data Protection Regulation, EU, retrieved August 23, 2018, https://gdpr-info.eu/art-83-gdpr/.
52. "Our Membership," British Medical Association, n.d., retrieved August 23, 2018, https://www.bma.org.uk/membership.
53. Valeria Fiore, "GPs Asked to Lobby MPs over 'Bombardment' of Patient Data Requests," Pulse Today, August 10, 2018, retrieved August 23, 2018, http://www.pulsetoday.co.uk/news/gp-topics/it/gps-asked-to-lobby-mps-over-bombardment-of-patient-data-requests/20037259.article.
54. Andrew Ross, "82% of Organisations Do Not Know Where All Their Critical Data Is Kept, Says Research," Information Age, August 21, 2018, retrieved August 23, 2018, https://www.information-age.com/critical-data-123474315/.
55. A. G. Tansley, "The Use and Abuse of Vegetational Concepts and Terms," *Ecology* 16, no. 3 (1935): 284–307.
56. F. Stuart Chapin, Pamela A. Matson, and Harold A. Mooney, *Principles of Terrestrial Ecosystem Ecology* (New York: Springer, 2002), 281–304.
57. Tim Herrera, "When You Track Oreos, Exercise and Everything Else," *New York Times*, August 15, 2018, retrieved August 17, 2018, https://www.nytimes.com/2018/08/15/technology/personaltech/track-your-life-app.html.
58. "UN Refugee Agency: Record 65.6 Million People Displaced Worldwide," BBC News, June 19, 2017, retrieved June 6, 2018, https://www.bbc.com/news/world-40321287.
59. "Figures at a Glance," United Nations Human Rights Commission, n.d., retrieved June 6, 2018, http://www.unhcr.org/en-au/figures-at-a-glance.html.
60. "Statelessness," US Department of State, n.d., retrieved June 6, 2018, https://www.state.gov/j/prm/policyissues/issues/c50242.htm.
61. Issie Lapowsky, "Trump Can't Block Critics on Twitter. Here's What This Means for You," *WIRED*, May 23, 2018, retrieved June 10, 2018, https://www.wired.com/story/donald-trump-blocking-on-twitter-unconstitutional/.
62. Masha Gessen, "'The Right to Have Rights' and the Plight of the Stateless," *New Yorker*, May 3, 2018, retrieved July 2, 2019, https://www.newyorker.com/news/our-columnists/the-right-to-have-rights-and-the-plight-of-the-stateless.
63. "The Meditation Room at United Nations Headquarters (1952)," United Nations Multimedia, n.d., retrieved August 19, 2018, https://www.unmultimedia.org/s/photo/detail/498/0049835.html.
64. "Dag Hammarskjöld: 'A Room of Light,'" United Nations, n.d., retrieved July 26, 2019, http://www.un.org/depts/dhl/dag/meditationroom.htm.
65. "Dag Hammarskjöld Designs United Nations Meditation Room," Read the Spirit, n.d., retrieved June 27, 2019, https://www.readthespirit.com/explore/dag-hammarskjolds-introduction-to-the-united-nations-meditaiton-room/.

CONCLUSION: THE NET STATE PATTERN

1. "Declassified Letter from N. Khrushchev to J. Kennedy," NSA Archive, George Washington University, October 24, 1962, retrieved September 21, 2018, https://nsarchive2.gwu.edu//IMG/cmc1ltr.pdf.
2. "The Cuban Missile Crisis, October 1962," Office of the Historian, US Department of State, n.d., retrieved September 21, 2018, https://history.state.gov/milestones/1961–1968/cuban-missile-crisis.
3. "Washington-Moscow Hotline," Cryptomuseum, n.d., retrieved September 21, 2018, http://www.cryptomuseum.com/crypto/hotline/index.htm.
4. "The Washington-Moscow Hotline," Electrospaces, October 28, 2012, retrieved September 21, 2018, https://electrospaces.blogspot.com/2012/10/the-washington-moscow-hot-line.html.
5. David Vergun, "Hotline, Now 50 Years Old, Continues to Promote Dialog with Russians," Army News Service, August 26, 2013, retrieved September 16, 2018, https://www

.army.mil/article/109986/Hotline__now_50_years_old__continues_to_promote_dialog_with_Russians/.

6. Allan Pease, "The Definitive Book of Body Language," *New York Times*, September 24, 2006, retrieved September 21, 2018, https://www.nytimes.com/2006/09/24/books/chapters/0924–1st-peas.html.
7. Alvin Snyder, "More History on a Murrow Obsession," USC Center on Public Diplomacy, July 28, 2009, retrieved July 2, 2019, https://www.uscpublicdiplomacy.org/blog/more-history-murrow-obsession.
8. Sean Hollister, "Here's How Facebook Defines Terrorism—and How It's Responding," CNET, April 23, 2018, retrieved August 26, 2018, https://www.cnet.com/news/facebook-shares-terrorism-definition-al-qaeda-isis/; Paul Cruikshank, "A View from the CT Foxhole: An Interview with Brian Fishman, Counterterrorism Policy Manager, Facebook," Combating Terrorism Center, West Point, September 2017, retrieved August 26, 2018, https://ctc.usma.edu/a-view-from-the-ct-foxhole-an-interview-with-brian-fishman-counterterrorism-policy-manager-facebook/.
9. Curt Mills, "Facebook Is Looking for a Counterterrorism Analyst," *US News & World Report*, November 14, 2016, retrieved August 26, 2018, https://www.usnews.com/news/national-news/articles/2016–11–14/facebook-is-looking-for-a-counterterrorism-analyst.
10. "Partnering to Help Curb Spread of Online Terrorist Content," Facebook Newsroom, December 5, 2016, retrieved August 26, 2018, https://newsroom.fb.com/news/2016/12/partnering-to-help-curb-spread-of-online-terrorist-content/.
11. "Global Internet Forum to Counter Terrorism," *Twitter Blog*, June 26, 2017, retrieved August 26, 2018, https://blog.twitter.com/official/en_us/topics/company/2017/Global-Internet-Forum-to-Counter-Terrorism.html.
12. Sheera Frenkel and Matthew Rosenberg, "Top Tech Companies Met with Intelligence Officials to Discuss Midterms," *New York Times*, June 25, 2018, retrieved August 26, 2018, https://www.nytimes.com/2018/06/25/technology/tech-meeting-midterm-elections.html.
13. Kevin Collier, "Tech Companies Are Gathering for a Secret Meeting to Prepare a 2018 Election Strategy," Buzzfeed News, August 23, 2018, retrieved August 26, 2018, https://www.buzzfeednews.com/article/kevincollier/tech-companies-are-gathering-for-a-secret-meeting-to.
14. Rana Dasgupta, "The Demise of the Nation State," *Guardian*, April 5, 2018, retrieved August 26, 2018, https://www.theguardian.com/news/2018/apr/05/demise-of-the-nation-state-rana-dasgupta.
15. Alex Tabarrok, "The Case for Getting Rid of Borders—Completely," *Atlantic*, October 10, 2015, retrieved August 26, 2018, https://www.theatlantic.com/business/archive/2015/10/get-rid-borders-completely/409501/.
16. "From Thomas Jefferson to Benjamin Rush, 23 September 1800," National Archives, n.d., retrieved August 17, 2018, https://founders.archives.gov/documents/Jefferson/01-32-02-0102.
17. James Madison, "The Federalist Papers, No. 46 (The Influence of the State and Federal Governments Compared)," Constitution (website), January 29, 1788, retrieved August 17, 2018, http://www.constitution.org/fed/federa46.htm.
18. Nelson Lund and Adam Winkler, "Common Interpretation: The Second Amendment," Constitution Center, n.d., retrieved August 17, 2018, https://constitutioncenter.org/interactive-constitution/amendments/amendment-ii.
19. James Madison, "The Federalist Papers, No. 51 (The Structure of the Government Must Furnish the Proper Checks and Balances Between the Different Departments)," Constitution (website), February 6, 1788, retrieved August 17, 2018, http://www.constitution.org/fed/federa51.htm.
20. G. A. Almond, "Political Theory and Political Science," *American Political Science Review* 60, no. 4 (1966): 869–79.

21. Abigail Geiger, "How Americans Have Viewed Government Surveillance and Privacy Since Snowden Leaks," Pew Research Center: Fact Tank, June 4, 2018, retrieved August 17, 2018, http://www.pewresearch.org/fact-tank/2018/06/04/how-americans-have-viewed-government-surveillance-and-privacy-since-snowden-leaks/.
22. "America in Crisis," Edelman, January 21, 2018, retrieved August 17, 2018, https://www.edelman.com/post/america-in-crisis.
23. James Madison, "Selected Quotes of James Madison," Constitution Society, n.d., retrieved June 27, 2019, https://www.constitution.org/jm/jm_quotes.htm.
24. James Madison, "Founders Online: From James Madison to Thomas Jefferson, 13 May 1798," National Archives, May 13, 1798, retrieved June 27, 2019, https://founders.archives.gov/documents/Madison/01-17-02-0088.
25. "America's Wars," US Department of Veterans' Affairs, n.d., retrieved August 17, 2018, https://www.va.gov/opa/publications/factsheets/fs_americas_wars.pdf; Ai Lei Tao, "Nation-State Actors Responsible for Most Cyber Attacks," *Computer Weekly*, September 22, 2017, retrieved August 17, 2018, https://www.computerweekly.com/news/450426775/Nation-state-actors-responsible-for-most-cyber-attacks.
26. *Encyclopædia Britannica Online*, s.v. "Widmanstätten Pattern," April 9, 2012, retrieved June 27, 2019, www.britannica.com/topic/Widmanstatten-pattern.
27. Vagn Fabritius Buchwald, *Iron and Steel in Ancient Times* (Copenhagen: Det Kongelige Danske Videnskabernes Selskab, 2005), 26.
28. Alice Walker, as quoted in William P. Martin, *The Best Liberal Quotes Ever: Why the Left Is Right* (Naperville, IL: Sourcebooks, 2004).

A GIFT OF JOY AND HOPE

A GIFT OF JOY AND HOPE

POPE FRANCIS

Translated from the Italian by Oonagh Stransky

New York • Nashville

Worthy
Hachette Book Group
1290 Avenue of the Americas, New York, NY 10104
worthypublishing.com
twitter.com/worthypub

First published in Great Britain in 2022 by Hodder & Stoughton,
a Hachette UK company.
First Worthy edition: September 2022

Worthy is a division of Hachette Book Group, Inc. The Worthy name and logo are trademarks of Hachette Book Group, Inc.

The publisher is not responsible for websites (or their content) that are not owned by the publisher.

The Hachette Speakers Bureau provides a wide range of authors for speaking events. To find out more, go to www.hachettespeakersbureau.com or call (866) 376-6591.

Library of Congress Control Number: 2022939621

ISBNs: 9781546003694 (hardcover), 9781546003717 (ebook)

Printed in the United States of America

LSC-C

Printing 1, 2022

CONTENTS

Don't hide away your dreams!
Don't drown them out;
give your dreams plenty of room,
dare to look out to the horizon,
dare to see what lies ahead
if you have the courage to build them.

Life is always a journey.
We become that which we move toward.
Let us choose the way of God, not of the self;
the way of *yes*, not *if*.
Together we shall discover that there is
no unexpected event,
no difficult climb,
no dark night
that cannot be faced with Jesus by our side.

MY WISH

My wish can be summed up in one word: *smile*.

Inspiration for this wish comes from Thailand, one of the last countries I visited, also known as "the country of the smile." The Thai people have a special kindness about them, a nobility and manner that are captured by this facial expression and present in all their gestures. My journey there had a profound impact on me and has led me to understand the smile as an expression of love, an expression of affection, a deeply human gesture.

When we look at a baby, we smile, and if a small smile blooms on the baby's face, we feel a deep, heartfelt emotion stir inside us. While the child may simply be responding to our gaze, their smile is incredibly powerful because it is new, pure, like water from a spring, and it awakens a true nostalgia for childhood in us adults.

This is what took place between Mary, Joseph, and Jesus in a special way. The Virgin and her husband, through their love, made a smile bloom on the lips of their newborn child. And when this happened, their hearts were filled with a new kind of joy, a joy that came from Heaven. And the little manger in Bethlehem was filled with light.

Jesus is God's smile. Jesus came to us to reveal the love of our heavenly Father, His goodness, and He did so by smiling at His parents, like every newborn child does. They, the Virgin Mary and St. Joseph, because of their great faith, were able to accept that message; in Jesus' smile they saw God's mercy, both for themselves and for all those who awaited His coming, the coming of the Messiah, the Son of God, King of Israel.

We too, my dear friends, can experience this feeling when we look down at the nativity scene; when we look at Baby Jesus, we can feel God smiling at us, smiling down upon the poor of the earth, upon all who await salvation, who hope for a kinder world, a place where war and violence do not exist, a place where men and women can live in dignity as the children of God.

Sometimes it is difficult to smile, and there is no lack of reasons for this. In these hard times, we especially need God's smile, and only Jesus can help us. Only He is the Savior. Sometimes we experience this concretely in our lives.

Sometimes things go well for us, but then there is the danger of feeling too safe. We must not forget that others are struggling. We need God's smile then so that we are stripped of our false sense of security and led back to simpler and easier ways of life.

Let us grant each other this wish, whatever the season: let us feel awe and wonder at God's smile, which Jesus brought to us. He Himself is this smile. Let us welcome Him and allow ourselves to be purified so that we can bring a humble, simple smile to others.

Share this wish with your loved ones at home, especially the sick and the elderly, so they can feel the gentle caress of your smile. Because that is what a smile is: a gentle caress, a caress of the heart, a caress of the soul. And, together, let us remain united in prayer.*

* The pages that follow outline a manifesto of rebirth, a new era of joy, a period made even more necessary because of these "times of illness." The thoughts and words included here represent a journey through the comments, speeches, homilies, apostolic exhortations, and encyclicals articulated by His Holiness.

ONE

CHANGE AND REBIRTH

HOPE NEVER LETS US DOWN

Optimism can let us down, but hope never does! These days our need for hope is great because we feel surrounded by darkness, disoriented by evil and violence, and distressed for the plight of so many of our brothers and sisters. We truly need hope! We feel disoriented, sometimes even discouraged and powerless, as though this darkness will never end.

We must not let hope slip away from us. God and His love walk alongside us. "I hope because God is with me." Each of us can say these words.

THE HAPPINESS OF A SHARED HUMANITY

In this fast-paced world, which has no shared roadmap, we increasingly sense that the gap between concern for one's personal well-being and the prosperity of the larger human family is widening; it feels as though there is a schism between the individual and the community of humankind. It is one thing to feel forced to live together, but it is another thing entirely to value the richness and beauty of the seeds of a shared life, which need to be sought out and sown. Yes, technology is constantly advancing, but how wonderful it would be if scientific and technological innovation brought greater equality and social inclusion with it. How wonderful it would be if, while discovering faraway planets, we could also rediscover the needs of our brothers and sisters who orbit around us!

THE NIGHTS OF OUR LIVES

We all have an appointment with God in the dark of night, in the night-time of our lives, in one of the many nights of our life: dark moments, moments of sin, moments of disorientation. We have a standing appointment with God on those nights, always. He will be there to surprise us at the moment we least expect Him, when we find ourselves truly alone. And on those nights, while we struggle with the unknown, we will realize that we are merely poor men and women; but then, in those very moments when we feel as though we are poor creatures, we discover that we need not fear, because God will give us a new name, one that will encompass the meaning of our entire lives. He will change our hearts and offer us the blessing reserved to those who have allowed themselves to be changed by Him. This invitation to let ourselves be changed by God is beautiful; God knows how to do it because He knows each one of us.

"Lord, you know me," each one of us can say. "Lord, you know me. Change me."

COME TO ME!

The words of Jesus in the Gospel of Matthew reach out to us: "Come to me, all you who labor and are overburdened, and I will give you rest" (Matthew 11:28). Often life is tiring; often life is tragic. Work is tiring, looking for work is tiring, and finding work these days is exhausting and requires a great deal of effort. But this is not the most burdensome aspect of life; what weighs more than all of this is the lack of love. Not being received with a smile, not being made to feel welcome—this weighs heavily on us. Some silences are even oppressive: those that exist within families, between husbands and wives, between parents and children, among siblings. Without love, our burdens become even heavier, intolerable. I think of elderly people living alone and families who receive no help in caring for someone at home who has special needs. "Come to me, all you who labor and are overburdened," Jesus tells us.

THE RIGHT SIDE OF THE TAPESTRY

The love that we show and spread often makes mistakes. Those who act and take risks often commit errors. Here the experience of Maria Gabriella Perin is precious witness for us. Losing her father shortly after her birth, she reflects on how this shaped her life, how it influenced her, how she got involved in a relationship that did not last but left her a mother, first, and now a grandmother.

> What I know is that God gives us stories. In his genius and mercy, he takes our triumphs and failures and

weaves beautiful tapestries that are full of irony. The reverse side of the fabric may look messy, with its tangle of threads—the events of our lives—and it may well be that this side is the one we dwell on in times of doubt. But the right side of the tapestry shows a magnificent story, and this is the side that God sees.*

WITH US EVERY DAY

Christ is alive! We need to remind ourselves of this constantly because otherwise we risk seeing Jesus Christ simply as a figure from the distant past, a memory, as someone who saved us two thousand years ago. But that is of no use to us; it would leave us unchanged and would not set us free. He who fills us with grace, He who liberates us, He who transforms us, He who heals and consoles us is a figure that is fully alive. He is the Christ arisen, filled with supernatural life and energy, and robed in boundless light. This is why St. Paul said, "if Christ has not been raised, your faith is pointless" (1 Corinthians 15:17).

If He is alive, He can be present in your life at any moment and fill it with light. With Him, there will never be sorrow and solitude. Even if everyone else leaves, He will be there, as He promised: "I am with you always; yes, to the end of time" (Matthew 28:20). He fills your life with His unseen presence and, wherever you go, He will be waiting there for you. He did not only come in the past, He comes to you each and every day, inviting you to set out toward ever-new horizons.

* Pope Francis and Friends, *Sharing the Wisdom of Time* (Loyola Press: Chicago, 2018).

BEYOND THE FAMILIAR

God is eternally new. He compels us to push forward and start over, to make changes, go beyond the familiar, move toward the fringes, and beyond. He leads us to places where humanity is most wounded, where men and women seek an answer to the meaning of life, where they look beyond shallow appearances and conformism. God is not afraid! He is fearless! He is always greater than all our plans and schemes. Unafraid of all that is marginal, He Himself became a marginal figure (cf. Philippians 2:6–8; John 1:14). If we dare make our way to the fringes, we will find Him there. Jesus is already there. Jesus precedes us in the hearts of our brothers and sisters, in their wounds, troubles, and profound desolation. He is already there.

WHERE IS MY HAND?

There is only one fair and correct way to look down on a person: if you are offering them a hand up. If one of us—and this includes me—looks down on a person, shows them contempt, it is of no use at all. But if we look down on a person in order to give them a hand up, to help them to their feet, then that man or woman beneath us is truly an extraordinary being. So, when you look down on someone, always ask yourself, "Where is my hand? Is it hidden away or is it offering help?" In this way you will find happiness.

This approach means developing a critical but underappreciated ability: the ability to learn how to make time for others, to listen, share, and understand. Only then will we be able to open our experiences and problems to the kind of love that can change us, thereby changing the world around us. If we do not give of our time, if we do not "waste time" with others, we will end up wasting it on things that, at the end of the day, will only leave us feeling empty and confused. Or, as they say in my country, we will "stuff ourselves" with so many things that we will end up having indigestion.

NOT ON OUR OWN

The Fifth Beatitude says: "Blessed are the merciful: they shall have mercy shown them" (Matthew 5:7). This Beatitude is particular: it is the only one where the action and the fruit coincide. Those who show mercy will find mercy; they will be "shown mercy." This reciprocity of forgiveness is found throughout the gospel. How could it be otherwise? Mercy is the very heart of God!

Two things cannot be separated: forgiveness granted and forgiveness received. And yet so many people struggle to forgive. Sometimes the harm experienced is so great that forgiving feels like climbing a very high mountain; it takes massive effort. A person thinks: "No, I cannot possibly do it." The reciprocity of mercy shows us that we have to turn our perspective upside down: we cannot do it on our own. We need the grace of God, we need to ask for it. Essentially, if the Fifth Beatitude promises mercy, and in the Lord's Prayer we ask for our trespasses to be forgiven, it means that we are essentially trespassers and that we need to find mercy!

PRAYER: A BULWARK AGAINST EVIL

In our daily life we feel the presence of evil: it happens every day. The first chapters of the Book of Genesis describe the gradual expansion of sin into human life. Adam and Eve doubt God's good intentions and believe they are dealing with an envious divinity that wants to impede their happiness. This is what leads to their rebellion. But their experience takes them in the opposite direction: their eyes are opened, they discover they are naked, that they have nothing. Do not forget this: that which tempts us does not pay well (cf. Genesis 3:1–7).

However, within these first pages of the Bible is another story, one that is less striking, a far humbler and more pious story, a story that represents redemption through hope. While almost everyone behaves wickedly, making hatred and conquest the great engine of human matters, some people are capable of praying to God with sincerity, capable of writing the destiny of humanity differently.

Prayer is a bulwark. It provides humans with refuge before the tidal wave of evil that floods the world. On looking closely, we understand that we actually pray to be saved from ourselves. It is important to pray: "Lord, please, save me from myself, from my ambitions, from my passions." The prayerful in the early pages of the Bible are peacemakers, for when prayer is true, it frees humanity from violent instincts. It is a gaze that is directed toward God, so that He might return to take care of humanity's heart. In the Catechism we read: "This kind of prayer is lived by many righteous people in all religions."* Prayer cultivates flowerbeds of rebirth in places

* Catechism of the Catholic Church 2569.

where human hatred has only enlarged the desert. Prayer is powerful because it attracts the power of God, and the power of God always gives life. Always. He is the God of life and brings about rebirth.

AN ANCHOR OF HOPE

Job was in darkness. He was at death's door. And in that moment of anguish, pain, and suffering, Job proclaimed hope: "I know that I have a living Defender and that he will rise up last, on the dust of the earth ... I shall look on God ... my eyes will be gazing on no stranger" (Job 19:25–7).

A cemetery is a sad place. It reminds us of our loved ones who have died before us and of our future, of death. But into this sadness we bring flowers as a sign of hope and eventually of celebration. Sorrow mingles with hope. We all feel this before the remains of our loved ones: both memories and hope. Hope helps us because we will have to make the same journey. Sooner or later, all of us. Some with more and some with less pain, but all of us. And then, there is the flower of the hope of rebirth, a powerful thread anchored in the hereafter. And it never disappoints. Jesus was the first to make the journey; we merely follow the journey that He took first. "I know that my Redeemer lives, and that he will be the last one standing on the ashes. I shall behold Him, with my eyes. My eyes shall behold Him, and no one else" (Job 19:25).

CHALLENGES

The evils of our world—and those of the Church—must not function as excuses for diminishing our commitment or our fervor. Let us look upon them as challenges that can help us grow.

THE HORSE AND THE RIVER

A moment of crisis is a moment of choice, which brings us face-to-face with the decisions we must make. All of us, in life, have had and will have moments of crisis: family crises, marriage crises, social crises, crises at work, so many crises. Even the pandemic that has struck us is a moment of social crisis.

How does one react in a moment of crisis? "After this, many of his disciples went away and accompanied him no more" (John 6:66). Jesus decides to question the apostles: "Then Jesus said to the Twelve, 'What about you, do you want to go away too?'" (John 6:67). Jesus urges them to make a decision. And it is Peter who makes the second confession: "Simon Peter answered, 'Lord, to whom shall we go? You have the message of eternal life, and we believe; we have come to know that you are the Holy One of God'" (John 6:68–9). In the name of the Twelve, Peter confesses that Jesus is the Holy One of God, the Son of God.

Earlier, Peter was the first to confess: "You are the Christ, the Son of the living God," he declares, but when Jesus began to explain the Passion, Peter stops Him and says, "No, no, Lord, not this"; Jesus then reproaches him, saying, "Get behind me, Satan! You are an obstacle in my path, because you are thinking not as God thinks but as human beings do" (Matthew 16:16–23).

In his second confession, Peter has matured a little and does not fight Jesus. He does not understand what Jesus is saying about eating His "flesh" and drinking His "blood"; he does not understand, but he trusts the Master (John 6:54–6). He trusts Him. And then he makes this second confession: "Lord, to whom shall we go? You have the message of eternal life" (John 6:68).

These words help us survive moments of crisis. In my country, a saying goes, "When you're riding a horse and have to cross a river, please, don't change horses in midstream." In moments of crisis, you need to be steadfast in your convictions of faith. Those who left "changed horses," sought out another teacher who was not so "hard on them," as they said to Him. Moments of crisis demand perseverance and silence; you need to stay put, be steadfast. It is not the right time for changes. It is the moment of faith, faithfulness to God, faithfulness to previously made decisions. It is also the moment of conversion because this faithfulness can inspire changes for the better. Such faith does not push us away from good. Moments of peace and moments of crisis: we Christians need to learn to manage both.

CHRIST LIVES!

He is our hope and brings all the beauty of youth to this world. Everything He touches becomes youthful, new, full of life. And so, the very first words that I say to each young Christian is this: Christ is alive and wants you alive!

He is in you, with you; He never abandons you. Wherever

you wander, the Risen One is by your side, He calls out to you and waits for you to return to Him and start over. When you feel old with sorrow, bitterness, fear, doubt, or failure, He will always be there to grant you strength and hope.

OUR SADNESS IS A SEED OF JOY

In times of suffering and pain, when we do not understand what is going on around us, and when we want to rebel, let us look for our mother, and let us be like the fearful little children we used to be and cling to her skirts and call to her in our hearts, "Mother!" We are not alone; we have a mother. We have our older brother, Jesus. We are not alone.

Let us look to the future with eyes filled with faith. Our sadness is a seed that will one day blossom into the joy that our Lord promised everyone who trusted His words: "Blessed are the gentle: they shall have the earth as inheritance" (Matthew 5:4). God's *compassion*, the way He suffers alongside us, gives eternal meaning and value to our struggles.

THOSE WHO MOVE FORWARD AND THOSE WHO TURN BACK

Many people, including Christians and Catholics in our community, do not walk ahead. They are tempted to stop. Many Christians among us have indeed stopped. Their hope is weak. Yes, they believe in Heaven but they do not seek it out. They follow the commandments, the precepts, they do it all, and yet are stuck. The Lord cannot draw leaven from them to grow His people.

Then there are the ones who go astray. All of us have sometimes taken the wrong turn. But the problem is not making a mistake; the problem is not turning back when we realize we have made a mistake. It is our condition as sinners that makes us take the wrong turn. We walk, but sometimes we take the wrong turn. And yet, we can always turn back. The Lord gives us the grace to turn back.

There is another group that is even more dangerous because they deceive themselves. These are the ones who walk in circles and never move forward. They are the wandering Christians: they go around and around as if life were an existential tour with no final destination, never taking their promises seriously. They walk in circles and deceive themselves by saying, "Look, I am walking." No, you're not walking; you're going around in circles! The Lord asks us to keep moving forward. He warns us against taking the wrong turn. He reminds us not to wander about aimlessly. He asks us to follow through on our promises and carry out what we promised.

YOUNG PEOPLE ARE THE PROMISE OF LIFE

Some time ago, a friend asked me what I see when I look at a young person. My response was that I see a person looking for their own path, someone who wants to fly with their own wings, who looks at the world and to the horizon with eyes full of the future, full of hope, as well as illusions. A young person stands on their two feet just like adults, but unlike adults, who keep their feet parallel, one next to the other, a young person always has one foot in front of the other, one foot forward, always ready to set out, to spring forward, always racing ahead. Talking about young people means talking about promises; and that means talking about joy. Young people have so much strength; they know how to look forward with hope. A young person is a promise of life that implies a certain degree of tenacity. They may be foolish enough to delude themselves but resilient enough to heal from the damage incurred by those delusions.

FALL IN LOVE!

Are you looking for passion? As the beautiful poem attributed to Fr. Pedro Arrupe advises, fall in love! Or, rather, let yourself fall in love, because:

> Nothing is more practical than
> finding God, than
> falling in Love
> in a quite absolute, final way.
> What you are in love with,
> what seizes your imagination,
> will affect everything.
> It will decide
> what will get you out of bed in the morning,
> what you do with your evenings,
> how you spend your weekends,
> what you read,
> whom you know,
> what breaks your heart,
> and what amazes you with joy and gratitude.
> Fall in Love,
> stay in love,
> and it will decide everything.

This love for God that embraces all of life with passion is only possible thanks to the Holy Spirit "because the love of God has been poured into our hearts by the Holy Spirit which has been given to us" (Romans 5:5).

TWO

THE DREAM OF BEAUTY

THE LACK OF POETRY IN THE WEST

Eastern wisdom is not merely the wisdom of knowledge, it is the wisdom of time, the wisdom of contemplation. It would be good for our Western society, always in too much of a hurry, to learn how to stop and contemplate things, even in a poetic manner. I believe the West lacks a little poetry. While there are many beautiful, poetic things in the West, the East always goes further. The East is capable of looking at things through eyes that see further. I will not use the word "transcendental" because some Eastern religions do not mention transcendence, but they certainly include a vision that goes beyond the limits of permanence. They go beyond. This is why I use the word "poetry" to describe something free. I think that slowing down and making time for wisdom would do us Westerners good. The culture of "rushing around" needs the culture of "stop a while." Stop and look at the world around you.

THE CREATION OF POETS

We cannot educate without introducing the heart to beauty. I would even go so far as to say that an education is not complete if it does not know how to shape people into poets. The path of beauty is a challenge that must be addressed.

A LIFE FULL OF GRACE

The prayers of humanity are closely linked to the sentiment of awe. The greatness of humankind is infinitesimal compared to the dimensions of the universe. While even an individual's greatest conquests may seem minor, the individual is not an insignificant being. In prayer, a sentiment of mercy is powerfully affirmed. Nothing exists by chance: the secret of the universe takes place when we exchange benevolent glances with another. Psalm 8 affirms that we are a little less than a god, that we are crowned "with glory and beauty" (Psalm 8:5). Our relationship with God is what makes us great, it is our enthronement. By nature, we are practically nothing, but by vocation, by calling, we are the children of the great King!

This is an experience that many of us have had. When the events of life, with all their hardships, risk suffocating the gift of prayer that lies within us, it is enough to look upon a starry sky, sunset, or flower to rekindle a spark of gratefulness. This experience is at the heart of the very first page of the Bible.

Prayer is the first weapon of hope. You pray and hope grows, it moves forward. I would even say that prayer opens the doors to hope. Hope is out there, but my prayers open the doors. People who pray hold fundamental truths; they are the ones who repeat over and over, first to themselves and then to everyone else, that life, despite all its trials and tribulations, despite the hard days, is full of awe-inspiring grace. As such, it must always be defended and protected.

IN HARMONY WITH CREATION

Jesus was able to invite others to look upon the beauty that exists in the world because He Himself was in constant touch with nature, because He gave it His full attention and awe. As He made His way throughout the land, He often stopped to contemplate the beauty sown by His Father, inviting His disciples to perceive the divine message within things: "look around you, look at the fields; already they are white, ready for harvest!" (John 4:35). "The kingdom of Heaven is like a mustard seed which a man took and sowed in his field. It is the smallest of all the seeds, but when it has grown it is the biggest of shrubs" (Matthew 13:31–2). Jesus lived in complete harmony with creation, and others were in awe of this.

BE ALERT TO BEAUTY

Let us be alert to beauty, to a sense of wonder, to the sense of awe that will open up new horizons and raise new questions. A consecrated life incapable of opening itself up to surprise is only a half-lived life. Let me repeat that. A consecrated life that is incapable of being surprised each day by joy or sadness is a consecrated life that is only half-lived. The Lord did not call us and send us forth into the world to impose obligations on people, or to lay heavier and more numerous burdens on people than they already have, but to share in joy, and look out to a horizon that is beautiful, new, and full of surprise.

THE SPLENDOR OF HUMANITY

Psalm 8 says: "I look up at your heavens, shaped by your fingers, at the moon and the stars you set firm—what are human beings that you spare a thought for them, or the child of Adam that you care for him?" (Psalm 8:3–4). The person who prays contemplates the mystery of life; they take in the starry skies above—those that astrophysicists explain to us in all their immensity—and wonder about the loving plan that exists behind such a formidable work! What is a human in this boundless vastness? A "pointless end," another Psalm states (Psalm 89:47). A being that is born and dies, an extremely fragile creature. And yet, throughout the universe, the only creature aware of such beauty is the human being. Yes, we are aware of this beauty!

THE FREEDOM OF PLAY

It is sad to see people who do not know how to daydream, who cannot dream. It is a great thing to be able to dream. "Don't be a fool, stop it, be real," some people say. But dreams are real, dreams help you look out to the horizon, they open up your life a little, they bring oxygen to your soul. Don't lose the ability to dream. A beautiful Italian song, "Nel blu dipinto di blu," is like a hymn to dreaming. And to playing! Play, too, speaks the language of freedom. One of the worst things that has happened to football today is that the game has lost all sense of play, it has become too commercial; but it is precisely the game, and for the love of the game, that makes you grow as a person. Play is both freedom and freeing. Never forget this idea.

THOSE WHO DO NOT KNOW HOW TO PLAY ARE NOT MATURE

A person who does not learn how to play as a child will never be fully mature and will always be a split person, a sterile person, incapable of creating a poem, incapable of dreaming. Play is important. Allow me to share something with you. When I offer confession to young married couples who mention they have two, or even three, children, I always say, "Let me ask you something: Do you play with your children?" Many of you will have children; don't forget to play with them. Fathers and mothers who know how to play with their children, who know how to get down on the floor and truly play with them—this is wisdom. This is what it means to raise a child well.

WE ALL RECEIVED A DREAM

Among the many things our parents save for us, it is good to find a few keepsakes that enable us to imagine what our grandmothers and grandfathers dreamed for us. Each one of us, long before our birth, received the gift of a blessing of a dream full of love and hope from our grandparents: the dream of a better life. And if not from our grandparents, then surely our great-grandparents had that happy dream for us while they looked down on their children in their cradles, and eventually their grandchildren. The original dream, the primordial one, is the dream of creation of God our Father, and it both precedes and accompanies the lives of all His children. Conserving the memory of this blessing, which extends from generation to generation, is a precious legacy that we need to keep alive so that we can pass it on.

This is why it is good to let older people tell their long stories, which often seem like myths or legends (they are, after all, the dreams of the elderly): they are often full of precious experiences, eloquent symbols, and hidden messages. These stories take time to tell. We should make ourselves available to listen to them and interpret them with patience. They are not messages like the kind you find on social media. We must learn to accept that the wisdom we need in order to live life cannot be squeezed into limits imposed on us by the current forms of communication.

LOVE TRUTH; SEEK BEAUTY

When someone sets out to find themselves, they are also seeking God. They may not find themselves, but, by seeking the truth, they set out on a path of honesty, goodness, and beauty. For me, a young person who loves and seeks the truth also loves goodness and is good. This person seeks out and loves beauty; this person is on a good path and will find God for sure! Sooner or later this young person will find Him! It is a long road, and some will not find Him in life, or at least not consciously. But those who are true and honest with themselves, good, and who deeply love beauty are also very mature and capable of an encounter with God, which is grace. An encounter with God is grace. You do not meet God because you heard about Him, nor can you pay to meet Him. It is a deeply personal journey, a path that must be taken. This is how we encounter Him.

THE HARMONY OF DIFFERENCES

As St. Paul writes: "Always be joyful . . . I repeat . . . The Lord is near" (Philippians 4:4–5). I would like to ask each of you a question that you could consider as a kind of "homework." Only you, on your own, can provide the answer to the question. With respect to joy, how are things at home? Is there joy in your family?

Dear families, you know well that the true joy we experience within our families is not superficial; it does not come from material objects, or from a life of comfort. True joy comes from profound harmony between people; it is something we feel in our hearts, something that lets us feel the beauty of being together, of helping each other along life's path. But at the base of this feeling of deep joy is the presence of God, the presence of God in our family, and His love, which is welcoming, merciful, and respectful toward all. Above all, it is patient. Patience is a virtue that God teaches us within the family, how to be loving and patient with each other. God alone knows how to create harmony from differences. When love for God is lacking, the family loses its sense of harmony; self-centeredness prevails, and joy fades. But families who live the joy of faith communicate naturally. They are the salt of the earth, the light of the world, and leaven for all of society.

THE BEAUTIFUL PATH OF LOVE

Spousal and familial love also reveal the vocation of a person to love in a unique way and forever; the trials, sacrifices, and crises of couples, and of families as a whole, represent pathways for growth in goodness, truth, and beauty. In marriage, we give ourselves completely, without calculation or reserve, and share everything, both gifts and hardship, always trusting in God's providence. This experience is one that young people can learn from both their parents and their grandparents. It is an experience of faith in God, mutual trust, profound freedom and holiness. Yes, because holiness means giving oneself with faith and sacrifice every single day! But problems in marriage can happen; people can have differing points of view, be jealous, have arguments. We need to tell young couples that they should never end the day without making peace. The sacrament of marriage is renewed in the act of making peace after an argument, a misunderstanding, a hidden jealousy, or even a sin.

Making peace brings unity to the family. It is important to tell young people and couples that this path is not easy, but it is a very beautiful path, very beautiful indeed. They need to know this!

IF THE ELDERLY DO NOT DREAM, THE YOUNG DO NOT SEE

The prophecy of Joel contains a verse that says, "I shall pour out my spirit on all humanity. Your sons and daughters shall prophesy, your old people shall dream dreams, and your young people see visions" (Joel 3:1; Acts 2:17). When young and old people alike open themselves to the Holy Spirit, a wonderful combination is produced. The old dream and the young have visions. How do the two complement each other?

The elderly have dreams woven through with memories and images of experiences they have lived through, marked by their experience and years. If young people sink their roots into those dreams, they can see into the future; they can have visions that broaden their horizons and show them new paths. But if the elderly do not dream, young people lose sight of the horizon.

THE HOLY SPIRIT IS HARMONY

At times it seems as though the Holy Spirit creates disorder in the Church because He differentiates and brings some people divine grace and gifts and not others. However, His actions are a source of wealth for all because the Holy Spirit is the Spirit of unity, which is not the same as uniformity, but leads back to harmony. In the Church, the Holy Spirit creates harmony. One of the Fathers of the Church has an expression that I love, "*ipse harmonia est*." He is harmony. Only He can bring about diversity, plurality, multiplicity, and at the same time unity.

GOOD SOLITUDE AND BAD SOLITUDE

In life, it is always important to find time to be on your own, without people around, to be by yourself so you can come face-to-face with your conscience. It is good for you. I do it sometimes. Thankfully, I do not feel lonely, in the sense of not having friends or people around me. I stay busy and this, too, is good—good solitude. It is important to be alone at some point during the day, every two or three days, and to know how to say, "I feel like being alone, I want to check in with myself and see what is happening in my life."

There is also a bad kind of solitude, which many people feel when they are not working or when their friends move far away. Bad solitude hurts, you slip into a state of melancholy and have bad thoughts, you might even feel jealousy or want revenge: "I feel lonely and want to hurt others." There is a lot of this kind of loneliness. But a little bit of good solitude, not too much of it, is good for everyone. It is important to check in on yourself, to look within. This leads to personal growth.

NEITHER ANXIETY NOR INSECURITY

We need to pursue our hopes and dreams constantly. And yet we also need to be careful of a certain temptation that can hold us back: anxiety. This can be a great enemy to us; it can lead us to give up, for example when we do not obtain instant results. Our best dreams are attained through hope, patience, and commitment, and never in haste. At the same time, we should not stop ourselves from doing something simply because we feel insecure; we must not be afraid to take chances or make mistakes. On the contrary, we should be fearful of remaining blocked, of becoming like the living dead, who do not live fully because they are afraid of taking risks, making mistakes, or following through on their commitments. If you make a mistake and fall, you can always get up and start again; no one has the right to rob you of hope.

STEPPING OUTSIDE OF OURSELVES

When an encounter with God is called an "ecstasy," it is because it takes us outside of ourselves and raises us up, overwhelming us with God's love and beauty. Yet we can also step outside of ourselves when we see the hidden beauty in other human beings, in their dignity and grandeur, how they have been made in the image of God and the Son. The Holy Spirit urges us to step outside of ourselves, embrace others with love, and seek out their goodness. This is why it is always better to live the faith together, to show our love in a community, to share our affection, time, and worries with other young people. The Church offers a wide range of opportunities for living one's faith in community; together, everything is easier.

DO NOT FORGET YOUR DIGNITY

While human beings are capable of degrading themselves in extreme ways, they are also capable of rising up, returning to the path of goodness, and starting anew, despite any and all psychological or social conditioning. We are capable of looking at ourselves honestly; we know how to acknowledge disgust and undertake new paths toward true freedom. No process can entirely suppress our openness to all that is good, true, and beautiful, or our capacity to react. This is an ability that God continues to cultivate deep in our hearts. I appeal to each one of you, all around the world: do not forget your dignity. No one can take it away from you.

THE BEAUTY OF SHARED BREAD

There is extraordinary beauty in the gathering of a family at the table, generously sharing what food it has, even if the provisions are meager. There is beauty in a slightly disheveled and no longer young wife who continues to care for her sick husband despite her own failing health and weakness. Although the springtime of their courtship has long since passed, there is beauty in the faithfulness of those couples who still love each other in the autumn of their lives, in those elderly couples who still hold hands when they take a walk. There is also a kind of beauty, one that is unrelated to appearances or fashion, in men and women who pursue their personal vocations with love, who act selflessly for their community, for their nation, who work hard at building a happy family, who advance social harmony with selfless and unrecognized efforts. To find, share, and highlight this beauty, which reminds us of Christ on the cross, is to lay the foundations for true social solidarity and the culture of encounter.

PAIN AND COMFORT

When someone is sick, we may think at times, "Let's call for the priest," but then "No!" we say. "It will bring bad luck" or "It will scare the sick person." Why do we think this? Because we believe that after the priest comes the undertaker. But this is not true. The priest comes to help the sick or elderly person; this is why the priest's visit is so important. Through the priest, Jesus arrives to assist the sick person, give him strength, give him hope, and help him, as well as to forgive his sins. And this is truly beautiful! One must not think that it is taboo. In times of pain and illness, it is good to know that we are not alone: the priest and those present during the anointing of the sick represent the entire Christian community, huddling around the suffering one and their family as one body, nurturing their faith and hope, supporting them through prayers and fraternal warmth. But the greatest comfort comes from the fact that the Lord Jesus Himself is present in the Sacrament. He takes us by the hand, He caresses us as He did the sick, and He reminds us that we already belong to Him, and that nothing—not even evil or death—will ever separate us from Him.

If we are in the habit of calling for the priest so that he can bring this Sacrament to the elderly and sick (and I am not talking about someone who merely has a cold, but a serious illness) and this act brings about comfort and the strength of Jesus to press on, let's do it!

A SHARED DREAM

To encounter another person is not the same as mimesis. It does not mean thinking the same way about things or repeating the same words that the other says: that's what parrots do. To encounter someone means something else. It means entering into the culture of encounter. It is a call and an invitation to have the courage to keep alive a shared dream. We have many differences, we speak so many different languages, we all wear different clothes, but, please, let us try and share a dream. We can do it! A shared dream does not erase us, it enriches us. We can have a great dream, a dream with a place for everyone. This is the dream for which Jesus gave His life on the cross, for which the Holy Spirit was poured out on the day of Pentecost, branding the heart of each and every person with the hope that it would find the room to grow and flourish. It is a dream named Jesus, sown by the Father with the confidence that it will grow and live in every heart. This is a concrete dream, for He was a man, and He runs through our veins, thrills our hearts, and makes us leap for joy whenever we hear the words: "love one another . . . as I have loved you. It is by your love for one another, that everyone will recognize you as my disciples" (John 13:34–5).

THREE

WHY GOD IS JOYFUL

LIKE A STAR THAT GLOWS INSIDE

These days there is an urgent need to recapture the light of faith, as once its flame dies out, all other lights begin to dim. The light of faith is unique; it has the capacity of illuminating each and every aspect of human existence. A light this powerful does not come only from us but from a more primordial source. It must come from God. Faith is born of an encounter with the living God who calls to us and reveals His love, a love that precedes us, one that we can lean on for strength, one that helps us build our lives. Transformed by this love, we see things with new eyes, we realize that this love contains a great promise of fulfillment, and a vision of the future opens up before us. Faith, a supernatural gift received from God, lights our way, guides us on our journey through time. On the one hand, it is a light that reaches us from the past, built on the foundations of the life of Jesus, which reveals His perfectly trustworthy love, a love capable of triumphing over death. On the other hand, since Christ has risen and leads us into the afterlife, faith is also a light from the future, illuminating vast horizons before us, taking us beyond our isolated individual selves and toward the expanse of communion. We come to see that faith does not dwell in shadow and gloom; it is a light for our darkness. Dante, in *The Divine Comedy*, after professing his faith to St. Peter, describes that light as the "spark that extends into a vivid flame and, like a star in heaven, glows in me."*

I would like for us to talk about this light of faith so that it will grow, enlighten our present, and become a star that will shine on the horizon of our journey, especially at this time, when humanity is particularly in need of light.

DO NOT BE A BAT IN THE SHADOWS

Some people—even among the clergy—cannot live in the light because they are too accustomed to the dark. The light is blinding; they cannot see. These are "bat-people," humans who can only move about at night. When we are in sin, all of us exist in this state: we cannot tolerate the light. "Though the light has come into the world people have preferred darkness to the light because their deeds were evil," Jesus says (John 3:19). It is more convenient for us to live in darkness. The light hits us in the face; it makes us see what we do not want to see. Even worse, the more the eyes of the soul live in darkness, the more they grow accustomed to it and become ignorant of what light actually is. Many human scandals and corruption show this to be true. Those who are corrupt do not know what the light is. They simply do not comprehend it. We, too, when we are in a state of sin, distance ourselves

* Paradise XXIV, line 145 (Mandelbaum translation).

from the Lord, become blind, and only feel at ease in the shadows. We move forward without seeing, like the blind do, as best we can.

Let us allow the love of God, who sent Jesus to save us, to enter into us. May the light of Jesus that "has come into the world" help us see things with God's light, the true light, and not with the shadows that the lord of darkness gives us (John 3:19).

The questions that we should ask ourselves every day is this: "Am I walking in the light or am I walking in the darkness? Am I a child of God, or have I become a poor bat?"

DO NOT GROW OLD BEFORE YOUR TIME

Sometimes all our youthful energy, dreams, and enthusiasm waver and we are tempted to dwell on ourselves and our problems, hurt feelings, and grievances. Don't let this happen to you! You will grow old inside before your time.

THE OUTSTRETCHED HAND OF GOD

Illness always brings big questions with it. Our first reaction to illness may be to rebel and even experience moments of bewilderment and desolation. We cry out in pain, and rightly so: Jesus suffered and in His own way. Through prayer, we can join His cry of pain. By uniting ourselves with Jesus in His passion, we discover the power of His closeness to our frailty and wounds. It is an invitation to cling to His life and sacrifice. If, at times, we feel "the bread of suffering and the water of distress," let us also pray that we find the guidance we need to feel the comfort that comes from "he who is your teacher" who "will hide no longer" but will remain by our side and accompany us every step of the way (Isaiah 30:20).

A TORCH IN THE DARKEST HOUR

The theme of caring for the sick during the most critical and terminal phases of life calls into question the Church's task of rewriting the "rules" for how we take care of the ailing person. The example of the Good Samaritan teaches us that it is necessary to rely on the gaze of the heart because often those who look do not see. Why? Because of a lack of compassion. It occurs to me that on numerous occasions when the gospel mentions how Jesus stood before a suffering person, it says, "He felt compassion." This is a refrain in the life of Jesus. Without compassion, the person who sees does not allow themselves to be involved in what they are observing and merely moves on; on the other hand, those who have a compassionate heart are touched and brought in, and Jesus takes care of them.

It is necessary to build a human platform of relationships around the ailing, a network that favors medical treatment while also remaining open to hope, especially in those extreme situations where physical illness is accompanied by emotional distress and spiritual anguish.

This relational—and not merely clinical—approach to caring for an ailing person, whom we regard as a unique and whole person, means that we should never abandon anyone with an incurable disease. Human life, because of its eternal destination, retains all its value and dignity in any condition whatsoever, even if fragile and in peril, and as such is always worthy of the highest consideration.

St. Teresa of Calcutta, who lived this way of closeness and

sharing, preserving until the very end an awareness and respect for human dignity, thus making dying more humane, used to say: "Whoever, on the journey of life, has lit even one torch in someone's darkest hour has not lived in vain."

HOW?

We are brought together in fear and uncertainty. So many troubled and broken hearts need comforting. I think of what Jesus said when He spoke about the Holy Spirit, the way He used a special word, *Paraclete* or "Consoler." Many of you have experienced His consolation, that inner peace which makes us feel loved, that gentle strength that always inspires courage, even in times of great pain. The Spirit reassures us that we are not alone, that God sustains us. Dear friends, that which we have received we must give as a gift to others: we are called upon to share the comforting Spirit and closeness of God.

How? Let us think about all the things that we long for: comfort, encouragement, someone to care for us, someone to pray for us, someone to weep with us and help us face our difficulties. "Always treat others as you would like them to treat you" (Matthew 7:12). Do we want to be heard? Let us listen. Do we need encouragement? Let us encourage. Do we want someone to care for us? Let us care for those who are alone. Do we need hope for tomorrow? Let us give hope today.

PRAY FOR THE GIFT OF HOPE

Hope is a gift of God. We must ask for it. It is placed deep within each human heart to shed light on this life, which is so often troubled and clouded over by a number of situations that bring sadness and pain. We need to nourish the roots of hope so that they can bear fruit; whatever evil we may have committed, we must be aware of the certainty of God's closeness and compassion. There is no corner of our heart that God's love cannot touch. Wherever someone has made a mistake, the Father's mercy is all the more present, awakening repentance, forgiveness, reconciliation, and peace.

THE SMILE THAT COMES FROM WITHIN

When we are in darkness and difficulty, we do not smile; hope teaches us that by smiling we can find the path that leads to God. One of the first things that happens to people who distance themselves from God is that they stop smiling. While they may laugh loudly and frequently, enjoying a joke or chuckle, their smile will be missing! Only hope brings a smile to our faces: the smile of hope that we will find God.

Life is often a vast desert. It is challenging to make our way through life, but if we place our trust in God, life becomes an easy journey along an open highway. Never lose hope. Keep believing, always, and in spite of everything. When a child stands before us, although we may have many problems and difficulties, a smile comes to us from within, we see hope in front of us: a child is hope! In this same way, we must discern the path of hope that will lead us to God, who became a Child for us. He will make us smile; He will give us everything!

THE VALUE OF WEEPING

Our world today lacks weeping. The marginalized weep, the neglected weep, the scorned weep, but those of us who lead a relatively comfortable life don't know how to weep. Certain realities of life can only be seen with eyes that have been cleansed by tears. I encourage each of you to ask yourself: Have I learned how to weep? Do I weep when I see a hungry child, a child on drugs living on the street, a homeless child, an abandoned child, an abused or exploited child? Or is my weeping only self-centered whining? This is the principal thing I would like to say: let us learn how to weep.

In the gospel, Jesus wept. He wept for His dead friend. He wept in His heart for the family that had lost its daughter. He wept in His heart when He saw the poor widowed mother burying her son. He was moved and wept in His heart when He saw crowds of people moving like sheep without a shepherd. If you don't learn how to weep, you are not a good Christian. This is my challenge to you: learn how to weep.

THROW OPEN THE DOORS OF CONSOLATION

If we want to experience His consolation, we must make room for the Lord in our lives. And for the Lord to reside within us, we must throw open the doors of our hearts and welcome Him in, not leave Him outside. The doors of consolation must always remain open, as Jesus loves to enter through them. Our daily readings of the gospel that we carry with us everywhere, our silent and holy prayers, confession, the Eucharist: these are the doors through which the Lord enters, bringing new colors to reality. But when the doors to our hearts are closed, His light cannot reach us, and we remain in the dark. Then we grow accustomed to pessimism, to things not working out, to a fixed reality. We end up closed in our sadness, in the depths of anguish, and isolated. On the other hand, if we throw open the doors of consolation, the light of the Lord shines in!

GOD ALWAYS TAKES THE FIRST STEP

Goodness attracts us, truth attracts us, life, happiness, and beauty all attract us. Jesus is at the juncture of this mutual attraction, of this dual movement. He is God and man, Jesus. God and man. Which of the two takes the initiative? God, always! God's love always comes before our own! He always takes the initiative. He waits for us, He invites us, the initiative is always His. Jesus is God made flesh, born for us. The new star that appeared to the Magi was a sign of the birth of Christ. Had they not seen the star, these men would not have set out on their journey. Their light leads and precedes us, truth and beauty precede us. God precedes us. The prophet Isaiah said that God is like the flower of the almond tree. Why? Because in those lands, the almond tree is the first to flower. And God always leads the way, He is always the one to seek us out first. He takes the first step.

IN TIMES OF ILLNESS, WE KNOW GOD "BY SIGHT"

Even when illness, loneliness, and our own inability make it hard for us to reach out to others, the experience of suffering can become a privileged means of transmitting grace and a source of gaining and growing *in sapientia cordis*, in the wisdom of the heart. We eventually understand what brought Job, at the end of his experience, to say to God: "Before, I knew you only by hearsay but now, having seen you with my own eyes, I retract what I have said" (Job 42:5–6). When people immersed in the mystery of suffering and pain accept their struggles with faith, they become living witnesses of a faith capable of embracing suffering, even though they may not be able to comprehend its full meaning.

HOPE IS A HELMET

Christian hope is a helmet. When we talk about hope, we often interpret the word in the common sense as referring to something beautiful that we desire but that may or may not be attained. We hope such-and-such will happen; hope in this instance is a desire. For example, people say, "I hope there will be good weather tomorrow!" And yet we know that there might be bad weather the following day. Christian hope is not like this. Christian hope is expecting something that has already taken place; the door is there, I hope to reach the door. What do I have to do? Walk toward the door! I am certain that I will reach the door. This is Christian hope: the certainty of walking toward something that exists, not something I hope might be there.

WHY IS GOD JOYFUL?

Chapter 15 of the Gospel of Luke contains the three parables of mercy: the lost sheep, the lost coin, and then the longest parable of all, typical of St. Luke, the parable of the father of two sons—the "prodigal" son and the son who believes he is "righteous," who believes he is saintly. All three of these parables speak of the joy of God. God is joyful. Remember this: God is joyful! And what constitutes the joy of God? The joy of God is forgiveness. Forgiveness! The joy of a shepherd who finds his little lamb; the joy of a woman who finds her coin; the joy of a father who welcomes home his lost son whom he had considered dead but who has come back to life, who has come home. Here is the entire gospel! Here! This is Christianity! And remember, this is not mere sentimentalism, this is not being a "do-gooder"! On the contrary, mercy is the true force that can save humanity and the world from the "cancer" which is sin, which is moral and spiritual evil. Only love fills that emptiness, those negative chasms that evil opens up in hearts and in history. Only love can do this, and this is God's joy!

FEAR IS A BAD COUNSELOR

A girl asks
I have my fears, but what are you afraid of?

Pope Francis replies
Of myself! Of fear! In the gospel, Jesus often says, "Don't be afraid!" He says it so many times. Why? Because He knows that fear is normal. We are afraid of life, we are afraid of facing challenges, we are afraid to come face-to-face with God. Everyone is afraid, everyone. You mustn't worry about being afraid. You have to accept that you will feel it and ask yourself, "Why do I feel afraid?" And then, facing God and yourself, try to clarify the situation or ask someone to help you do so. Fear is not a good counselor; it gives you bad advice. It pushes you down a road that is not the right one.

This is why Jesus often said, "Don't be afraid!" Then, we also have to try and know ourselves: each of us has to try and know ourselves and try to understand where we will make the most mistakes, and we need to be afraid of that area. Because there is bad fear and good fear. Good fear is like caution: it's about being cautious. "In this and this I am weak; be careful and don't fall." Bad fear is the one that cancels you out, erases you, annihilates you, and doesn't let you do a single thing. This is bad fear, and you should cast it far away.

FROM *IF* TO *YES*

The road to Emmaus is a story that begins and ends *on the move* (cf. Luke 24:13–35). On the outbound journey, the disciples, saddened by the "epilogue" of Jesus' life, leave Jerusalem to return home to Emmaus—a walk of about eleven kilometers, the journey taking place during the day, much of it downhill. And then there is the return journey back to Jerusalem—another eleven kilometers, which takes place as night falls, is partially uphill, and is tiring after both the fatigue of the outbound journey and the events of the day. Two trips: an easy, daytime one and a tiring, night-time one. And yet the first trip takes place under the veil of sadness, the second one in joy. During their outbound trip, the Lord walks beside them, but they do not recognize Him; on the return trip, they do not see Him any longer but they feel Him nearby. On the way, they are discouraged and hopeless; on the way back, they hurry to bring the good news of their encounter with the Risen Jesus to others so they can share it.

The two journeys of these early disciples tell us—Jesus' disciples today—that two opposing directions lie before us.

There is the outbound path taken by people who let themselves be paralyzed by life's disappointments and who walk in sadness. And then there is the path of those who do not put themselves and their problems first, but whom Jesus visits, whose brothers await their return so they can be cared for. This is the turning point: we must stop orbiting only around ourselves and the disappointments of the past, unmet dreams, and many bad things that have happened in our lives. So very often we tend to keep revolving around and around our own problems. We have to leave this cycle behind and move forward by accepting the greatest and truest reality of life: Jesus lives, Jesus loves me. I can do something for others. This is the greatest reality. It is a beautiful, positive, and bright reality! This is the U-turn we need to make: to go from *thoughts about me* to the *reality of my God*; we need to go from *if* to *yes*.

THE HOLY SPIRIT MAKES MIRACLES

"You must be born from above" means to be born with the strength of the Holy Spirit (John 3:7). We cannot just receive the Holy Spirit, but we can let Him transform us. Our docility opens the door to the Holy Spirit. He is the one who changes us, transforms us, brings about this rebirth from on high. This is the promise that Jesus makes, to send the Holy Spirit (cf. Acts 1:8). The Holy Spirit is capable of making miracles that we cannot even begin to imagine.

One example is the first Christian community described in the Acts of the Apostles. What is described there is no fantasy. It is a model to be followed when we meekly allow the Holy Spirit to enter into us and transform us—an ideal community, we might say. It is true that problems arise immediately afterward, but the Lord helps us see what we can obtain when we are open to the Holy Spirit, if we are docile. There is harmony in this community (cf. Acts 4:32–7). The Holy Spirit is the master of harmony: He is capable of making it happen and has made it happen. Now it needs to happen in our hearts. Many things inside us need to change to create harmony, and they can, because He Himself is harmony. And He, with harmony, creates things such as this community. As history and the Book of the Acts of the Apostles tell us, although there are many problems in the community, nevertheless this is a model for us: the Lord has permitted this model of an almost "heavenly" community to exist so that we can see the direction in which we need to go.

BEYOND NARROW SPACES

"You are light for the world" (Matthew 5:14). Light disperses darkness and enables us to see. Jesus is the light that has dispelled the darkness, yet darkness still remains, both in the world and within individuals. It is the task of Christians to disperse the darkness by radiating out the light of Christ and proclaiming the gospel. It is a radiance that stems from our words, but especially from our "good works" (Matthew 5:16). A disciple and a Christian community are lights to the world when they direct others to God, helping people experience His goodness and mercy. A disciple of Jesus is light when they manage to show their faith beyond narrow spaces, when they help eliminate prejudice and slander, when they bring the light of truth into situations spoiled by hypocrisy and lies. When they shed light. But it is not *my* light, it is the light *of Jesus*: we are merely instruments, so that the light of Jesus can reach everyone.

WITHOUT YOU, THE DARK OF NIGHT

"If He had managed to free us, if God had listened to me, if life had gone the way I wanted it to, if I had this or that . . . ," all said in a whiny tone. These "ifs" are not helpful; they are not fruitful. They help neither us nor others. They are similar to those two disciples walking to Emmaus who then shifted their doubt to certainty, from *if* to *yes* (cf. Luke 24:13–35). It is as if they are saying: "Yes, the Lord is alive, He walks with us; yes, we will continue our journey so we can announce it now, not tomorrow; yes, I can do this for the people so that they will be happier, so that people can better themselves, so I can help many people; yes, yes I can." They go from *if* to *yes*, from complaints to joy and peace, because when we complain, we are not joyful; we live in the gray, gray world of sadness. And this does not help us, nor does it allow us to grow.

How did this change of pace, going from *I* to *God*, from *if* to *yes*, take place within the disciples? By *encountering Jesus*. The two disciples of Emmaus first open their hearts to Him, then they listen to Him explain the Scriptures, and then they invite Him into their home. These are the three steps that we, too, can take. First, we need to open our hearts to Jesus, entrust Him with our burdens, hardships, and disappointments of life, entrust Him with the "ifs." Then, the second step: we need to listen to Jesus, pick up the gospel and read Chapter 24 of Luke. Third: we need to pray to Jesus with the same words used by those disciples when they invited Him in: "they pressed him to stay with them saying, 'It is nearly evening, and the day is almost over' " (Luke 24:29).

CHRISTIANS WITHOUT EASTER

There are Christians who live as if it were always Lent and never Easter. I realize, of course, that joy is not always expressed in the same way throughout one's life, especially during times of great difficulty. Joy adapts and changes; but when all is said and done, there is always a flicker of light that lives inside us. We know that we are loved infinitely. I understand the grief of people who have to endure great suffering, yet slowly and surely, we have to let the joy of faith revive us. Our faith has to be like a firm secret, even amid the greatest of distress. "I have forgotten what happiness is... This is what I shall keep in mind and so regain some hope: surely Yahweh's mercies are not over, his deeds of faithful love not exhausted; every morning they are renewed; great is his faithfulness!... It is good to wait in silence for Yahweh to save" (Lamentations 3:17, 21–23, 26).

"THANK YOU" IS A BEAUTIFUL PRAYER

We all carry joy. Have you ever thought about this? That you carry joy? Or would you prefer to be a bearer of only bad news, of things that make people sad? We are all capable of carrying joy. This life is the gift that God gave us: and it is too short to be consumed by sadness and bitterness. Let us praise God and be happy simply to exist. Let us look at the universe, let us look at beauty, and let us also look at the crosses we bear and say, "You exist, you made us like this, for you." It is important that we feel the disquiet of the heart that leads to thanks and praise of God. We are children of the great King, of the Creator, we can see His signature throughout all of creation. These days we do not safeguard creation enough, but creation is signed by God, who made it out of love. May the Lord help us understand this ever more deeply and lead us to say "Thank you." The prayer of "Thank you" is a beautiful one.

THE CHURCH, HOUSE OF CONSOLATION

When we are united in communion, the consolation of God is at work in us. We can find consolation in the Church which is *the house of consolation*. It is here that God wishes to console us. We would do well to ask ourselves: "Do I, as part of the Church, bring about God's consolation? Do I welcome others as guests and comfort those who appear tired and disillusioned?" Even when enduring affliction and exclusion, a Christian is always called on to bring hope to the hearts of those who have given up, encourage the downhearted, offer the light of Jesus, the warmth of His presence, and the restorative nature of His forgiveness. So many people are suffering, there are so many victims of injustice, and all these people live in a state of anxiety. Our hearts need anointing with God's comfort; it might not solve our problems, but it gives us the power of love, allowing us to bear pain in peace.

CONTINUE TO SURPRISE US

A fundamental element of Pentecost is astonishment. We know our God is a God of astonishment. No one expected anything further from the disciples after Jesus' death; they were a small, insignificant group of defeated orphans, lost without their Master. Instead, an unexpected event took place that astounded people: they were astonished to hear each of the disciples speaking in their own tongues and all talking about the marvels of God (cf. Acts 2:6–7, 11). The Church born at Pentecost is an astonishing community because a new message is proclaimed with God-given strength—the resurrection of Christ—and in a new language, the universal language of love. The disciples are adorned with power from on high and speak with courage; a few minutes earlier they huddled meekly, but now they speak with courage and candor, with the freedom of the Holy Spirit.

This is how the Church is meant to be: capable of astounding while proclaiming to all that Jesus Christ has conquered death, that God's arms are always open, that His patience is always there, waiting for us in order to heal and forgive us. Jesus arose and bestowed His Spirit on the Church for this very mission.

Remember: if the Church is alive, she must always surprise. The living Church must astound us. A Church that cannot surprise us is weak, sick, and dying, and needs to be taken to the emergency room as quickly as possible!

VITAL COMPONENTS

God has a large population of grandparents around the world. Nowadays, in the secularized societies of many countries, most parents do not have a Christian upbringing and living faith that they can pass on to their children in the way that grandparents can pass on to their grandchildren. Grandparents are the vital link for teaching our children and youth about faith. We must get used to including them in our pastoral outreach and relying on them as regular and vital components of our communities. They are not merely people on whom we are called to help and protect; they can be key players in our pastoral evangelical work, as privileged witnesses of God's faithful love.

A "LIST" OF DISEASES

I believe that a "list" of diseases, like those drawn up by the Desert Fathers, might serve us all.

1. The disease of thinking we are "immortal," "immune," or downright "indispensable," causing us to neglect our regular check-ups. A Curia that is not self-reflective, does not keep up with the times, and does not make the effort to be fit is a sick body. A simple visit to the cemetery will show us the names of many people who thought they were immortal, immune, and indispensable! It is the disease of the rich fool in the gospel who thought he would live forever, but it also strikes those who become lords and masters, who feel that they are above everyone else and not at the service of others (cf. Luke 12:13–21). This disease often arises from a pathological need for power or a superiority complex, stemming from a form of narcissism that lovingly gazes at its own image but does not see the image of God in the face of others, especially the weakest and those most in need. The antidote to this epidemic is the grace of realizing that we are sinners, being able to say wholeheartedly, "We are useless servants: we have done no more than our duty" (Luke 17:10).
2. Another disease is the "Martha complex," also known as being excessively busy. It is found in people who immerse themselves in work and inevitably neglect "the better

part," that of sitting at the feet of Jesus (cf. Luke 10:38–42). This is why Jesus called out to His disciples and told them to "rest for a while" (Mark 6:31). A neglect of rest leads to stress and agitation. A time of rest after completing one's work is necessary, vital, and should be taken seriously: spend time with your family, enjoy a holiday so you can recharge both spiritually and physically. We need to take a lesson from Ecclesiastes, which says that "there is a season for everything" (Ecclesiastes 3:1–15).

3 Then there is the disease of mental and spiritual "petrification." It is found in those who have a heart of stone, "stubborn people" (cf. Acts 7:51–60). It is found in those who, over time, lose their inner serenity, alertness, and daring, who hide under a pile of papers, become paper pushers, and turn "away from the living God" (Hebrews 3:12). It is dangerous to lose the sensitivity that allows us to weep with those who weep and to rejoice with those who rejoice! This is the disease of those who lose the sentiments of Jesus because, over time, their hearts grow hard and become incapable of loving the Father and our neighbor unconditionally. To be a Christian, you need to "Make your own the mind of Christ Jesus" and show sentiments of humility and generosity (Philippians 2:5).

4 The disease of excessive planning and practicality. When an apostle organizes everything down to the last detail, believing that everything will fall into place thanks to perfect planning, he merely becomes an accountant or office manager. It's true that we need to be well prepared for things, but we must not fall into the temptation of trying to box in or direct the freedom of the Holy Spirit,

which is always greater and more generous than any human plan. We contract this disease because it is always easier and more comfortable to rely on fixed, predictable, and unchanging ways when, actually, the Church shows her fidelity to the Holy Spirit to the degree that she does not try to control or tame Him. Tame the Holy Spirit! He is freshness, imagination, and newness.

5 The disease of poor coordination. When the body loses coordination and communion between its parts, it no longer functions in harmony and with balance. It becomes an orchestra that only makes noise. Its parts do not collaborate; they lose the spirit of fellowship and teamwork. When the foot says to the arm, "I don't need you," or the hand says to the head, "I'm in charge here," awkwardness and scandal are the result.

6 There is also the disease known as "spiritual Alzheimer's." This consists in losing all memory of our personal "salvation history," our personal history with the Lord, our "first love." It involves a progressive decline in the spiritual faculties and sooner or later becomes a great handicap in a person, making them incapable of doing anything on their own, except for living in a state of absolute dependence on their often imaginary perceptions. We see this at work in those who have lost the memory of their first encounter with the Lord. We see it in those who no longer understand life's meaning in "Deuteronomic" terms. We see it in those who are completely caught up in the present moment—their passions, whims, and obsessions—or in those who build walls and routines around themselves and become more and more enslaved to the idols they carved with their own hands.

7 The disease of rivalry and vainglory. When appearances, the color of our clothes, and honorific titles become our primary object, we forget the words of St. Paul: "Nothing is to be done out of jealousy or vanity; instead, out of humility of mind everyone should give preference to others, everyone pursuing not selfish interests but those of others" (Philippians 2:3–4). This is a disease that leads us to become men and women of deceit, living a false "mysticism" and false "quietism." St. Paul himself defines such people as "enemies of Christ's cross" because "they glory in what they should think shameful, since their minds are set on earthly things" (Philippians 3:18–19).

8 The disease of existential schizophrenia. This is the disease of those who live a double life, the fruit of a hypocrisy typical of a mediocre and progressive spiritual emptiness which no kind or number of academic titles can assuage. It is a disease that often strikes those who abandon pastoral service and restrict themselves to bureaucratic matters, thus losing all contact with reality and actual people. In this way they create a parallel world, one where they put aside everything they teach others, and begin to live a hidden and often dissolute life. This serious disease requires urgent and indeed emergency care.

9 The disease of gossiping, whispering, and rumor-mongering. I have already spoken about this disease a number of times, but it is never enough. This grave illness starts out simply enough, perhaps only with some small talk, but eventually takes over and transforms a person into a "sower of weeds" (like Satan) or even "a cold-blooded killer" of the reputations of colleagues and

friends. This disease is typical of cowards who lack the courage to speak directly and, instead, talk behind other people's backs. St. Paul admonishes us to do all things "free of murmuring or complaining so that you remain faultless and pure" (Philippians 2:14–15). Brothers and sisters, stay alert to the terrorism of gossip!

10 The disease of idolizing superiors. This is the disease of those who court their superiors in the hope of earning favor. They are victims of careerism and opportunism; they honor people, not God. They think only about what they can get out of things and not what they can give. They are small-minded people, unhappy and inspired only by their own lethal selfishness. Even "higher-ups" can be affected by this disease when they court their junior colleagues in order to gain their submission, loyalty, and psychological dependency; the end result is complicity.

11 The disease of indifference toward others. This happens when individuals think only of themselves and lose the sincerity and warmth of human relationships. When knowledgeable people do not share their knowledge with less knowledgeable colleagues. When we learn something and then keep it to ourselves rather than sharing it in a helpful way with others. When, out of jealousy or deceit, we take joy in seeing others fall instead of helping them up and encouraging them.

12 The disease of glumness, when grim and dour people think that to be serious or to be taken seriously they have to put on a face of melancholy and severity and treat others—especially those considered their inferiors—with firmness, brusqueness, and arrogance. In actual fact, this show of

severity and sterile pessimism is often a symptom of fear and insecurity. An apostle must make the effort to be courteous, serene, enthusiastic, and joyful, someone who transmits joy everywhere she or he goes. A heart filled with God is a happy heart and radiates infectious joy: it is so clear! Let's not lose that joyful, humorous, and even self-deprecating spirit that makes people likable even in difficult situations. A good dose of humor is healthy! We would all do well to make a habit of reciting the prayer of St. Thomas More (see p. 175). I say it every day and it does me a great deal of good.

13 The disease of hoarding, when an apostle tries to fill an existential void in their heart by accumulating material goods, not out of need, but merely to feel secure. The fact is that we cannot bring material goods with us; "the shroud has no pockets." All our earthly treasures—even if they were gifts—will never be able to fill up the void; on the contrary, they will only make it deeper and more insistent. To these people the Lord says, "You say to yourself: I am rich, I have made a fortune and have everything I want, never realizing that you are wretchedly and pitiably poor, and blind and naked too...so repent in real earnest" (Revelation 3:17,19). Accumulating goods only burdens and inexorably slows down our journeys! This reminds me of an anecdote. The Spanish Jesuits used to describe the Society of Jesus as the "light brigade of the Church," and I remember once when a young Jesuit was moving and loading up a truck with his many possessions—suitcases, books, objects, and gifts—an old Jesuit standing nearby was heard to say with a smile, "So, this is the so-called light brigade of the Church?" Moving house is a good time to check for this disease.

14 The disease of closed circles: where belonging to a group becomes more powerful than belonging to a Body and, in some circumstances, even to Christ Himself. This disease always begins with good intentions but, over time, it enslaves its members and becomes a cancer that threatens the harmony of the Body and causes immense evil—and scandals—especially to our weaker brothers and sisters. Self-destruction or "friendly fire" from fellow soldiers are the most insidious dangers. This evil strikes from within. As Christ says: "Any kingdom which is divided against itself is heading for ruin" (Luke 11:17).

15 Lastly, there is the disease of worldly profit and ostentation: when an apostle transforms his service into power and his power into a commodity in order to gain worldly profit or an even greater grasp on power. This is the disease of people who insatiably try to accumulate power and to this end are always ready to slander, defame, and discredit others, even in newspapers and magazines. Their goal is to put themselves on display and show that they are more capable than others. This disease does great harm to the Body because it leads people to justify the use of any means whatsoever to attain their end goal, often relying on the name of justice and transparency! This reminds me of a priest who used to telephone journalists just to tell them—and often to invent—private and confidential matters involving his confrères and parishioners. He was only concerned about appearing on the front page; this made him feel powerful and glamourous and caused great harm to others and to the Church. Poor soul!

Brothers and sisters, these diseases and these temptations are a danger to each Christian and every Curia, community, congregation, parish, and ecclesial movement. Do not forget that they can strike at both the individual and community level.

DO NOT BE EMPTY DOLLS!

How can young people make time for God in our frenetic society, where people are focused exclusively on being competitive and productive? More and more frequently we see people, communities, or even whole societies, with highly developed exteriors but impoverished and underdeveloped inner lives, lacking in real life and vitality. They seem like dolls, empty inside. Everything bores them. Many young people these days do not dream. A young person who does not dream, who does not take the time or make the space to dream, is a terrible thing; it means God will not be able to enter into their dreams, and that person will not live a fruitful life. Men and women have forgotten how to laugh, do not play, and have lost all sense of wonder and surprise. They are like zombies; their hearts have stopped beating. Why? Because of their inability to celebrate life with others. Listen: you will find happiness and you will be fruitful if you keep alive your ability to celebrate life with others. So many people in our world are materially rich but live like slaves to unparalleled loneliness! I often think about how many young and old people feel loneliness in our prosperous but often anonymous societies. Mother Teresa, who worked with the poorest of the poor, once said something deeply prophetic and precious: "The most terrible poverty is the loneliness and feeling of being unloved."

THE JOY THAT COMES FROM COMPASSION

The Church appreciates how God operates in other religions and "rejects nothing that is true and holy in these religions. She regards with sincere reverence those ways of conduct and of life, those precepts and teaching which...often reflect a ray of that Truth which enlightens all."* And yet we Christians are very much aware that

> if the music of the Gospel ceases to resonate in our very being, we will lose the joy born of compassion, the tender love born of trust, the capacity for reconciliation that has its source in our knowledge that we have been forgiven and sent forth. If the music of the Gospel ceases to sound in our homes, our public squares, our workplaces, our political and financial life, then we will no longer hear the melodies that challenge us to defend the dignity of every man and woman, whatever his or her origin.†

Others drink from other sources. For us, the wellspring of human dignity and fraternity lies in the gospel of Jesus Christ. It gives rise to "Christian thought and the actions of the Church, to the primacy of relationships, to the encounter with the sacred mystery of the other, and to universal communion with the entire human family, as a vocation of all."‡

* Pope Paul VI, *Nostra Aetate* (28 October 1965).

† Address of the Holy Father, 24 September 2018.

‡ Pope Francis, "Lectio Divina," Pontifical Lateran University (26 March 2019).

FOUR

FREEDOM FROM SADNESS

A WHIRLWIND OF THOUGHTS

After Jacob helps his many people and livestock to cross the river, the patriarch remains alone on the distant bank (cf. Genesis 32:23–33). He ponders what awaits him the following day. How will his brother Esau, whose birthright he had stolen, behave? Jacob's mind is a whirlwind of thoughts. And just as it is growing dark, a stranger suddenly grabs him and begins to wrestle with him. The Catechism explains this to us: "The spiritual tradition of the Church has retained the symbol of prayer as a battle of faith and as the triumph of perseverance."*

Jacob wrestles the entire night and never lets go of his adversary. In the end he is defeated when his opponent strikes his sciatic nerve, and thereafter he will always walk with a limp. The mysterious wrestler asks the patriarch Jacob for his name and then says, "No longer are you to be called

* Catechism of the Catholic Church 2573.

Jacob, but Israel since you have shown your strength against God and men and have prevailed" (Genesis 32:28–9). The wrestler changes Jacob's name, life, and attitude. Jacob then says to the wrestler, "Please tell me your name," but the wrestler does not reveal it to him; he blesses him instead. Jacob understands that he has encountered God "face to face" (Genesis 32:30–31).

Wrestling with God is a metaphor for prayer. On other occasions, Jacob had shown that he could dialogue with God. He had sensed that He was a friendly and close presence. But that night, after the lengthy struggle that nearly overwhelms him, the patriarch emerges altered. He changes in name, way of life, and personality: he becomes a different man. For once he is no longer the master of the situation, no longer a shrewd and calculating individual. God brings him back to being a true mortal, someone who trembles and fears, because during the struggle, Jacob was afraid. For once Jacob has nothing to present to God but frailty and powerlessness, in addition to his sins. And it is *this* Jacob who receives God's blessing, who goes on to limp into the promised land, vulnerable and wounded, but with a new heart.

At first, Jacob was a confident person and relied on shrewdness. He was impervious to grace and resistant to mercy; he did not know what mercy was and never thought he needed it. "I am in command here!" But God saved what had been lost. He revealed to Jacob his limits, that he was a sinner who needed mercy, and He saved him.

THE DESERT AND THE SEEDS OF GOODNESS

How often the seeds of goodness and hope that we try to sow get choked by the weeds of selfishness, hostility, and injustice! And not only around us, but within our own hearts too. We are troubled by the growing gap between rich and poor in our societies. We see signs of idolatry of wealth, power, and pleasure that come at a high cost to human lives. Closer to home, so many of our friends and contemporaries, even if surrounded by immense material prosperity, suffer from spiritual poverty, loneliness, and quiet despair. It is as if God has been left out of the picture entirely. It is as if a spiritual desert were spreading through the world. It affects the young, robbing them of hope and, in all too many cases, of life itself. And yet this is the world into which you have been called to go forth and bear witness to the gospel of hope, the gospel of Jesus Christ, and the promise of His kingdom.

WHAT FREES US FROM SADNESS?

Being of service to one another and serving each other frees us from sadness! By so doing, the Lord will free us of ambition and rivalry, which undermine the unity of communion. He frees us from mistrust and sadness, the latter being dangerous because it casts us down. It is dangerous! Be careful! He frees us from fear, from inner emptiness, isolation, regret, and complaints. Even in our communities, there is no shortage of negative attitudes that make people self-referential, more concerned with defending themselves than giving of themselves. But Christ frees us from this existential grayness, for as the responsorial Psalm 40 puts it, God is "my helper, my

Savior" (Psalm 40:17). For this reason, we disciples of the Lord, although weak and sinners—all of us!—are called to live our faith with joy and courage, enter into communion with God and with our brothers, adore God, and face life's labors and trials with strength.

WITHIN OUR STRUGGLES

God Himself is the One who takes the initiative and chooses, as He did with Mary, to enter into our homes and daily struggles, so full of anxiety and desire. And it is precisely within our cities, schools, universities, in our city squares and hospitals, that the most beautiful proclamation that we could possibly hear is made: "Rejoice, the Lord is with you!" This joy generates life, hope, and becomes flesh in the way we face the future, in our approach to others. And this joy becomes solidarity, hospitality, and mercy toward all.

CONSTANT BATTLE

A Christian life is a constant battle. It takes strength and courage to withstand the devil's temptations and proclaim the gospel. And yet this battle is sweet, for it allows us to rejoice each time the Lord triumphs in our lives.

STRONG ROOTS SO AS NOT TO FLY AWAY

I have seen young, beautiful trees, with branches reaching up to the sky, reaching as high as they can, and they look to me like a hymn to hope. Later, following a storm, I sometimes find them fallen, lifeless. The trees had spread out their branches without being firmly rooted in the earth and they fell as soon as nature unleashed her power. This is why it pains me to see young people being encouraged to build a future without first putting down roots, as if the world has just begun. It is impossible to grow without having strong roots that help us stand tall and firmly grounded. It is easy to fly away when there is nothing firm for us to hold on to, nothing to ground us.

LIVE BOLDLY

Holiness is *parrhesia*, or boldness, an impulse to spread the word and leave a mark on this world. In order to be able to do this, Jesus comes toward us and calmly and firmly says, "Don't be afraid" (Mark 6:50). "I am with you always; yes, to the end of time" (Matthew 28:20). These words enable us to go forth and serve with the same courage that the Holy Spirit stirred in the apostles, compelling them to proclaim the name of Jesus Christ. Boldness, enthusiasm, the freedom to speak out, apostolic fervor—all of these are part of the word *parrhesia*. The Bible includes this word to describe the freedom of a life open to God and others.

PEOPLE WHO MOVE MOUNTAINS

Let us, like the apostles, say to the Lord: "Increase our faith!" (Luke 17:5). Yes, Lord, our faith is small, our faith is weak and fragile, but we offer it to you as it is, so that you can make it grow. And how does the Lord reply? He says, "If you had faith like a mustard seed you could say to this mulberry tree, 'Be uprooted and planted in the sea,' and it would obey you" (Luke 17:6). A mustard seed is tiny, but Jesus says that even faith the size of a mustard seed, as long as it is true and sincere, is enough to achieve what is humanly impossible or unthinkable. And it is true! We all know simple and humble people, whose faith is so strong they can move mountains! There are mothers and fathers, for example, who face very difficult situations; or people who become ill, even gravely ill, who still manage to transmit serenity to those who visit them. These people, precisely because of their faith, do not boast about what they do, but instead they say (as Jesus instructs them to do in the gospel), "We are useless servants: we have done no more than our duty" (Luke 17:10). Many people out there have this strong and humble faith, and they do such good for all of us!

NEVER GIVE UP

One of the more serious temptations that stifles boldness and fervor is a sense of defeatism that turns us into querulous and disillusioned pessimists or "bellyachers." You cannot go into battle if you are not convinced of victory. If we start out lacking in confidence, we have already lost half the battle, and buried our talents. Even if we are painfully aware of our frailties, we have to press on without giving up, remembering what the Lord said to St. Paul: "My grace is enough for you: for power is at full stretch in weakness" (2 Corinthians 12:9). Christian triumph is always a cross, yet a cross that is also a victorious banner carried with a forceful tenderness against the assaults of evil. The evil spirit of defeatism is brother to the temptation to separate, before its time, the wheat from the weeds, and is the fruit of an anxious and self-centered lack of trust.

A LIGHT THAT NEVER FADES

There are many stars, lights that twinkle and guide us, in our lives. It is up to us to choose which ones to follow. For example, there are *flashing lights* that come and go, like the small pleasures of life: although they may be good, they are not sufficient, because they do not last long and they do not leave us the peace that we seek. Then there are the *dazzling lights* of money and success, that promise so much and so fast: they are seductive, but their intensity is blinding and causes our dreams of glory to fade into the most obscure darkness. The Magi, instead, invite us to follow a *steady* light, a *gentle* light that does not wane, because it is not of this world: it comes from Heaven and it shines. Where does it shine? In the heart.

This true light belongs to the Lord, or rather, it is the Lord Himself. He is our light: a light that does not dazzle but accompanies us and bestows on us its unique joy.

TO BE TRULY FREE

Faced with the pressure of events and trends, we would never be able to find the right path on our own, and even if we were to find it, we would never have the strength to persevere, to face the climb and the unexpected obstacles in our path. And this is where the invitation from the Lord Jesus comes in. He tells us that we can follow Him if we want to. He asks us if we want to travel with Him on the journey, not to exploit us, not to make slaves of us, but to free us. Only together with Jesus, by praying to Him and following Him, do we find clarity of vision and the strength to move forward. He loves us, He has chosen us, He gave Himself to each of us. He is our defender and big brother and will be our only judge. How beautiful it is to be able to face life's ups and downs in the company of Jesus, to have Him and His message with us! He does not take away our autonomy or freedom. On the contrary, by fortifying our fragility, He allows us to be truly free, to have the freedom to do good, the strength to continue doing it, and the capability of forgiving and asking for forgiveness. This is Jesus who accompanies us, this is the Lord!

FAMINE OF HOPE

Today our world is experiencing a tragic famine of hope. How much pain, emptiness, and inconsolable grief there is! Let us become messengers of the comfort bestowed on us by the Spirit. Let us transmit hope, and the Lord will open new paths as we journey toward the future.

I would like to share with you something about this journey. In these difficult times, I would very much like us, as Christians, to be even more closely connected, as witnesses of mercy to humanity. Let us ask the Spirit for the gift of unity; we can only spread the spirit of fraternity if we live as brothers and sisters. We cannot ask others to be united if we ourselves take different paths. So let us pray for one another; let us take responsibility for each other.

THROW YOURSELVES INTO THE ARMS OF GOD

God loves you. You may have heard this already but it makes no difference, I want to remind you of it. God loves you. Never doubt this, whatever may happen to you in life. You are, in every moment, infinitely loved.

Perhaps your experience of fatherhood was not the best. Perhaps your earthly father was distant or absent, harsh and domineering, or maybe he just was not the father you needed. I do not know. What I can tell you with absolute certainty is that you can, with total security, throw yourself into the arms of your heavenly Father, the God who gave you life and who continues to give you life at every moment of the day, and He will be there for you, while also fully respecting your freedom.

For Him, you are precious; you are not insignificant. You are important to Him.

STRENGTH MEANS NOT LOSING HOPE

We need to be strong every day of our lives, to move forward with our lives, families, faith. The apostle Paul said something we do well to heed: "There is nothing I cannot do in the One who strengthens me" (Philippians 4:13). When we face the trials of daily life, when difficulties arise, let us remember that we can do all things in Him who strengthens us. The Lord always strengthens us, He never lets our strength waver. The Lord does not try us beyond our possibilities. He is always with us. Say to yourself: "I can do all things in Him who strengthens me."

Sometimes we may be tempted to give in to laziness, or worse, discouragement, especially when faced with the hardships and trials of life. In these cases, let us not lose heart, let us invoke the Holy Spirit so that through the gift of fortitude He may lift our hearts and communicate new strength and enthusiasm to our life and our faith in Jesus.

HE WHO WANTS LIGHT SHOULD STEP OUTSIDE

He who wants the light should step out of himself and seek it, not stay inside, withdrawn, watching what happens around him, but rather, by entering into the fray. Christian life is a continuous journey made of hope, a quest, a journey which, like that of the Magi, continues even when the star momentarily fades from view. On the journey there are traps to watch out for: superficial and mundane gossip, which slows us down; paralyzing selfish whims; and the pit of pessimism that ensnares hope.

TAKE AWAY THE STONE

Still to this day, Jesus says to us: “Take the stone away” (John 11:39). God did not create us for the tomb, He created us for life, which is beautiful, good, joyful. But as the Book of Wisdom says, “Death came into the world only through the Devil’s envy” and Jesus Christ came to free us from its bonds (Wisdom 2:24).

We are thus called to remove the stones from the path of everything that hints at death. For example, the hypocrisy with which some people live out their faith is death; the destructive criticism of others is death; insults and slander are death; the marginalization of the poor is death. The Lord asks us to remove these stones from our hearts, and life will then flourish again around us. Christ lives, and those who welcome Him and follow Him come into contact with life. Without Christ, or outside of Christ, life is not just absent, it also leads to death.

THE TIME FOR COURAGE

Now is the time for courage! We need courage to steady our faltering steps, to recapture our enthusiasm that allows us to devote ourselves to the gospel, to reacquire confidence in the strength of our mission. Now is the time for courage, although having courage doesn't guarantee success. Courage is required of us in order to fight, but it does not necessarily mean we will win; to proclaim, although we shall not necessarily convert. Courage is needed to offer an alternative to the world, without ever becoming polemical or aggressive. Courage is needed so we can open up to everyone, without ever diminishing the absoluteness and uniqueness of Christ, the one Savior of all. Courage is required so that we can resist incredulity, without ever becoming arrogant. We need the courage of the tax collector in the gospel, "not daring even to raise his eyes to heaven . . . he beat his breast and said, 'God, be merciful to me, a sinner!' " (Luke 18:13). Now is the time for courage! Courage is what we need today!

SHAME CAN BE GOOD

Shame can be good. It is healthy to feel a little shame. Being ashamed can be healthy for you. In my country, when a person feels no shame, we say that he is "shameless," *sin verguenza*. But shame can do good because it makes us humbler. The priest receives confession and our shame with love and tenderness and forgives us on God's behalf. It is good to talk with a priest from a human point of view, to unburden ourselves of things that weigh heavily on our hearts. It is a way of unburdening before God, before the Church, with someone like a brother. Do not be afraid of confession! When you stand in line to go to confession, you may feel many things, even shame, but when confession is over, you feel free, strong, beautiful, forgiven, candid, happy. This is the beauty of confession! I would like to ask you to think about the last time you went to confession. When was it? Was it two days ago or two weeks ago? Two years, twenty years, forty years ago? If a lot of time has passed, do not waste another day. Go! The priest will be kind with you. You will find Jesus there, and Jesus is the kindest priest there is. Jesus welcomes you; He welcomes you with so much love. Be courageous and go to confession!

HE WHO RISKS, NEVER LOSES

The Lord does not disappoint those who take risks, and when someone takes a step toward Jesus, they discover that He is already there, waiting with open arms. Now is the time to say to Jesus: "Lord, I have let myself be deceived; in a thousand ways I have shunned your love, yet here I am once more, to renew my covenant with you. I need you. Save me once again, Lord, take me once more into your redeeming embrace."* It is

* Evangelii Gaudium, 3.

so beneficial to us to find our way back to Him after we have been lost!

PRAYER SOWS LIFE

I recall the story of an important government leader, not from recent times but from the past. He was an atheist who had no religious direction but, as a child, he had heard his grandmother pray and her words remained in his heart. At a very difficult time in his life, he recalled this: "My grandmother used to pray." And so, he began to recite the same words that his grandmother had used, and he found Jesus. Prayer is a chain of life: many men and women who pray sow life. Prayers, even small prayers, sow life. This is why it is so important to teach children to pray. I suffer when I encounter children who do not know how to make the sign of the cross. They have to be taught to make the sign of the cross properly, because it is the first prayer. Then, perhaps over time they will forget or take another path, but the first prayers a child learns remain in the heart, because they are a seed of life, the seed of dialogue with God.

KNOWING HOW TO SEE WITH THE HEART

Jesus had friends. He loved them all but He had friends with whom He had a special relationship, as one does with great friends, where there is a stronger love, a deeper trust. And very often, He spent time at the house of brethren, such as Lazarus, Martha, and Mary.

Jesus felt pain at the illness and death of His friend Lazarus. He arrived at the tomb and was deeply moved. Troubled, He asked, "Where have you put him?" and Jesus broke out in tears (John 11:34–5). Jesus, God made flesh, wept. Another occasion in the gospel tells of how Jesus wept: He wept over Jerusalem (cf. Luke 19:41–2). With what tenderness Jesus weeps! He weeps deep from the heart. He weeps with love. He weeps alongside His friends who weep. He may have wept at other times in His life, we do not know, but surely on the Mount of Olives. Jesus always weeps out of love.

He wept when He was deeply moved or troubled. Jesus' emotions are often mentioned in the gospel: "he felt sorry for them" (Matthew 9:36). Jesus cannot look at people and not feel compassion. His eyes see with His heart. Jesus sees with His eyes, but He also sees with His heart and knows how to weep.

Today, faced with a world where so many people are suffering, and not simply because of the effects of the pandemic, I have to ask myself, "Am I capable of weeping as Jesus would certainly have done and is doing now? Is my heart like that of Jesus? And if it is hard for me to weep, even if I can speak to many and do good and help, if I cannot weep, I must ask the Lord for this grace. Lord, let me weep alongside you while you weep with your people who are suffering." Many people weep today. From this altar, from the sacrifice of Jesus, a Jesus who was not ashamed of weeping, let us ask for the grace to weep.

NOW IS NOT THE TIME TO SLEEP!

Ten bridesmaids await the arrival of the Bridegroom, but He is late and they fall asleep. When it is suddenly announced that the Bridegroom is arriving, they prepare to welcome Him but only five of them, who were wise, have oil to burn in their lamps while the others, who were foolish, have none, and so their lamps remain dark. While they go to get some oil, the Bridegroom arrives, and when they return, the foolish virgins find that the door to the hall of the marriage feast is shut.

They knock on it again and again, but it is too late. The Bridegroom replies: I do not know you. The Bridegroom is the Lord, and the time of waiting for His arrival is the time of mercy and patience that He grants us before His Final Coming. It is a time of watchfulness, a time when we must keep the lamps of faith, hope, and charity lit, a time for keeping our hearts open to goodness, beauty, and truth. It is a time for living according to God, because we know neither the day nor the hour of Christ's return. What He asks of us is to be ready for the encounter, ready for the beautiful encounter with Jesus, which means watching for the signs of His presence, keeping our faith alive with prayer, taking the sacraments, and being careful not to fall asleep, not to forget about God. The life of slumbering Christians is a sad one, not a happy one. Christians must be happy. Do not doze off!

MASTERS OF OUR FEELINGS

What do I think about before I go to bed? Sometimes there's no time to think: boom! you fall asleep because you're so tired. I am surely not alone in this. I usually try to go to bed a little earlier so that I can take the time to look back at what has been going through my heart during the day. What kinds of feelings have I had? Why did I have that feeling in that particular situation? Why did I have other feelings during other situations? Why did I feel anger toward this person? Why did I feel kindly toward that other person? I try to see what is going on in my heart. This helps me a great deal. Sometimes I find that my feelings are not good, I find roots of selfishness and envy. Yes, even me! We all have so many bad things inside! But I also find good things, too, good roots. I don't want my heart to be a road where feelings come and go without my trying to understand them. No, you have to take the time. This is what I do: I take a little bit of time, just a little, less than ten minutes, and try to analyze what has gone through my heart and the significance of the feelings I experienced. I take the good things and thank life and God for them, while I try to examine what the bad things and bad feelings mean so they don't happen again. In other words, I try to look at what happened inside me during the day, to look at my feelings. This is very important. So often we focus on thoughts and say to ourselves, "I thought this or that." But what did you feel? It does so much good to be a master of your feelings—not to micro-manage them, but to understand what they mean and the messages they hold.

WHAT THE CHURCH NEEDS

I will tell you what the Church needs: it needs you, your collaboration, and even before that, your communion with me and with each other. The Church needs your courage of proclaiming the gospel in every opportune and inopportune moment, to bear witness to the truth. The Church needs your prayer for the safe journey of Christ's flock, the prayer that—let us not forget!—is the proclamation of the Word and the first task of the bishop. The Church needs your compassion especially in this time of great pain and suffering. Let us together express our spiritual closeness with the ecclesial communities, with all Christians who suffer discrimination and persecution. We must fight against discrimination! The Church needs our prayers so that it is bolstered by faith and can counter evil with goodness. This prayer of ours extends to every man and woman who suffers injustice because of his or her religious convictions.

The Church needs us to be men and women of peace and to make peace through our works, desires, and prayers. We must be artisans of peace! This is why we invoke peace and reconciliation for the people who are afflicted by violence, exclusion, and war.

A EULOGY TO DISQUIET

Our love of God and our relationship with the living Christ do not refrain us from dreaming. Nor do they require us to limit our horizons. On the contrary, this love elevates us, encourages us, and inspires us to live a better and more beautiful life. Much of the longing present in the hearts of young people can be summed up in the word "dissatisfaction." As St. Paul VI said, "In the very dissatisfaction that torments you... there is a ray of light."* Restless discontent, combined with

* Pope Paul VI, Mass to the Young People, Sydney, 2 December 1970.

the awe we experience when faced with new horizons, generates the boldness that leads you to take responsibility for your life, for your mission. This healthy restlessness, which is so typical of youth, continues to dwell in every heart that remains young, open, and generous. True inner peace coexists with this profound discontent. As St. Augustine says, "You have made us for yourself, and our hearts are restless until they rest in you."*

* St. Augustine, *Confessions*, Chapter 11.

FIVE

JOY HAS THE FINAL WORD

REJOICE!

“Rejoice and be glad,” Jesus says to those who have been persecuted or humiliated for his sake (Matthew 5:12). The Lord asks everything of us, and in return He offers us true life, the happiness for which we were created. He wants us to be saints and not to settle for a bland or mediocre existence.

CONTEMPLATE JESUS, OVERFLOWING WITH JOY!

Contemplate the happy Jesus, overflowing with joy. Rejoice with Him as with a friend who has triumphed. They killed Him, the holy one, the just one, the innocent one, and yet He triumphed in the end. Evil does not have the last word. And it will not have the last word in your lives, for you have a friend who loves you and wants to triumph in you. Your Savior lives.

If He lives, there can be no doubt that goodness will have the upper hand in our lives, and all our struggles will prove worthwhile. And if this is so, we can stop complaining and look to the future, for this is always possible with Him. This is our certainty. Jesus is eternally alive. If we hold fast to Him, we will have life and be protected from the death and violence that lie in wait for us along the way.

All other solutions will prove to be inadequate and temporary. They may be helpful for a short amount of time, but soon we will find ourselves defenseless, abandoned, and exposed to the storms of life. With Him, on the other hand, our hearts are rooted in security, which endures all.

ARE YOU HAPPY, HOLY FATHER?

A student asks
Everyone in this world seeks happiness. But I was wondering, are you happy? And why?

Pope Francis replies
I am happy! Definitely! I don't know why I am happy, maybe because I have a job, because I am not unemployed. I have an important job, the job of shepherding! I am happy because I found my path in life and this path makes me happy. And it is an easy happiness, because at this age it is not like being young, there is a difference. I have an inner peace, a profound peace, a happiness that comes with age. And despite following a path that is filled with difficulties (because there are always problems, even now), this happiness doesn't go away. This happiness sees the problems, it suffers them, and then it moves forward, it does something to resolve them. Peace and

happiness live in the deepest part of the heart. It is a grace of God, truly. A grace. I did nothing to deserve this happiness.

WHAT FILLS THE HEART?

The joy of the gospel fills the hearts and lives of all those who encounter Jesus. Those who allow themselves to be saved by Him are set free from sin, sorrow, inner emptiness, and loneliness.

POVERTY CAN BE HAPPINESS

The First Beatitude states that the poor in spirit are blessed for theirs is the kingdom of Heaven. When so many people suffer as a result of the financial crisis, it might seem strange to link poverty and happiness. How can we consider poverty a blessing?

First of all, let us understand what it means to be "poor in spirit." When the Son of God became flesh, He chose the path of poverty and self-emptying. As St. Paul said in his Letter to the Philippians: "Make your own the mind of Christ Jesus: Who, being in the form of God, did not count equality with God something to be grasped. But he emptied himself, taking the form of a slave, becoming as human beings are; and being in every way like a human being" (Philippians 2:5–7). Jesus is God who strips Himself of His glory. Here we see God's choice to be poor—"although he was rich, he became poor"—in order to enrich us through His poverty (2 Corinthians 8:9). This is the mystery that we contemplate in the crib when we see the Son of God lying in a manger, and later on the cross, where His self-emptying reaches its culmination.

The Greek adjective *ptochós*, as well as meaning materially "poor," also means "beggar." It should be seen as linked to the Jewish notion of the *anawim*, "God's poor," suggesting humility, an awareness of one's limitations, of one's existential poverty. The *anawim* trust in the Lord; they know they can count on Him.

THE UNIMAGINABLE PATH OF GOD

What does the word "blessed" really mean? Why does each of the eight Beatitudes begin with the word "blessed"? The original term does not suggest that someone has a full belly or that someone is doing well, but that a person lives in a state of grace, that they progress in God's grace and along God's path: patience, poverty, service to others, comfort. Those who move forward in these things are happy and shall be blessed.

In order to give Himself to us, God often chooses unimaginable paths, ones that might lead to our limitations, to tears, to a sense of defeat. This is the paschal joy that our oriental brothers and sisters speak of, the joy of the stigmata, the joy of He who has been through death and has experienced the power of God, and is alive. The Beatitudes always lead you to joy; they are the paths we need to take to reach joy.

ARE WE CAPABLE OF EXPERIENCING THE ESSENTIAL?

The happiness of the poor—the poor in spirit, that is—is twofold, and relates to possessions and God. In terms of material possessions, this poverty in spirit is sobriety; it is not necessarily sacrifice, but the ability to savor the essence of things and share, the ability to experience each day anew and with a sense of wonder at the goodness of things, without being weighed down by dark and voracious consumerism. (The more I have, the more I want; the more I have, the more I want: this is voracious consumerism.) This kills the soul. And the men and women who live this way, the people who embrace this attitude of "the more I have, the more I want," are not happy and will never attain happiness.

With regards to God, the happiness of the poor is a form of praise, a way of recognizing that the world is a blessing, and that at its origin lies the creative love of the Father. But it is also a way of being open, a docility toward His Lordship. He, the Lord, is the Great One, He who desired to create the world for humankind, He who desired it so that men and women could find happiness.

BAD SPIRITS (AND THE GOOD SPIRIT OF A BLIND MAN)

In Chapter 5 of the Book of John, Jesus, on arriving in Jerusalem, goes to a pool where the sick went to heal themselves. It was said that every now and then an angel came down and stirred the waters, as if it were a river, and the first person to throw themselves into the waters would be healed. Many unwell people lay nearby waiting to be healed, waiting for the water to stir: "crowds of sick people, blind, lame, paralyzed" (John 5:3).

A man was there who had been ill for thirty-eight years. Thirty-eight years, waiting to be healed! That certainly makes you think, doesn't it? It sounds a little too long to me. When someone is sick and they want to be healed, they do what they can to get the help they need, they make an effort, they try and be smart, and do things fast. But this person waited there for thirty-eight years, to the extent that he didn't know if he was ill or dead. "When Jesus saw him lying there" and knew that he had been ill for a long time, He said to him, "Do you want to be well again?" (John 5:6). The man's answer is interesting. He does not say "Yes." Instead, he complains. Does he complain about the illness? No. " 'Sir,' replied the sick man, 'I have no one to put me into the pool when the water is disturbed; and while I am still on the way, someone else gets down there before me' " (John 5:7). This is the kind of person who is always late and always has an excuse for it. "Jesus said, 'Get up, pick up your sleeping-mat and walk around.' The man was cured at once" (John 5:8–9).

Sadness is the seed of the devil

This man's attitude makes us think. Was he sick? Yes, perhaps he had some form of paralysis, but it appears that he could walk, at least a little. But mainly, this man's heart was sick, his soul was sick, he was sick with pessimism, sadness, and apathy. These were his illnesses. "Yes, I want to live, but..." he just stayed there. His reply was not "Yes, I want to be healed!" No, he wanted to complain about the others, who got there before him. His reply to Jesus' offer to heal him is actually a complaint about others. He spent thirty-eight years complaining about others... and did nothing to heal himself.

The key is his encounter with Jesus after he has been healed. "After a while Jesus met him in the Temple and said, 'Now you are well again, do not sin any more, or something worse may happen to you' " (John 5:14). The man was living in sin but not because he had done something bad. No, his sin was surviving and complaining about others. The sin of sadness is the seed of the devil, the incapability of deciding about one's own life, and instead looking at the life of others and complaining. The man does not criticize the others, he complains, as if to say, "They always get there first, I am the victim here, in this life." People like this live and breathe complaints.

Gray lives

Let us compare this man with the person who was blind from birth from Chapter 9 of the Gospel of John: with what joy and enthusiasm the latter welcomed his healing, and with what determination he went to speak with the Doctors of the Law in defense of Jesus. However, in Chapter 5, the ill man merely went to inform the Doctors of the Law: "The man went back and told the Jews that it was Jesus who had cured him" (John 5:15). This reminds me of just how many of us Christians live in a state of apathy, incapable of doing anything, always complaining about everything. Apathy is a poison; it is a fog that surrounds the soul and suffocates it. It is also a drug, because if you use it often enough, you eventually like it. And you become addicted to sadness, addicted to apathy. It is like the air you breathe. And these are rather common sins in our world today: sadness, apathy, a kind of melancholy.

It would do us good to re-read Chapter 5 of John, to better understand this disease to which we can all fall prey. Water saves us. But I cannot save myself, the paralyzed man says. Why not? Jesus wants to know. Because other people are at fault, the man says, and goes on to wait for thirty-eight years. Then Jesus heals him. We do not see the reaction of the other people around him who were also healed, but they probably take up their mats and dance, sing, give thanks, and rejoice to the world! This person, meanwhile, just keeps complaining. When the Jews say that he cannot move his mat because it is the Sabbath, the healed man retorts that the one who cured him told him that he could do so. He does not defend the one who healed him. And instead of going to Jesus to thank Him, the healed man merely points out Jesus to the Doctors of the Law and "told the Jews that it was Jesus who had cured him" (John 5:15). This individual leads a gray life, gray from the evil spirits of apathy, sadness, and melancholy.

Water and apathy
Let us think about water as a symbol of strength and life. Jesus uses water to regenerate us through baptism. And let us think about ourselves. Let us be careful not to slip into that apathy, into that so-called "neutral" sin. Because it is neither white nor black, it's hard to say exactly what it is, but it is a sin that the devil uses to annihilate our spiritual and personal life. May the Lord help us understand just how evil this sin is.

WHO TEACHES US TO LAUGH AND CRY?

Children know how to laugh and cry. When I pick some of them up and embrace them, they smile. Others see me all dressed in white and think I am a doctor and that I will give them a jab, so they cry, for no reason at all! That's the way children are: they laugh and cry, two actions that grown-ups no longer know how to do, that we "block" ourselves from doing. We forget how to laugh. Our smiles grow rigid, like cardboard, unnatural, lifeless, artificial, like that of a clown. Children, however, laugh and cry so easily. It all depends on our hearts, and often our hearts are blocked, and we have lost this capability. Children can teach us how to laugh and cry all over again. We need to ask ourselves: Do I smile naturally, easily, with sincerity and love, or is my smile artificial? Do I still cry or have I forgotten how? These are two very human questions that children teach us.

WHEN I ENCOUNTER A HOMELESS PERSON . . .

When I encounter a person sleeping outdoors on a cold night, I have many options. I could look at this bundle of clothes as an annoyance, an idler, an obstacle in my path, a troubling sight, a problem for politicians to sort out, or even like rubbish that clutters a public space. Or I could respond with faith and charity, and see in this person a human being with a dignity identical to my own, a creature infinitely loved by the Father, an image of God, a brother or sister redeemed by Jesus Christ. This is what it means to be a Christian! How else can holiness be understood except through the startling recognition of the dignity of each human being?

JOY IS NOT IN THINGS BUT IN THE ENCOUNTER

We know that material things can satisfy certain desires, make us feel certain emotions, but in the end they are superficial joys, they are not profound, intimate joys: they're a momentary tipsiness that does not really make us happy. But joy is not a brief tipsiness: it is something else entirely!

True joy does not come from owning things. No! It is born of an encounter, from having a relationship with others, from feeling accepted, understood, and loved, from accepting, understanding, and loving. It is not because of a passing fancy, but because we understand that the other is a complete person. Joy is born from the freedom of an encounter! It is hearing someone say, and not necessarily with words, "You are important to me." This is beauty. And this is precisely what God leads us to understand. He calls out to you and says, "You are important to me, I love you, I count on you." Jesus says this to each one of us! Joy is born from this! Joy comes from the feeling of Jesus looking upon us. Understanding and hearing, this is the secret of our joy. Feeling loved by God, knowing that we are not numbers for Him, but people, and hearing Him call out to us.

Becoming a priest or person of the cloth is not always our own decision. I do not trust the seminarian or novice who says, "I have chosen this path." No, I do not like that at all! It just won't do! Rather, we need to respond to a call of love. I hear something inside that moves me and I reply, "Yes." The Lord makes us understand this love through prayer, but also through signs that we encounter in our lives and the people He sets on our path. The joy of the encounter with Him and with His call lead to an opening up, not to a shutting down. The encounter leads to service in the Church.

NO GLOOMY FACES

Let us not have gloomy faces. Let us not be discontent or dissatisfied, for, as they say, "a gloomy disciple is a disciple of gloom." Just as everyone does, we too have our troubles, we have our dark nights of the soul, disappointments, infirmities; we too slow down as we grow older. But it is precisely in these things that we should find "perfect joy" and learn how to recognize the face of Christ, who became like us in all things. We should rejoice in the knowledge that we are like Him, and that He, out of love for us, did not refuse the sufferings of the cross.

THE ROAD OF THE SAINTS (AND YOUR OWN)

If there is one thing that all the saints have in common, it is that they are genuinely happy. They discovered the secret of authentic happiness, which lies deep within the soul and has its source in the love of God. This is why we say that the saints are blessed.

WITH A LITTLE HUMOR

Christian joy is generally accompanied by a sense of humor. We see this clearly, for example, in the writings of St. Thomas More, St. Vincent de Paul, and St. Philip Neri. Being ill-humored is not a sign of holiness. "Rid your heart of indignation" (Ecclesiastes 11:10). God "gives us richly all that we need for our happiness" (1 Timothy 6:17). Sadness can be seen as a sign of ingratitude, as an indication that we are so caught up in ourselves that we are unable to recognize God's gifts.

FAITH IS A STRENGTH THAT BELONGS TO THOSE WHO KNOW THEY ARE NOT ALONE

No one has ever heard of a sad saint with a mournful face. It's unheard of! It would be a contradiction. The heart of a Christian is filled with peace because they know how to place their joy in the Lord even when going through difficult moments in life. Having faith does not mean never having difficult moments, but having the strength to face those moments knowing that we are not alone. This is the peace that God grants His children.

PIETY, NOT PIETISM

If the gift of piety makes us grow in relation to and in communion with God, leading us to live as His children, at the same time it helps us to share this love with others and recognize them as our brothers and sisters. This will usher forth feelings of piety—but not pietism!—toward those around us and those we encounter every day. Why do I say "not pietism"? Because some people think that to be pious is to close one's eyes, strike a pose, and pretend to be a saint. In Piedmont we say: to play the *mugna quacia*.* This is not piety. The gift of piety means to be able to rejoice with those who are rejoicing, and weep with those who are weeping, to be able to stand by those who are lonely or in anguish, help those who have made a mistake, comfort the afflicted, and welcome and assist those in need. The gift of piety is closely linked to gentleness. The gift of piety given to us by the Holy Spirit

* Literally, a peaceful nun; figuratively, a wolf in sheep's clothing or dissembler.

makes us gentle, calm, patient, at peace with God, and places us at the service of others.

A JOY SO GREAT THAT NO ONE CAN TAKE IT AWAY

Joy! It was the first word that the archangel Gabriel addressed to the Virgin: "Rejoice, you who enjoy God's favor! The Lord is with you" (Luke 1:28). The life of those who have discovered Jesus is filled with such great inner joy that nothing and no one can take it away from them. Christ gives His followers the strength they need not to be sad or discouraged, or feel as though their problems have no solution. Sustained by this truth, Christians never doubt that actions done with love go on to generate joy, that joy is the sister to hope and is capable of breaking down the barriers of fear and opening doors on to a promising future.

BE CUNNING

One aspect of the light that guides us on the journey of faith is holy "cunning." This, too, is a virtue. It consists of a spiritual shrewdness that enables us to recognize dangers and avoid them. The Magi used this light of "cunning" when, on their way back, they decided not to pass through Herod's shadowy lands but to take another route. These wise men from the East teach us how not to fall into the traps of darkness and how to defend ourselves from the shadows that seek to invade our life. The Magi and their holy cunning protected the faith.

We also need to protect our faith and guard it from

darkness. Sometimes darkness appears to be light. This is because the devil, as St. Paul says, sometimes disguises himself as an angel of light. Holy cunning is necessary in order to protect faith, to guard it from those mesmerizing voices that exclaim, "Listen to me, this is what we are going to do."

Faith is a grace, a gift. We are entrusted with the task of guarding it through prayer, love, and charity. We need to welcome the light of God into our hearts and, at the same time, cultivate that spiritual cunning that combines simplicity with astuteness. As Jesus told His disciples, "be cunning as snakes and yet innocent as doves" (Matthew 10:16).

SISTER COMPLAINT

The apostle St. Paul says to the Thessalonians, "Always be joyful." How exactly should I be joyful? He goes on to explain: "pray constantly; and for all things give thanks" (1 Thessalonians 5:16–18). We find Christian joy in prayer and in giving thanks to God: "Thank you, Lord, for so many beautiful things!" But some don't know how to give thanks to God, people who always have something to complain about.

Once I knew a Sister who was a good woman and worked hard, but her life was one long complaint. In fact, in the convent, they called her "Sister Complaint." A Christian cannot always look for something to complain about—"That person has something that I don't!" or "Did you see what they just did?" This is not the Christian way! It is unpleasant to meet Christians with embittered, wry faces, who are not at peace. No saint ever had a mournful face! Saints always show joy on their faces, or in times of suffering, peace. The greatest suffering was the martyrdom of Jesus, and yet He always had an expression of peace on His face. He was always concerned for others: His mother, John, the thief. He was always concerned about others.

THE DEPTH OF LIFE

The person who sees things as they truly are and empathizes with pain and sorrow is capable of touching life's depths and finding authentic happiness. They find the comfort of Jesus, not of the world. Such people are unafraid of sharing the suffering of others and do not flee from painful situations. They discover that the meaning of life lies in aiding those who suffer, understanding other people's anguish, and bringing them relief. These people sense that the other is flesh of our flesh; they are not afraid of drawing near and even touching their wounds. They feel so much compassion for others that all distances vanish. In this way they can embrace St. Paul's exhortation: "be sad with those in sorrow" (Romans 12:15). Holiness is knowing how to weep with others.

SIX

ASK THE RIGHT QUESTIONS AND YOU WILL FIND THE ANSWERS

A WISE TEACHER

A wise teacher once said that the key to gaining wisdom was not so much finding the right answers but asking the right questions. You can start by asking yourself: "Do I know how to respond to things? Do I respond well to things? Do I have the right answers?" If you can answer "Yes," then well done! Bravo! But then you should ask yourself the following question: "Do I know how to ask the right questions? Do I have a restless heart that urges me to ask questions about life, about myself, about others, about God?" The right answers may help you pass an exam, but you won't pass the exam of life without the right questions!

WHO DO I EXIST FOR?

I ask myself, what is the worst form of poverty? If we are being honest, we will admit that the worst kind of poverty we could possibly face is loneliness and the feeling of being unloved. While combating this spiritual poverty is a task that belongs to all of us, it falls especially on the shoulders of the younger generations. It is your task because it requires a change in priorities and the choices we make. It implies that we recognize that the most important thing is not what we can have or acquire, but who we share things with. It is less important to focus on what I live for but for whom I live. Learn to discern among the questions. Do not ask: "What do I live for?" but instead: "Who do I live for? Who do I share my life with?" Things are important but people are essential. Without them we grow dehumanized, lose our individuality, lose our names, and become just another object, maybe stronger or better than others, but still nothing more than an object—and we are not objects! We are people. The Book of Sirach says, "A loyal friend is a powerful defense: whoever finds one has indeed found a treasure" (Ecclesiasticus 6:14). That is why it is always important to ask: "Who do I live for?" Of course, you live for God. But He wants you to live for others, too, and He has given you many qualities, skills, gifts, and charisms that are not just for you, but to share with those around you, with others. Don't just live life. Share life.

HOW DO WE FACE THE TIMES AHEAD?

I would like the essential Christian virtue of hope to inspire the way ahead.

Of course, hope requires a sense of realism. We must acknowledge the many troubling issues that confront our world and the challenges that lurk on the horizon. We must call the problems that face us by their proper names and find the courage to resolve them. We must keep in mind that our human family is scarred and wounded by an ongoing succession of increasingly destructive wars that particularly affect the poor and vulnerable. Sadly, the new year does not seem to show as many encouraging signs as additional tensions and acts of violence.

It is precisely because of this situation that we must hold on tightly to hope. And it takes courage to have hope. It means acknowledging that evil, suffering, and death will not have the final word, and that even the most complex questions can and must be faced and resolved. Hope is the virtue that inspires us and keeps us moving forward, that gives us wings with which to fly, even when obstacles seem insurmountable.

ALL RESTLESS, ALL SEEKING

The search for happiness belongs to people of all eras and ages. God has placed an irrepressible desire for happiness and fulfillment in the heart of every man and woman. Have you not noticed how restless your hearts are? How they are always searching for a treasure that might quench your thirst for the infinite?

WHO ARE THE RIGHTEOUS?

The vision of Heaven described in the Book of the Apocalypse is so very beautiful, with the Lord God, beauty, goodness, truth, tenderness, and love in all its fullness. This is what awaits us. Those who have gone ahead and died in the Lord are there. They proclaim that they have been saved not by their works, though they surely did good works, but by the Lord: "Salvation to our God, who sits on the throne, and to the Lamb!" (Revelation 7:10). It is He who saves us, it is He who at the end of our lives takes us by the hand, like a father, and leads us to the Heaven where our ancestors rest. One of the elders asks: "Who are these people, dressed in white robes, and where have they come from?" (Revelation 7:13). The elder wants to know who these righteous ones are, who are these saints in Heaven? And the reply is: "These are the people who have been through the great trial; they have washed their robes white again in the blood of the Lamb" (Revelation 7:14).

We can enter Heaven thanks to the blood of the Lamb, the blood of Christ. Christ's own blood has justified us and opened the gates of Heaven for us. And if, on the first day of November, we remember our brothers and sisters who have gone before us in life and who are in Heaven, it is because they have been washed in the blood of Christ. This is our hope, Christ's blood! It is a hope that does not disappoint. If we walk with the Lord in life, He will never disappoint us!

HOW CAN WE MAKE HOPE A PART OF OUR DAILY LIVES?

The dizzying pace in which we are forced to live might seem as though it would rob us of hope and joy. The pressures and powerlessness we experience when faced with so many difficult situations might seem as if they would dry up our souls and make us insensitive to the countless challenges we face. Paradoxically, although we are rushing about building a better society, at least in theory, in the end we have no time for anything or anyone. We lose time that we could spend with our family, communities, friends, in acts of solidarity and remembering.

We would do well to ask ourselves this: "How can we live the joy of the gospel in our cities today? Is Christian hope possible in the here and now?"

These two questions lie at the wellspring of our identities, at the hearts of our families, at the crux of our towns and cities.

HERE AM I, SEND ME!

Illness, suffering, fear, and isolation are challenges we frequently face. Poverty challenges us: the poverty of those who die alone, who have been abandoned, who have lost their jobs and income, who have neither home nor food. Being forced to observe social distancing and stay at home offers us the chance to rediscover our need for social relationships and our need for a communal relationship with God. Far from increasing mistrust and indifference, this particular situation we are living in should make us even more attentive to how we relate to others. Prayer, through which God moves and touches our hearts, opens us up to our brothers' and sisters' needs for dignity and freedom, as well as to the needs of creation.

The impossibility of gathering together as a Church to celebrate the Eucharist has led us to share the experience of the many Christian communities that cannot celebrate Mass every Sunday. Throughout all of this, we hear anew God's question: "Whom shall I send?" He awaits our generous and convincing reply: "Here am I, send me" (Isaiah 6:8). God continues to look for those whom He can send forth into the world to bear witness to His love, His deliverance from sin and death, and His liberation from evil (cf. Matthew 9:35–8; Luke 10:1–12).

WHAT AWAITS ME?

None of us knows what life will bring us. Sometimes we do bad things, even terrible things, but please, do not despair: the Father is always waiting for us! "Come back!" These are His words: "Come back!" I will go home because my Father is waiting for me. And the greater the sinner I am, the greater the celebration will be. To all the priests who read these words: please embrace the sinners and be merciful. God never gets tired of forgiving; He never gets tired of waiting for us.

GOD CHANGES FEAR INTO HOPE

When is the identity of Jesus revealed in the gospel? When the Roman centurion says, "In truth this man was Son of God" (Mark 15:39). He says it right after Jesus has given His life on the cross, there are no two ways about it: he sees that God is *love omnipotent* and nothing further. Love is His nature. This is how He is made. He is Love.

You might object: "What can I do with such a weak God, one who dies? I'd rather have a strong God, a powerful God." But as you know, power fades while love remains. Only love safeguards our life, embraces our fragility, and transforms it. At Easter, it is God's love that heals our sin with His forgiveness, He who turned death into a passage of life, who changed our fear into trust, our anxiety into hope. Easter shows us that God can transform everything into goodness. With Him, we can trust that everything will work out in the end. And this is not an illusion. Jesus' death and resurrection is not an illusion. It was truth! This is why on Easter morning we are told: "There is no need for you to be afraid" (Matthew 28:5). And the anxiety-inducing questions about evil do not fade immediately but find in the Risen One a solid foundation that saves us from shipwreck and drowning.

WHAT ABOUT ME? WHICH TABLE DO I WANT TO SIT AT?

Looking around, we see so many offers of food that do not come from the Lord, many of which appear to be much more satisfying. Some people nourish themselves with money, others with success and vanity, others with power and pride. But the food that truly nourishes and satiates us is that which the Lord gives us! The food the Lord offers us is different from other kinds, and because it might not seem as tasty as other dishes the world offers us, we dream of different dishes, just as the Hebrews in the desert did when they longed for the meat and onions they used to eat in Egypt. But they forgot that when they consumed those meals, they sat at the table as slaves. In moments of temptation, their memory felt real, but it was an unhealthy one, a selective memory. A memory of enslavement, not a free one.

Today, each of us can ask: "What about me? Where do I want to eat? Which table do I want to sit at to be nourished? At the Lord's table? Or shall I eat more flavorful foods but as a slave?" We may also ask ourselves: "What memories do I have? Those of the Lord who saves me, or those of the meat and onions of slavery? Which recollection quenches my soul?"

The Father tells us that He fed us with manna, which our ancestors did not know (cf. Deuteronomy 8:16). Let us reawaken this memory. This is our task, to reawaken that memory. And let us learn to recognize the false bread that comes from selfishness, self-reliance, and sin, and see how it deceives and corrupts us.

WHY DO YOU LIVE THIS WAY?

When faced with a non-believer, the last thing I want to do is try and convince them of my faith. The last thing I will do is talk. I must live in a manner that is consistent with my faith. And by bearing witness, the curiosity of the other person will awaken. Eventually, they will question: "Why do you live this way?" And then I can begin to talk, only then I will speak. But never, ever proselytize the gospel. If someone claims to be a disciple of Jesus and wants to attract you through proselytizing, he is not a true disciple of Jesus. We do not proselytize. The Church does not grow through proselytizing. As Pope Benedict once said, the Church grows through attraction, by bearing witness. Soccer teams or political parties can proselytize, but faith does not proselytize. If someone asks me: "But why do you live this way?" I reply, "Read, read the gospel; this is my faith." And I say this without any pressure.

COME BACK TO THE FATHER'S EMBRACE

When I read or listen to Chapter 14 of the prophet Hosea, where it says "Israel, come back to Yahweh your God," I am reminded of a song by Carlo Buti from seventy-five years ago (Hosea 14:2). All the Italian families in Buenos Aires used to sing this song: "Come back to your Father. He'll sing you a lullaby." Come back: your Father is calling you to come back. God is your Father, He is not a judge, He is your Father. Come back home, listen. And the memory of that song (I was still just a boy) leads me straight to the father in Chapter 15 of Luke: "While he was still a long way off, his father saw him and was moved with pity" (Luke 15:20). This was the son who had taken everything and "left for a distant country where he squandered his money on a life of debauchery" (Luke 15:13). For the father to see him coming from a distance means that he was waiting for him: he went up onto the roof a number of times each day! For weeks, months, even years. He had been waiting for his son and now saw him from afar. Come back to your Father, He is waiting for you. This is God's tenderness that speaks to us, especially during Lent. This is the time for returning to ourselves, for remembering our Father, and for going back to Him.

We might say things like, "I'm ashamed to come back, Father," or, "I've done too many bad things." What does the Lord say? "I shall cure them of their disloyalty, I shall love them with all my heart, for my anger has turned away from them. I shall fall like dew on Israel, he will bloom like the lily and thrust out roots like the cedar of Lebanon" (Hosea 14:5–6). Come back to your Father, the God of tenderness will heal us. He will heal so very many of life's wounds and all the trouble we have made. Everyone has done something!

Let us think about going back to God as if we were returning to a warm embrace with our Father.

FAITH AND PERSEVERANCE

Faith and perseverance go together: if you have faith, the Lord will certainly grant you what you ask. And if the Lord makes you wait, well then, keep knocking at the door and in the end the Lord will grant you grace. But remember, the Lord does not do it to make Himself desired or because He thinks it's "better to wait." No, He does it for our own good, so that we take it seriously. He wants us to take prayer seriously and not be like parrots who simply repeat words: blah blah blah, not feeling anything. Jesus Himself reprimands us: "In your prayers do not babble as the gentiles do, for they think that by using many words they will make themselves heard" (Matthew 6:7). No, we must persevere; this is faith.

CAN YOU REALLY LOVE SELFLESSLY?

A student asks

I would like to ask you how to deal with the kind of suffering that comes after performing a good deed, a noble deed. In my soul, I always desire some kind of recognition. Why is it that I cannot give something without wanting something in return? Do you think I will ever be able to love without ulterior motives? Does selfless love really exist?

Pope Francis replies

It's true that when we do good acts there is always the danger that we do them for some kind of personal interest, either for recognition or reward. It is very difficult to travel the road of selflessness and perform selfless acts. The only way to accomplish them is through love. Those who love do not love out of self-interest. Love is its own reward. Love is a great word. I would almost say that loving is the greatest thing we can do. Think, for example, of parents who sacrifice themselves for their children and do not ask for a reward, or children who sacrifice themselves for their elderly parents, so they can feel affection and have everything they need. They do this with tenderness. Friends sacrifice themselves for other friends out of love. The path of love is the only thing that can assure us that we are not being selfish. But it takes work to get there; it takes maturity and generosity. Generally, we always show a little self-interest; we are always looking for something. But the very fact that you asked this question means that you feel this concern, which is already a great thing. It means that you are examining yourself: "Am I doing this in order to achieve something for myself or not? Am I trying to get someone else's slice of pie?" Just asking this question shows maturity.

Ask yourself these questions as they are the questions of life. All of us—each and every one of us, me included!—are a little selfish and do things out of self-interest. I used to know someone who was very selfish. Her classmates called her, in Spanish, *Yo, me, mi, conmigo, para mi*, "I, me, mine, with me, for me." In other words, self-centered. People who have a self-centered character will never do well in life. They are bitter people. They look out only for themselves and, if someone else has greater success than they do, they become bitter and want more. This is why the path of love is a difficult one. It requires us to cut back our bad attitudes, not the good ones.

The path of generosity assists us; we need to let other people speak and listen. Some people begin to talk even before the other has finished. Instead, we need to let people talk, to give other people room. We may meet someone who is boring. We all know boring people. It can be tiring listening to some people, but we have to listen patiently. All these things cut back on our tendency to take away from others and guide us toward generosity. This is the way of love. And it is paved with small sacrifices.

LET US BE COMFORTED

The Lord comforts with closeness, truth, and hope. These are the three ways that the Lord offers His consolation.

He is always close, never distant: "I am here." These are beautiful words: "I am here." "I am here, with you." And very often He is with us in silence. We know that He is there. He is always there. That closeness is God's style. Even in the incarnation, He makes Himself close by. The Lord comforts us with His closeness. He does not use empty words. On the contrary, He prefers silence. This is the strength of His closeness, His presence. He speaks little, but is close at hand.

A second way that Jesus comforts us is through truth: Jesus is truthful. He does not say things that turn out to be lies. He does not say things like, "Stay calm, it will pass, it will be all right, it will work out." No. He tells the truth. He does not hide the truth. He Himself says, "I am the Way; I am Truth" (John 14:6). And the truth is that He will leave, which is to say, He will die. "I am going now to prepare a place for you" (John 14:2). We are all faced with death; this is the truth. And yet He says it so simply, so gently, without wounding us, without hurting us. But we are faced with death. He does not hide the truth.

And this is the third way that Jesus comforts us: with hope. Yes, these are difficult times, but He also says, "Do not let your hearts be troubled. You trust in God, trust also in me" (John 14:1). Allow me to tell you something: Jesus says, "In my Father's house there are many places to live in; otherwise, I would have told you. I am going now to prepare a place for you" (John 14:2). He is the first to go. He opens the doors for us, the doors through which we will all pass, or at least I hope we will. "I shall return to take you to myself, so that you may be with me where I am" (John 14:3). The Lord returns each time one of us leaves this world behind. "I will come and take you with me." This is hope. He will come and take us by the hand and bring us with Him. He does not say things like, "No, you will not suffer, it's nothing." No, He tells the truth. "I am near you. Here is the truth: it is a bad time, one of danger and death. But do not let your heart be troubled, be at peace, for that peace is at the base of all consolation; I will come and I will take you by the hand and lead you to where I will be."

It is not easy to allow ourselves to be comforted by the Lord. In difficult times, we often get angry with the Lord and do not allow Him to come to speak to us like this, with this tenderness, closeness, gentleness, truth, and hope.

TWO RICHES THAT NEVER FADE

God questions us about the meaning of our lives. Using an image to help us understand, He offers us a "strainer" through which our life can be poured, reminding us that almost everything passes in this world, like running water. But certain treasured realities are left behind, like gems in a strainer. What endures? What has value in life? What riches do not disappear? Two, for sure: the Lord and our neighbor. These two riches never fade! These are the greatest goods and are meant to be loved. Everything else—the heavens, earth, beauties of all kind—passes; but we must never forget about God or our neighbor.

A NEW AGE FOR SALVATION

The resurrection is not simply an event of past history to be remembered and celebrated; it is much more. It is the saving proclamation of a new age that resounds and bursts onto the scene: "now it emerges; can you not see it?" (Isaiah 43:19). The future is the advent that the Lord calls on us to build. Faith grants us a realistic and creative imagination, capable of leaving behind all forms of repetition, substitution, and maintenance. An imagination that calls on us to bring about an ever new time: the time of the Lord. Although the presence of an invisible, silent, far-reaching virus has thrown us into crisis and turmoil, let this other discreet, respectful, and non-invasive presence teach us anew how to face reality without fear. If an impalpable presence has managed to disrupt and upset our priorities and the apparently overpowering global agendas that are suffocating and devastating our communities and our sister earth, let us not be afraid to let the presence of the Risen Lord point out our path, open new horizons, and grant us the courage to live this unique moment of history to the fullest. A handful of fearful folk were able to change the course of history by courageously proclaiming the God who is with us today. Do not be afraid now!

A LIFE IMMERSED IN LIGHT

We live in a provisional reality, one that ends. On the other hand, in the afterlife, after the resurrection, death will no longer be at the horizon and we will experience everything, even human bonds, through God's dimension, in a transfigured way. Even marriage, a sign and instrument of God in this world, will shine brightly, transformed in the full light of the glorious communion of saints in Paradise.

The "children of the resurrection" are not just a few privileged ones, but all men and women, because the salvation brought by Jesus is for all of us (Luke 20:36). And the life of the risen shall be "the same as the angels," meaning wholly immersed in the light of God, completely devoted to His praise, in an eternity filled with joy and peace (Luke 20:36). But watch out! Resurrection not only means rising up after death; it is a new genre of life that we can already experience today. It is the victory over nothingness, and it is at our door. Resurrection is the foundation of faith and Christian hope!

WAIT FOR THE FUTURE BY LIVING THE PRESENT

I would like to remind you that when Cardinal Francis Xavier Nguyen Van Thuan was imprisoned in a concentration camp, he did not spend his days merely waiting to be freed. He chose to "live the present moment, filling it to the brim with love" and proclaimed, "I will seize the occasions that present themselves every day; I must accomplish ordinary actions, in an extraordinary way."* As you work to achieve your dreams, make the most of each day, and do your best to let each moment overflow with love. While this youthful day may well be your last, it is worth making the effort to live it as enthusiastically and fully as possible.

* Venerable Francis Xavier Nguyen Van Thuan († 2002) was imprisoned by the Vietnamese government for thirteen years. This prayer is from *Five Loaves, Two Fish*, Tinvui Media, Tr., copyright © 1997 (Edizioni San Paolo, Pauline Books & Media/The Daughters of St. Paul, Boston, MA).

SEVEN

BE HOPE

EVERYTHING IS CONNECTED

When our hearts are truly open to universal communion, nothing and no one is excluded from this sense of fraternity. Consequently, all indifference or cruelty toward fellow creatures of this world eventually affects how we treat other human beings. We have but one heart, and the wretchedness that leads us to mistreat an animal will reveal itself in how we interact with other people. All acts of cruelty toward creatures are considered "contrary to human dignity."* We can hardly consider ourselves to be loving individuals if we are cruel or ignore a certain element of reality: "Peace, justice, and the preservation of creation are three interconnected themes which cannot be separated and treated individually without once again falling into reductionism."† Everything is

* Catechism of the Catholic Church 2418, later quoted in Encyclical Letter *Laudato Sì*.

† From a Pastoral Letter written in 1987 at the Conference of Dominican Bishops, later quoted in Encyclical Letter *Laudato Sì*.

connected; human beings are like brothers and sisters on a wonderful pilgrimage, united by God's love, which extends out to all His creatures and links us affectionately to Brother Sun, Sister Moon, Brother River, and Mother Earth.

HOW CAN WE *NOT* WORK TOGETHER?

As Christians, we are united by faith in God, the Father who gives us life and loves us so deeply. We are united by faith in Jesus Christ, the one Savior, who set us free with His precious blood and glorious resurrection. We are united by our desire to hear His word and be guided in our steps. We are united by the fire of the Spirit that sends us forth. We are united by the new commandment that Jesus left us, to build a civilization based on love, and by a shared passion for this kingdom that the Lord calls on us to build with Him. We are united by the struggle for peace and justice. We are united by the conviction that not everything ends with this life, but that we will be called to the heavenly banquet where God will wipe away our tears and see all that we did for those who suffer.

All this unites us. How can we *not* struggle together? How can we *not* pray and work side by side to reveal the sacred countenance of the Lord and care for His work of creation?

BUILDING A SHARED HOME

The world is a joyful mystery to be contemplated with gladness and praise. Our urgent challenge is to protect our shared home as it concerns the entire family of humanity. We need to work together to seek a sustainable and integrated path forward, as we know that things can change. The Creator does not abandon us; He never abandons His loving plan or regrets creating us. Humanity can still collaborate. We can still build our shared home together.

TRANSFORM YOURSELVES INTO HOPE

While to the eyes of the world a boy or girl may count for little or nothing, in God's eyes they are an apostle of the kingdom, a sign of hope! I would like to ask all young people one thing: Do you want to be a sign of hope for God? Do you want to be hope for the Church?

A young heart that welcomes Christ's love becomes hope for others and is a powerful force! But young people must learn to transform both us and themselves into hope! Open the doors to a new world of hope. This is the task of young people.

Let us reflect on the significance of the many young people who have encountered the Risen Christ, bring His love to everyday life, live it, and communicate it. They won't appear on the front pages of newspapers because they won't be the ones to perpetrate acts of violence. They won't cause scandals and so they won't make news. However, if they remain united with Jesus, they will build His kingdom. Their unity will lead to brotherhood, acts of sharing, works of mercy. They are a powerful force and can make the world a more just and more beautiful place. They can transform it! I would like to ask all the young men and women of the world: Do you have the courage to take this challenge?

Are you ready to be the force of love and mercy that will be brave enough to want to transform the world?

Dear friends, this is the real news we want to read. This is the Good News, even if it will not appear in the papers or on the television: God loves us, He is our Father, and He sent His Son Jesus to stand by each of us and offer us salvation.

THOSE WHO SUFFER CARRY THE LIGHT

The light of faith does not erase the sufferings of this world. Many men and women found their faith through suffering: the suffering have always borne the light! So it was with St. Francis of Assisi and the leper or with Blessed Mother Teresa of Calcutta and her poor. These two figures understood the mystery of suffering. By standing by people who suffered, they did not take away all their pain, they were not able to explain all acts of evil. Faith is not a light that eliminates all darkness but a lamp that guides us on our way during the night, one that suffices for the journey. God does not provide those who suffer with a reason for everything; His response is His presence, a history of goodness that touches each instance of suffering and brings a ray of light. Through Christ, God Himself wishes to share this path with us and offer us His gaze so that we might see the light that lies within it. Christ is the one who, having endured suffering, "leads us in our faith and brings it to perfection" (Hebrews 12:2).

Suffering reminds us that faith brings hope to the common good. This is its service and this hope looks to the future. It knows that through God, through the Risen Jesus, our society will find a solid and lasting foundation. And in this way, faith is linked to hope. Even if earth, our home, is slowly falling into decay and being destroyed, we have an eternal dwelling place that God has already prepared for us in Christ, in His body (cf. 2 Corinthians 4:16; 5:5). The dynamic of faith, hope, and love (cf. 1 Thessalonians 1:3; 1 Corinthians 13:13) leads us to embrace the worries of all human beings undertaking the journey toward "the well-founded city, designed and built by God" (Hebrews 11:10). For "perseverance... gives us hope, and a hope which will not let us down" (Romans 5:4–5).

WILL WE COME OUT OF THE PANDEMIC BETTER PEOPLE?

Before us lies the duty to build a new reality. We can help, but really, the Lord will do it for us: "I am making the whole of creation new" (Revelation 21:5).

When we leave behind this pandemic, will we pick up where we left off? Continue doing what we used to do and how we used to do it? No, everything will be different. All this suffering will be for nothing if we cannot build a society that is more just, more equitable, more Christian—and not only in name, but in fact. We need to build a world that is based on Christian conduct. If we do not fight to put an end to the pandemic of poverty that exists across the globe, in each of our countries, in our cities, then this era will have been in vain.

From the great trials of humanity, including this pandemic, we either come out better or worse. We certainly will not come out unchanged.

I ask you: How do you want to come out of this? Better or worse? This is why we open our hearts to the Holy Spirit, so that He can change them and help us come out of it better.

If we do not live in the way that Jesus teaches us—"For I was hungry and you gave me food... I was a stranger and you made me welcome ... in prison and you came to see me" (Matthew 25:35–6)—we will not come out of this better.

ONLY LOVE CAN EXTINGUISH HATRED

While Jesus was experiencing the agony of the cross, just before dying, while living the supreme moment of suffering and love, many people around Him taunted Him cruelly with the words: "save yourself; come down from the cross!" (Mark 15:30). Saving oneself is a great temptation and spares no one, including us as Christians. It is the temptation to think only of saving ourselves and our circle, to focus on our problems and interests, as if nothing else mattered. It is a very human instinct, but a wrong one. This was the final temptation of the crucified God.

Save yourself. "The passers-by" said these words first, they "jeered at him" (Mark 15:29). They were ordinary people. They had heard Jesus preach and witnessed His miracles. Now they taunt Him as if to say, "Show us how you get down from the cross." They have no pity, they just want miracles; they want to see Jesus make His way down from the cross. Sometimes we wish we could see a god that works wonders rather than one who is compassionate, a god who is powerful in the eyes of the world, who shows his might and scatters to the wind those who wish us ill. But this is not our God; these feelings come from within ourselves. How often we wish that God would become like us, rather than us having to become like God. We prefer to worship at the altar of ourselves rather than that of God. Such worship grows and feeds on *indifference toward others*. Those passers-by were only interested in Jesus for their own satisfaction. When He was merely an outcast hanging on a cross, He was no longer of interest to them. They could see Him with their eyes, but not with their hearts. Indifference kept them far from the true face of God.

Save yourself. The chief priests and the scribes were the next people to utter those words. They were the ones who condemned Jesus, for they saw Him as dangerous. It is easy to crucify others when it comes to saving ourselves, we all know how. Jesus allowed Himself to be crucified to teach us not to blame others. "The chief priests and the scribes mocked him among themselves in the same way with the words, 'He saved others, he cannot save himself'" (Mark 15:31). They knew Jesus and remembered how He had healed others and performed miracles, coming to a cruel conclusion: they see saving others as useless, they think that Jesus, who gave Himself unreservedly for others, was lost! The mocking tone of their accusation is garbed in religious language, twice using the verb "to save." But the "gospel of save yourself" is not the gospel of salvation. Making others bear the cross is the most false of the apocryphal gospels. The true gospel bids us to take up the cross of others.

Save yourself. Even those who were crucified alongside Jesus taunted Him. How easy it is to criticize others, speak badly about others, point out the evil in others but not in ourselves, and even blame the weak and the outcast! But why did the other crucified people get angry with Jesus? Because He did not take them down from the cross. They said to Him: "Are you not the Christ? Save yourself and us as well" (Luke 23:39). They wanted Jesus to resolve their problems and that's all. But God does not come to free us from our ever-present daily problems, which will always re-present themselves, but rather to liberate us from the real problem, which is our lack of love. This is the primary cause of our personal, social, international, and environmental ills; thinking only of ourselves is the father of all evils. But one of the criminals crucified alongside Jesus looks over at Him and sees humble love. This man was granted access to paradise for thinking not of himself but of Jesus, the man on the cross next to him (cf. Luke 23:42).

Calvary was the site of a great "duel" between God, who came to save us, and humans, who only want to save themselves. It was a battle between faith in God and worship of self, between the person who accuses and God who excuses. In the end, God's victory was revealed. His mercy descended over earth. From the cross came forgiveness and fraternal love was reborn. "The Cross makes us brothers and sisters."* Jesus' open arms on the cross are a symbol of this change. God does not point a finger at any one person but embraces all. Love alone extinguishes hatred. Love alone can triumph over injustice. Love alone makes room for others. Love alone is the path toward our complete communion.

BEGIN THE JOURNEY!

This message is for our youth, whose gentle age and open-minded vision of the future make them truly generous. Their uncertainties, worries about the future, and the problems they have to encounter on a daily basis risk paralyzing their enthusiasm and shattering their dreams, to the extent that they begin to believe it cannot be worth the effort of getting involved, that the God of the Christian faith is somehow limiting their freedom. Dear young friends, do not be afraid

* Benedict XVI, Address at the Way of the Cross at the Colosseum, 21 March 2008.

to step outside yourselves and begin the journey! The gospel is the message that brings freedom to our lives, transforming them, and making them even more beautiful.

CHOOSE THE HORIZON

"The women came quickly away from the tomb and ran to tell his disciples" (Matthew 28:8). God always begins with women, always. They lead the way. They do not doubt: they know. They saw Him, they touched Him. They also saw the empty tomb. It is also true that the disciples did not believe the women. The men probably made comments about the women, that their imaginations were too vivid, I do not know; but in any case, the men had their doubts. And yet the women were certain. Thanks to them we are on this path today: Jesus is risen, He is alive among us (cf. Matthew 28:9–10)!

And then there were other issues. Some people thought, "This empty tomb will bring us many problems," so they tried to hide it. As always, when we do not serve the Lord God, we serve that other god—money. Let us remember what Jesus said: there are two masters, the Lord God and the lord money. A person cannot serve two masters. And to distance themselves from this fact, this reality, the priests and the Doctors of the Law chose the other path, the one offered to them by the lord of money. Their silence was bought. They were paid to be silent witnesses (cf. Matthew 28:12–13).

One of the centurions confessed as soon as Jesus died: "In truth this man was Son of God" (Mark 15:39).These poor men did not understand and were afraid because their lives were at stake. So they went to the priests and the Doctors of the Law. And they received money: they were paid to keep silent. The guards were paid so that they would say that the disciples had stolen the body of Jesus. They were paid to be silent. This, my dear brothers and sisters, is not just a bribe, this is corruption, pure and simple. If you do not confess to Jesus Christ the Lord, ask yourself why: ask about the seal on your own sepulcher, if there is corruption. Many people do not confess to Jesus because they do not know Him. This is because we have not talked about Him consistently, which is our fault. But when one takes this road in the presence of evidence, it is the road of the devil, the road of corruption.

May the Lord, both in our personal and social lives, help us choose proclamation, for it is like a horizon, always open. May He lead us to choose the well-being of the people and never fall into the tomb of the god of money.

A REGENERATING FIRE

A person needs to mature into an adult without losing the values of youth. Each stage of life offers its own permanent grace, each has its own enduring value. A well-lived youth always remains in our heart, continuing to grow and bear fruit throughout adulthood. Young people are naturally attracted by the infinite horizon that opens up before them, and a risk of becoming an adult, with all its securities and comforts, is that this horizon will shrink and lose all the excitement it triggered in youth. Instead, the very opposite should happen. As we mature, grow older, and build our lives, we should strive to retain that enthusiasm of youth; we should always feel the amazement of that ever-greater sense of openness. We can rediscover and increase our sense of youthfulness at any point in our lives.

When I began my ministry as Pope, the Lord broadened my horizons and granted me a renewed sense of youth. The same thing can happen to a couple that has been married for many years or to a monk in his monastery.

Some things need to settle as we get older, but maturity can coexist with a youthful heart, with that regenerating fire.

FAILURE AND AWARENESS IN THE FAMILY

How much pain exists in families when one of their members—often a young person—is hooked on alcohol, drugs, gambling, or pornography! So many people have lost the meaning of life, their prospects for the future, or hope! So many people are plunged into this poverty by unjust social conditions or unemployment, which takes away the dignity to care for their families, or by lack of equal access to education and health care. In cases like these, such moral destitution is tantamount to impending suicide. This type of poverty, which also causes financial ruin, is invariably linked to the spiritual poverty that we experience when we turn away from God and reject His love.

If we believe we do not need God, who reaches out to us through Christ, because we think we can make do on our own, we are headed for failure. God alone can truly save and free us.

A DIFFICULT PATH TOWARD CONVIVIALITY

Acceptance and a dignified integration into society are stages of a difficult process, but it is unthinkable that we address this problem by putting up walls. I grow fearful when I hear certain speeches by new populist leaders; they remind me of speeches that spread fear and hatred back in the 1930s. It really is unthinkable that the processes of acceptance and integration be accomplished through the building of walls. In so doing, we immediately cut one group off from the wealth of the other, denying both the opportunity for growth. When we reject the desire for fellowship that is written in our hearts and which is part of the history of humanity, we block the way that leads to the unification of the family of humankind, which continues to advance despite all challenges. Last week, an artist from Turin sent me a picture created with pyrography that portrays the flight to Egypt. It shows a St. Joseph, but not the peaceful one we are all used to seeing on holy cards. It shows him dressed like a Syrian refugee, carrying a child on his shoulders. It portrays the pain and the bitter tragedy of the Child Jesus on His flight to Egypt, the same drama happening today.

Dialogue enables us to come together, overcome prejudices and stereotypes, tell our stories, and learn more about ourselves. Dialogue and conviviality.

To this end, when they are provided with resources and opportunities to take charge of their own futures, young people represent a special opportunity: they show us that they are capable of generating a promising and hope-filled future. This will only happen when there is full acceptance, one that is not superficial but heartfelt and benevolent, when it is practiced by people at all levels, both on the everyday level in terms of interpersonal relationships and on a political and institutional level, when encouraged by those who shape culture and bear greater responsibility in the area of public opinion.

LOVE IS GREATER THAN DEGRADATION

The extreme poverty experienced in places that lack harmony, open spaces, or the potential for integration, can lead to brutality and exploitation by criminal organizations. In unstable neighborhoods in large cities, where people experience a combination of overcrowding and social anonymity, a sense of uprootedness often exists, spawning antisocial behavior and violence. I would like to stress that love is always stronger. People who live in these conditions can weave bonds of belonging and togetherness, and transform overcrowding into a sense of community, where walls of the ego are torn down and the barriers of selfishness are overcome. This experience of shared salvation often generates creative ideas of ways of improving buildings and neighborhoods.

EVERYTHING IS IN TRANSFORMATION

Everything is transformed: the desert blooms, comfort and joy permeate hearts. These signs are fulfilled in Jesus. He Himself affirms them by responding to the messengers sent by John the Baptist. What does Jesus say to these messengers? "The blind see again, and the lame walk, those suffering from virulent skin-diseases are cleansed, and the deaf hear, the dead are raised to life" (Matthew 11:5). They are not words but facts showing how salvation brought by Jesus absorbs the human being and regenerates them. God came to us to free us from the slavery of sin; He set His tent in our midst in order to share our existence, to heal our lesions, to bind our wounds, and give us new life. Joy is the fruit of this intervention of God's salvation and love.

We are called to let ourselves be drawn in by the feeling of exultation, by feelings of joy, a heartfelt joy, a joy within that leads us forward and gives us courage. A Christian who is not joyful is either lacking something or not a Christian! The Lord comes into our life as a liberator; He comes to free us from all forms of interior and exterior slavery. It is He who shows us the path of faithfulness, patience, and perseverance because, upon His return, our hearts will overflow with joy.

GOD HEALS OUR "MEMORIES"

We must remember all the kindness we have received. If we do not remember it, we become strangers to ourselves, and we merely stroll casually through life. Without memory, we uproot ourselves from the soil that nourishes us and allow ourselves to be carried off like leaves in the wind. However, if we do remember, we connect with our strongest bonds, feel part of a living history, and breathe together as one people. Memory is not something private; it is the path that unites us to God and others.

But there is a problem: What if the chain of transmission of memories is interrupted? How can we remember something we have only heard about if we did not experience it? God knows how difficult it is, He knows how weak our memory is, and He did something remarkable for us: He left us *a memorial*. He did not just leave us words, for it is easy to forget things that we hear. He did not just leave us the Scriptures, for it is easy to forget things that we read. He did not just leave us signs, for we can even forget things that we see. He gave us Food, and it is not easy to forget something we have eaten! He left us a kind of Bread in which He is truly present, alive and true, which expresses the flavor of His love. Receiving this Bread, we can say to ourselves that He is the Lord and He remembers me! This is why Jesus told us: "do this in remembrance of me" (1 Corinthians 11:24).

Orphaned memories

The Lord heals first and foremost our *orphaned memories*. We are living in an era of much abandonment. So many people have memories marked by a lack of affection or bitter disappointments, caused by those who should have given them love and instead orphaned them. We wish we could go back and change the past but cannot. God, however, can heal these wounds by placing within our memory a greater love: His own. The Eucharist brings us the Father's faithful love, which heals our sense of being orphans. It gives us Jesus' love, which transformed a tomb from a place of ending to a place of beginning, and in the same way can transform our lives. He fills our hearts with the comforting love of the Holy Spirit, who never leaves us, and always heals our wounds.

Negative memories

Through the Eucharist, the Lord also heals our negative memories, that negativity that often finds its way into our hearts. The Lord heals the negativity that brings back memories of how things went wrong, the kind that leave us feeling useless, as though all we ever do is make mistakes; that we, ourselves, are a mistake. Jesus comes to tell us that this is not so. He wants to be close and each time we receive Him, He reminds us that we are precious, we are the guests He has invited to His banquet, friends with whom He wants to dine. And not only because He is generous, but because He is truly in love with us. He sees and loves the beauty and goodness that we are. The Lord knows that evil and sins do not define us; they are diseases, infections. And He comes to heal them with the Eucharist, which contains the "antibodies" to our negative memories. With Jesus, we can become immune to sadness. We will always recall our failures, troubles, our problems at home and at work, our unrealized dreams. But their weight will not crush us because Jesus is present even more deeply, encouraging us with His love. This is the strength of the Eucharist, which transforms us into carriers of God, carriers of joy, and not of negativity. We who go to Mass can ask, "What do we bring to the world—sadness and bitterness or the joy of the Lord? Do we receive Holy Communion and then carry on complaining, criticizing, and feeling sorry for ourselves?" This does not improve anything, whereas the joy of the Lord is a life-changer.

Closed memories

Finally, the Eucharist heals our *closed memories*. The wounds we keep deep inside create problems not only for us, but also for others. They make us fearful and suspicious. We start off by being closed and end up becoming cynical and indifferent. Our wounds lead us to react to others with detachment and arrogance, believing that this behavior will help us control situations. Yet this is an illusion, for only love can heal fear at its root and free us from the self-centeredness that imprisons us. This is what Jesus does. He approaches us gently, in the disarming simplicity of the Host. He is like Bread being broken, He splits open the hard shell of our selfishness. He gives of Himself to teach us that by opening our hearts we can be freed from our inner barriers, from the paralysis of the heart.

LESS IS MORE

Christian spirituality offers an alternative way of understanding the notion of "quality of life" by encouraging a prophetic and contemplative lifestyle, one that is capable of deep pleasure without being obsessed with a need to consume things. We need to relearn an ancient lesson that is found in many different religious traditions, and even in the Bible. It is the conviction that "less is more." A constant flood of new consumer goods can confuse the heart and prevent us from cherishing everything we have, at all moments of the day. On the other hand, if we are serenely present in all realities, however small they may be, we leave ourselves open to greater depth of understanding and personal fulfillment. Christian spirituality suggests both moderation and the capacity to be happy with little: a return to simpler ways, a chance to stop and appreciate the smaller things, to be grateful for the opportunities that life affords us, to be spiritually detached from our possessions, and not to succumb to sadness for what we lack.

WALKING TOGETHER WE GO FAR

These are the things that our society ought to invest in: health, employment, and the elimination of inequalities, and poverty. As never before, our vision needs to include all of humanity: we cannot go back to our selfish pursuit of success without caring about those left behind. And even if many people will do precisely that, the Lord asks us to change course. On the day of Pentecost, Peter spoke with a certain boldness (known as *parrhesia*), prompted by the Spirit. "You must repent," he urged, be converted, change the direction of your lives (Acts 2:38).

That is what we need to do: go back to walking toward God and our neighbor. We must not be isolated or anesthetized when faced with the cry of the poor and the devastation of our planet. We need to be united in facing the many pandemics that are spreading around us: the virus, of course, but also the pandemics of hunger, war, contempt for life, and indifference to others. Only by walking together will we be able to go far.

JUSTICE AND OTHER VIRTUES

I invite each of you to feel engaged not only in an external commitment toward others but also as a personal task within yourselves: let us think of our conversions. This is the only justice that generates justice!

It must be said, however, that justice alone is not enough. It needs to be accompanied by other virtues, especially the cardinal virtues, which act as hinges: prudence, temperance, and fortitude.

Prudence gives us the ability to distinguish true from false and allows us to attribute to each his own. *Temperance*, the element of moderation and balance regarding facts and situations, allows us freely to decide according to our conscience. *Fortitude* allows us to overcome the difficulties we encounter and resist pressures as well as passions.

WHAT WORLDLY PEOPLE IGNORE

The worldly person ignores problems of sickness and sorrow, both within their own family and around them; they avert their gaze. The world has no desire to weep; it would rather disregard painful situations, cover them up, or hide them. So much energy is expended on fleeing from situations of suffering in the belief that reality can be concealed.

A person who sees things as they truly are, a person who sympathizes with pain and sorrow, is capable of touching life's depths and finding authentic happiness. They are comforted not by the world, but by Jesus. People such as these are unafraid to share in the suffering of others. They do not flee from painful situations. In so doing, they discover that the meaning of life lies in aiding those who suffer, understanding their anguish, bringing relief. They sense that the other is flesh of our flesh; they are not afraid to draw near and even touch their wounds. They feel compassion for others in such a way that all distance vanishes. In this way they embrace St. Paul's exhortation: "be sad with those in sorrow" (Romans 12:15).

Knowing how to weep with others: this is holiness.

THE REVOLUTION OF KINDNESS

When we think of the elderly and talk about them, especially in pastoral terms, we must learn to alter our verb tenses slightly. It is not only the past that exists for the elderly, as if they only have a life behind them, a mildewed archive. No. The Lord can and wants to write new pages with the elderly, pages of holiness, service, and prayer. *The elderly are both the present and future of the Church*. Yes, they are also the future of a Church that prophesies and dreams, together with young people! This is why it is so important that the elderly and young speak to each other, so very important.

The prophecies of the elderly take place when the light of the gospel enters fully into their lives. When, like Simeon and

Anna, they take Jesus in their arms and announce the *revolution of tenderness*: the Good News of the one who came into this world to bring the light of the Father.

NATURE IS FILLED WITH WORDS OF LOVE

Nature is filled with words of love, but how can we listen to them amid the constant noise and interminable and nerve-wracking distractions, or when masked by the cult of appearances? Many people today feel a profound imbalance that drives them to undertake frenetic activity just so they feel occupied. They live in a constant hurry, which in turn leads them to ride roughshod over everything around them. This, in turn, even affects how they treat the environment. Integrated ecology means taking the time to recover a serene harmony with creation, reflecting on our lifestyle and our ideals, and contemplating the Creator who lives among us and surrounds us, whose presence must not be contrived but found, unearthed.

This heartfelt attitude looks upon life with serene attentiveness; we need to be fully present without thinking of what comes next, and accept each moment as a gift from God to be lived to the fullest. Jesus taught us this attitude when He invited us to contemplate the lilies of the field and the birds in the air, and when He saw the rich young man and understood his restless feelings, "Jesus looked steadily at him and he was filled with love for him" (Mark 10:21). He knew how to be completely present with everyone and every creature. In so doing, He showed us how to overcome the unhealthy anxiety that makes us superficial, aggressive, and compulsive consumers.

EIGHT

THE GIFT OF A SMILE

APPRECIATE EVERYTHING!

When sobriety is lived freely and consciously, it is liberating. It is not a lesser life, or one that is lived with any less intensity. On the contrary, the people who enjoy life the most are those who have given up poking around here and there, always looking for things they do not have; they experience what it means to appreciate each person and thing, they become more familiar with the simpler things in life and know how to enjoy them. In this way, they cast off unsatisfied wants and reduce their fatigue and anxiety. A person needs little to live well, especially if they cultivate additional pleasures and find joy in encounters with others, in service, in developing their gifts, in music and art, in contact with nature, in prayer. Happiness means knowing how to limit certain needs that only serve to confuse us, and by so doing, we remain open to the many different possibilities that life offers.

WE ALL NEED MERCY

We are all debtors. All of us. We are in debt to God, who is so generous, and to our brothers and sisters. We all know that we are not the father or mother, husband or wife, brother or sister that we ought to be. We are all "in deficit" in life. And we need mercy. We all know that we have done wrong, that we could have done something better, that we could have been better.

This poverty becomes the strength of forgiveness! We are debtors, and "because the standard you use will be the standard used for you," we would do well to open up and cancel others' debts and forgive (Luke 6:38). Every single person should try to remember that they need to forgive, that they need forgiveness, that they need patience; this is the secret to mercy. *By forgiving one is forgiven.* "Christ died for us while we were still sinners" (Romans 5:8). By receiving His forgiveness, we become capable of forgiving. Our poverty and the lack of justice become opportunities to open ourselves up to the kingdom of Heaven, to the generosity of God, who is Himself mercy.

SO MANY SAINTS HIDDEN AMONG US

Let us remember those men and women who lead difficult lives, who fight to feed their family or educate their children: they do all this because the spirit of fortitude helps them. So many men and women—whose names remain unknown—honor our people, honor our Church, because of their strength. Their strength lies in how they move forward with their lives, families, jobs, faith. These brothers and sisters of

ours are saints, everyday saints, saints hiding in th
gift of fortitude enables them to carry on with th
fathers, mothers, brothers, sisters, citizens. There
of them among us! Let us thank the Lord for these Christians and their hidden holiness: the Holy Spirit is at work within them and helps them move forward! It will benefit us deeply to think about these people: If they manage to do all this, why can't we? It will do us good to ask the Lord to give us the gift of strength.

AN INCONCEIVABLE JOY

After the resurrection, the Lord went to His disciples. They knew that He had risen, and even Peter knew because he had spoken with Him that morning. The two who came back from Emmaus knew it, but when the Lord appeared they were afraid, "In a state of alarm and fright" (Luke 24:37). They had the same feelings on the lake, when Jesus walked on water. But then Peter found courage and made a deal with the Lord, saying: "if it is you, tell me to come to you across the water" (Matthew 14:28). But on the day of resurrection, Peter was quiet; he had spoken with the Lord that morning—nobody knows what they said to each other—and he remained silent. They were filled with such fear and upset, they thought they were seeing a ghost. But Jesus said: "Why are you so agitated, and why are these doubts stirring in your hearts? See by my hands and my feet that it is I myself" and He showed them His wounds (Luke 24:38–9). Those wounds were the treasure that Jesus took to Heaven to show to the Father on our behalf. "Touch me and see for yourselves; a ghost has no flesh and bones" (Luke 24:39). And then there is a line that gives me personally a great deal of comfort, one of my favorite passages in the gospel: "Their joy was so great that they still could not believe it" (Luke 24:41). They still did not believe it; they were overwhelmed with astonishment. Joy prevented them from believing. Their joy was so great that they said to themselves: "No, this cannot be true. This joy is not real, it is too joyful." They were overwhelmed by joy, paralyzed by joy.

LIMPING MY WAY IN

Once I heard an elderly man (a good man, a good Christian, but a sinner) say, "God will help me; He will not leave me alone. I will enter the kingdom of Heaven; I might have to limp my way in, but I will enter."

HOPE MUST NOT BE STOLEN FROM US

Together with faith and charity, hope propels us toward a certain future, one that has a different horizon than that offered to us by the idols of this world, one that grants us a new kind of momentum and strength with which we can face our daily lives. Let's not allow hope to be stolen from us, or let it be dimmed by easy solutions and replies that ultimately block our progress, "fragmenting" time and leaving behind only empty space. Time is always superior to space. Space blocks our processes, whereas time propels us toward the future and encourages us to move forward with hope.

HE WHO ASKS FOR ALL, GIVES ALL

When, in God's presence, we examine our life's journey, no areas should be off limits. We can continue to grow in all aspects of life, constantly offering more to God, even in the areas that are most difficult for us. But we need to ask the Holy Spirit to free us and expel that fear that makes us ban Him from certain parts of our lives. He who asks for all from us, grants us all. He does not want to enter our lives in order to cripple or diminish them, but to bring fulfillment. We come to understand that discernment is not a solipsistic self-analysis or a form of egotistical introspection, but an authentic way to step outside ourselves and toward the mystery of God, who in turn assists us in carrying out the mission of helping our brothers and sisters that He assigned to us.

A TWO-WAY MOVEMENT

In order to stay alive, we need to breathe, and this is an action we do without realizing it; we all breathe automatically. To stay alive in the fullest sense of the word, we need to learn how to breathe spiritually, through prayer and meditation, through an inner movement that allows us to hear God speak to us in the depths of our hearts. At the same time, we also need to make outward movements to reach out to others through acts of love and service. This two-way movement allows us to grow and discover not only that God loved us, but that He entrusted each of us with a unique mission and vocation that we will come to discover to the extent that we give ourselves to others, to people.

NO MORE FEAR

The secret to a successful life is to love and to give oneself over to love. Only in this way will we be able to find the strength to "sacrifice with joy." Even the most difficult trials will become a source of great joy. To this end, certain life choices we make will no longer evoke fear but will appear in their true light, as a way of fully realizing one's freedom.

LORD, GRANT ME THE GRACE TO IMPROVE

Conversion, a change of heart, is a process that clears us of moral encrustation. At times it is a painful process because no path of holiness exists without some sacrifice or spiritual battle. It is a battle for good, a battle not to fall into temptation; we do what we can to reach the peace and joy of the Beatitudes. A Christian life is not made up of dreams and beautiful aspirations, but of concrete commitments so that we can open ourselves ever further to the will of God and love for our brethren. But even the smallest concrete commitment cannot be achieved without grace. Conversion is a grace that we must constantly ask for: "Lord, give me the grace to improve. Give me the grace to be a good Christian."

KEEP THE LAMP OF HOPE BURNING

Christian hope is not just a desire or wish or sense of optimism: for a Christian, hope is fervent expectation, impassioned by the ultimate and definitive fulfillment of a mystery, the mystery of God's love, in which we are born again and which we already experience. And it is the sense of waiting for someone who we know will arrive: Christ the Lord is approaching, ever closer, day by day. He is coming to bring us, at long last, into the fullness of His communion and peace. The Church has the task of keeping the lamp of hope burning brightly so that it shines like a sign of salvation, illuminating for all of humanity the path that leads to an encounter with the merciful face of God.

THIS IS THE PACT

Education is not merely about transmitting concepts; this is a holdover from the Age of Enlightenment that we need to surpass. Education transmits much more than just concepts: it is an enterprise that demands that everyone involved—families, schools, social, cultural, and religious institutions—be active participants. To this degree, in some countries, it is said that the educational pact is broken because the social aspect of education is lacking. In order to educate fully, one needs to speak the language of the brain, the language of the heart, and the language of the hands, through actions. Students need to think about what they feel and do, they need to feel what they think and do, and they need to do what they think and feel. It should all be integrated. By encouraging this comprehensive training of the mind, heart, and hands, by granting young people an intellectual and a socio-emotional education, by transmitting individual and societal values and virtues, by teaching committed citizens how to be concerned for justice, and by imparting the skills and knowledge that will prepare young people for the world of work and society, families, schools, and institutions become essential vehicles for the *empowerment* of future generations. When this exists, the educational pact is not broken. This is the pact.

WHOEVER YOU ARE, YOU MAY BE A SAINT

Some people think that sanctity means closing your eyes and putting on a blissful expression. No! This is not sanctity! Sanctity is something greater, deeper, something God gives us. It is actually by living with love and bearing witness as a Christian in everyday matters that we are called to become saints. Each one of us has to do this in the conditions in which we find ourselves.

Are you consecrated? Be saintly by living out your ministry with joy.

Are you married? Be saintly by loving and taking care of your spouse, the way Christ took care of the Church.

Are you an unmarried baptized person? Be saintly by carrying out your work with honesty and skill and by offering time in the service of your brothers and sisters. But, Father, you might say, I work in a factory, I work as an accountant, I work with numbers; it's hard to be a saint there. You can! You can be a saint in your place of work. God gives you the grace to become holy. God communicates with you. Everywhere, always, one can be holy, one can open oneself up to this grace, which is at work inside us and leads us to holiness.

Are you a parent or a grandparent? Be saintly by teaching your children or grandchildren about Jesus, and how to follow Him. It takes so much patience to do this: to be a good parent, a good grandparent. It takes so much patience and with this patience comes holiness, just by showing patience.

Are you a Sunday school teacher or volunteer? Be a saint by becoming a visible sign of God's love and His presence alongside us.

All walks of life lead to holiness, always! In your home, on the street, at work, in church, in every moment and in all states of life, the path to sainthood is open to you. Don't be discouraged. Pursue this path. God alone grants us this grace. This is all the Lord ever asks of us: that we be in communion with Him and at the service of our brethren.

WHAT IS SALVATION AFTER ALL?

Salvation is an encounter with Jesus, who loves and forgives us by sending us the Spirit who comforts and defends us. Salvation is not the consequence of our missionary work or the result of talking about how the Word was made flesh. For each of us, salvation can take place only through the lens of an encounter with the one who calls to us. For this reason, the mystery of predilection can only begin with an outburst of joy and gratitude. The joy of the gospel is that "great joy" of the poor women who, on Easter morning, went to the tomb of Christ, found it empty, then encountered the Risen Jesus and raced home to tell the others (cf. Matthew 28:8–10). Only because we have been chosen and singled out can we bear witness to the glory of the Risen Christ before the entire world.

GESTURES THAT SOW PEACE

St. Thérèse of Lisieux invites us to practice the gentle way of love and not miss out on the opportunity to offer a kind word, smile, or small gesture that nurtures peace and friendship. Integrated ecology is made up of simple daily gestures that break with the logic of violence, exploitation, and selfishness. A world of unchecked consumption is one and the same as a world that mistreats life in all its forms.

DO NOT BE AFRAID OF HOLINESS

Do not be afraid of holiness. It will not strip you of energy, vitality, or joy. On the contrary, you will become what the Father had in mind when He created you, and you will be faithful to your deepest self. To depend on God sets us free from all forms of enslavement and leads us to become aware of our great dignity.

THE TINY DETAILS OF LOVE

Let us recall how Jesus asked His disciples to pay attention to details. The tiny detail of wine running out at a party. The tiny detail of a missing sheep. The tiny detail of the widow who gave away her two coins. The tiny detail of having spare oil for a lamp in case the bridegroom is late. The tiny detail of asking the disciples how many loaves of bread they had. The tiny detail of having a fire lit and a meal of fish cooking while He waited for the disciples at dawn.

A community that cherishes the small details of love, whose members care for one another and create an open and evangelizing environment, is a place where the Risen Lord is present, sanctifying it in accordance with the Father's plan.

THE PRAYER OF INTERCESSION IS LOVE FOR YOUR NEIGHBOR

A supplication is an expression of a heart that trusts in God by a person who realizes that they cannot manage on their own. The lives of God's faithful are marked by supplications born of faith and filled with love. Let us not diminish the importance of the prayer of petition, which can calm our hearts and help us persevere with hope. Prayer of intercession has a particular value, for it is both an act of trust in God and, at the same time, an expression of love for our neighbor. Some people, spiritually prejudiced people, believe that prayer should be an undiluted contemplation of God, free from distraction, as if the names and faces of others were somehow an intrusion to be avoided. But actually our prayer will be all the more pleasing to God and more holy if, through intercession, we attempt to practice the twofold commandment that Jesus left us. Intercessory prayer is an expression of our fraternal concern for others, since through it we can embrace their lives, their deepest troubles, and their loftiest dreams. About those who commit themselves generously to intercessory prayer, we can quote the words of Scripture: "This is a man... who loves his brothers and prays much for the people" (2 Maccabees 15:14).

LET US WALK IN JOY

At the end, we will find ourselves face-to-face with the infinite beauty of God; we will be able to read with admiration and happiness the mystery of the universe, which will share unending plenitude with us (cf. 1 Corinthians 13:12). Even now we are journeying toward the Sabbath of eternity, the New Jerusalem, toward our common home in Heaven. Jesus says, "I am making the whole of creation new" (Revelation 21:5). Eternal life will be a shared experience of awe, where all creatures, resplendently transfigured, will take their rightful place, and will have something to give those poor men and women who will have been liberated once and for all.

In the meantime, we come together to take care of this home that has been entrusted to us, knowing that all the good that exists here will be taken up into the heavenly feast. In union with all creatures, we journey through this land seeking God, for "if the world has a beginning and if it has been created, we must inquire who gave it this beginning, and who was its Creator."* Let us sing as we go! May our struggles and concern for this planet never take away the joy of our hope.

God, who calls on us to commit with generosity to Him, offers us the light and the strength needed to continue on our way. In the heart of this world, the Lord of life, who loves us so much, is always present. He does not abandon us, He does not leave us alone, for He has united Himself with our earth. His love compels us constantly to find new ways forward. Praise be the Lord!

IF YOU LET YOURSELF BE LOVED

If, in your heart, you can learn to appreciate the beauty of this message, if you are willing to encounter the Lord, if you are

* Quotation by Basil the Great included in Encyclical Letter *Laudato Sì*.

willing to let Him love you and save you, if you can become friends with Him and start to talk to the living Christ about the real problems in your life, then you will have a profound experience that will be capable of sustaining your entire Christian life.

THE SAINTS NEXT DOOR

I like to see holiness in the patience of God's people: in parents who raise their children with immense love, in the men and women who work hard to support their families, in the sick or elderly who never lose their smile. In their daily perseverance I see the holiness of the Church Militant. Frequently this is a case of holiness belonging to "the folks next door," people who live nearby, who reflect God's presence, or to use another term, "the middle class of holiness."

Let us be spurred on by the signs of holiness that the Lord shows us through the humblest members of those people who share in Christ's prophetic office, spreading their living witness of Him, especially through their lives of faith and charity. As St. Teresa Benedicta of the Cross suggests, they are the ones who make real history. As she writes:

> The greatest figures of prophecy and sanctity step forth out of the darkest night. But for the most part, the formative stream of the mystical life remains invisible. Certainly, the most decisive turning points in world history are substantially co-determined by souls no history book ever mentions. And we will only find out about those souls to whom we owe the decisive turning

points in our personal lives on the day when all that is hidden is revealed.*

TAKE CARE OF ALL THAT EXISTS

St. Francis is an example par excellence of someone who cared for the weak and who joyfully and authentically lived a life of integral ecology. He is the patron saint of those who study and work in the area of ecology, and he is much loved even by non-Christians. He was particularly concerned for God's creation and for the poor and outcast. He loved deeply and was deeply loved for his joy, generosity, and openheartedness. He was a mystic and a pilgrim who lived in simplicity and in wonderful harmony with God, others, nature, and himself. He shows us just how inseparable the bond is between concern for nature, justice for the poor, commitment to society, and inner peace.

His life helps us see that integral ecology calls for an openness to transcend the limiting categories of mathematics and biology, and leads us to the heart of what it means to be human. Just as when we fall in love with someone, whenever Francis admired the sun, the moon, or the smallest of animals, he would burst into song and include all creatures in his praise. He communed with all of creation, preaching even to the flowers, and inviting them "to praise the Lord, just as if they were endowed with reason."† His response to the world around him was so much more than an intellectual appreciation or mathematical equation because, for him, all creatures

* St. Benedicta of the Cross (Edith Stein), 1891–1942.
† From Encyclical Letter *Laudato Sì*.

were his brothers and sisters, and linked together through bonds of affection. This is why he felt compelled to care for all that exists.

TO GIVE THE GIFT OF A SMILE

Kindness frees us from the cruelty that sometimes infects human relationships. It frees us from the anxiety that prevents us from thinking about others. It frees us from the frantic flurry of activity that ignores the fact that other people have the right to be happy. Nowadays, we rarely find the time or energy to stop and be kind to others, to say "excuse me," "pardon me," or "thank you." Yet, every now and then, miraculously, a kind person appears, someone who sets everything else aside and shows interest, gives the gift of a smile, speaks a word of encouragement, and listens amid general indifference. Doing this on a daily basis could create a healthy social atmosphere where misunderstandings would be overcome and conflict forestalled. Kindness is not a secondary detail, nor is it a superficial or bourgeois virtue. Precisely because it entails esteem and respect for others, once kindness becomes the culture of a society, it can transform lifestyles, relationships, and the ways ideas are discussed and compared. Kindness facilitates consensus and opens up roads, while exasperation burns bridges.

HOPE TRANSFORMS A DESERT INTO A GARDEN

True history is not made by the powerful, but made by God together with His little ones. True history, the one that will remain eternal, is written by God with His little ones. His little ones are those who surround Jesus when He is born: Zechariah and Elizabeth, old and barren; Mary, the young virgin maiden betrothed to Joseph; the shepherds, who were scorned and ignored. The little ones, made great by their faith, *know how to keep hope alive*. Hope is the virtue of the little ones. The great ones, satisfied with life, do not know what hope is. Let me repeat that. They do not know what hope is.

Let us be taught hope. While we wait for the coming of the Lord, let us be confident that the desert of our lives (and each of us knows what desert he or she is walking in) will become a garden in bloom. Hope does not disappoint!

THE PATH OF TRUE HAPPINESS

It is always beneficial for us to read and reflect on the Beatitudes! Jesus proclaimed them in His first great sermon on the shore of the sea of Galilee. There was a very large crowd, so Jesus went up on the mountain to teach His disciples. This is why it is known as the "Sermon on the Mount." In the Bible, a mountain is regarded as a place where God reveals Himself, and Jesus, by preaching on the mount, reveals Himself to be a divine teacher, a new Moses. What does He tell us? He shows us the way of life, the way that He Himself has taken, the way that He is, and He proposes that this is the way to the path of true happiness.

By proclaiming the Beatitudes, Jesus asks us to follow Him and travel with Him along the path of love, the only path leading to eternal life. It is not an easy journey, yet the Lord promises us His grace and tells us that He will never abandon us. We face so many challenges in life: poverty, distress, humiliation, the struggle for justice, persecutions, the difficulty of daily conversion, the effort to remain faithful to our call to holiness, and many others. But if we open the door to Jesus and allow Him to be part of our lives, if we share our joys and sorrows with Him, then we will experience the peace and joy that only God, who is infinite Love, can give.

NINE

MY PRAYERS

MOTHER, HELP OUR FAITH!

Mother, help our faith!
Open our ears to hear God's word
and to recognize His voice and call.
Awaken in us a desire to follow in His footsteps,
to go forth from our own land and to receive His promise.
Help us to be touched by His love,
that we may touch Him in faith.
Help us to entrust ourselves fully to Him
and to believe in His love, especially at times of trial,
beneath the shadow of the cross, when our faith is called to mature.
Sow in our faith the joy of the Risen One.
Remind us that those who believe are never alone.
Teach us to see all things with the eyes of Jesus,
that He may be light for our path.
And may this light of faith always increase in us,
until the dawn of that undying day
which is Christ Himself, your Son, our Lord! *Amen.*

HEALTH OF THE SICK

O Mary,
you shine continuously on our journey
as a sign of salvation and hope.
We entrust ourselves to you, Health of the Sick,
who at the cross
united with Jesus' pain,
keeping your faith firm.
You, Salvation of the Roman People,
know what we need,
and we trust that you will provide for those needs
so that, as at Cana of Galilee,
joy and celebration may return
after this moment of trial.
Help us, Mother of Divine Love,
to conform ourselves to the will of the Father
and to do what Jesus tells us.
He who took our suffering upon Himself,
and burdened Himself with our sorrows
to bring us, through the cross,
to the joy of resurrection. *Amen.*

MOTHER OF LIFE

Mother of Life,
in your maternal womb Jesus took flesh,
the Lord of all that exists.
Risen, He transfigured you by His light
and made you the Queen of all creation.
For that reason, we ask you, Mary,

to reign in the beating heart of Amazonia.
Show yourself the Mother of all creatures,
in the beauty of the flowers, the rivers,
the great river that courses through it
and all the life pulsing in its forests.
Tenderly care for this explosion of beauty.
Ask Jesus to pour out all His love
on the men and women who dwell there,
that they may know how to appreciate and care for it.
Bring your Son to birth in their hearts,
so that He can shine forth in the Amazon region,
in its peoples and in its cultures,
by the light of His word,
by His consoling love,
by His message of fraternity and justice.
And, at every Eucharist,
may all this awe and wonder be lifted up
to the glory of the Father.
Mother, look upon the poor of the Amazon region,
for their home is being destroyed by petty interests.
How much pain and misery,
how much neglect and abuse there is
in this blessed land overflowing with life!
Touch the hearts of the powerful,
for, even though we sense that the hour is late,
you call us to save what is still alive.
Mother whose heart is pierced,
who yourself suffer in your mistreated sons and daughters,
and in the wounds inflicted on nature,
reign in the Amazon,
together with your Son.
Reign so that no one else can claim lordship

over the handiwork of God.
We trust in you, Mother of life.
Do not abandon us in this dark hour. *Amen.*

COMMON PRAYER FOR EARTH AND FOR HUMANITY

Loving God, Creator of Heaven, earth, and all therein contained. Open our minds and touch our hearts, so that we can be part of creation, your gift. Be present to those in need in these difficult times, especially the poorest and most vulnerable. Help us to show creative solidarity as we confront the consequences of the global pandemic. Make us courageous in embracing the changes required to seek the common good.

Now more than ever, may we all feel interconnected and interdependent. Enable us to succeed in listening and responding to the cry of the earth and the cry of the poor. May their current sufferings become the birth-pangs of a more fraternal and sustainable world.

We pray through Christ our Lord, under the loving gaze of Mary Help of Christians. *Amen.*

SHAME, REMORSE, AND HOPE

Lord Jesus, our gaze is turned to you, full of shame, remorse, and hope.

Before your supreme love shame pervades us for having left you alone to suffer for our sins:

shame for having fled when tested despite having said

to you a thousand times: "*even if all leave you, I will never leave you*";

shame for having chosen Barabbas and not you, power and not you, appearances and not you, the god of money and not you, worldliness and not eternity;

shame for having tempted you with mouth and heart, each time that we faced a trial, saying to you: "*you are the Messiah, save yourself and we will believe*!";

shame because many people, and even some of your ministers, have allowed themselves to be deceived by ambition and vainglory, losing their dignity and the love they had at first;

shame because our generations are leaving young people a world fractured by divisions and war; a world devoured by selfishness in which the young, the children, the sick, the elderly are marginalized;

shame for having lost our shame.

Lord Jesus, grant us always the grace of holy shame!

Our gaze is also filled with remorse which before your eloquent silence implores your mercy:

remorse which germinates in the certainty that you alone can save us from evil; you alone can heal us from our leprosy of hate, selfishness, pride, greed, vengeance, avarice, idolatry; you alone can embrace us again, restoring our filial dignity and rejoicing in our return to home, to life;

remorse which blossoms from feeling our pettiness, our nothingness, our vanity, and which allows itself to be caressed by your pleasing and powerful call to conversion;

remorse of David, who, from the abyss of his misery, found in you his unique strength;

remorse which arises from our shame, born from the certainty that our heart will always be unsettled until we find you and in you its sole source of fulfillment and calm;

the remorse of Peter, who, in meeting your gaze, wept bitterly for having denied you before others.

Lord Jesus, grant us always the grace of holy remorse!

Before your supreme majesty, in the obscurity of our despair, a glimmer of hope ignites because we know that your unique measure of loving us is that of loving us without measure:

hope because your message continues to inspire, still today, many people and peoples for whom good alone can conquer evil and cruelty; forgiveness alone can destroy rancor and vengeance; fraternal embrace alone can dispel hostility and fear of the other;

hope because your sacrifice continues, still today, to emit the perfume of divine love which caresses the hearts of many young people who continue to consecrate their lives to you, becoming living examples of charity and gratuitousness in this world of ours, devoured by the logic of profit and easy earnings;

hope because so many missionaries continue, still today, to challenge humanity's dormant conscience, risking their lives to serve you in the poor, the rejected, the immigrants, the invisible, the exploited, the hungry, and the imprisoned;

hope because your Church, holy and comprised of sinners, continues, still today, despite attempts to discredit her, to be a light that illuminates, encourages, comforts, and witnesses to your boundless love for humankind, a model of altruism, an ark of salvation, and font of certainty and truth;

hope because the resurrection has sprung from your cross, fruit of the greed and cowardice of many Doctors of the Law and hypocrites, transforming the darkness of the tomb into the splendor of the dawn of the Sunday whose sun never sets, teaching us that your love is our hope.

Lord Jesus, grant us always the grace of holy hope!

Help us, Son of Man, to strip ourselves of the arrogance of the robber placed at your left and of the short-sightedness of the corrupt, who have seen in you an opportunity to exploit, a condemned man to criticize, a defeated man to deride, another occasion to ascribe to others, and even to God, their own faults.

We ask you instead, Son of God, to identify us with the good robber who looked at you with eyes full of shame, remorse, and hope; who, with the eyes of faith, saw divine victory in your seeming defeat and thus knelt before your mercy, and with honesty, stole paradise! *Amen.*

THE CROSSES OF THE WORLD

Lord Jesus, help us to see in your cross all the crosses of the world:
the cross of people hungry for bread and for love;
the cross of people alone and abandoned even by their children and kin;
the cross of people thirsty for justice and for peace;

the cross of people who lack the comfort of faith;
the cross of the elderly who struggle under the weight of years and of loneliness;
the cross of migrants who find doors closed in fear
and hearts armored by political calculations;
the cross of little ones, wounded in their innocence and their purity;
the cross of humanity that wanders in the darkness of uncertainty
and in the obscurity of temporary culture;
the cross of families split by betrayal, by the seductions of the evil one
or by homicidal levity and selfishness;
the cross of consecrated people who tirelessly seek to bring your light into the world
and feel rejected, derided, and humiliated;
the cross of consecrated people who, along the way, have forgotten their first love;
the cross of your children who, while believing in you
and seeking to live according to your word,
find themselves marginalized and rejected even by their families and their peers;
the cross of our weaknesses, of our hypocrisy, of our betrayals, of our sins
and of our many broken promises;
the cross of your Church that, faithful to your gospel, struggles to spread your love
even among the baptized themselves;
the cross of the Church, your Bride, that feels constantly assailed from within and without;
the cross of our common home that is gravely withering before our selfish eyes,

blinded by greed and by power.
Lord Jesus, revive in us the hope of resurrection and of your definitive victory
over all evil and all death. *Amen.*

THANK YOU

Thank you, Lord, for being with us here today.
Thank you, Lord, for sharing our sorrows.
Thank you, Lord, for giving us hope.
Thank you, Lord, for your great mercy.
Thank you, Lord, because you wanted to be like one of us.
Thank you, Lord, because you keep ever close to us,
even when we carry our crosses.
Thank you, Lord, for giving us hope.
Lord, may no one rob us of hope!
Thank you, Lord, because in the darkest moment of your own life, on the cross,
you thought of us and you left us a mother, your Mother.
Thank you, Lord, for not leaving us orphans! *Amen.*

A PRAYER FOR OUR EARTH

All-powerful God, you are present in the whole universe
and in the smallest of your creatures.
You embrace with your tenderness all that exists.
Pour out upon us the power of your love,
that we may protect life and beauty.
Fill us with peace, that we may live
as brothers and sisters, harming no one.

O God of the poor,
help us to rescue the abandoned and forgotten of this earth,
so precious in your eyes.
Bring healing to our lives,
that we may protect the world and not prey upon it,
that we may sow beauty, not pollution and destruction.
Touch the hearts
of those who look only for gain
at the expense of the poor and the earth.
Teach us to discover the worth of each thing,
to be filled with awe and contemplation,
to recognize that we are profoundly united
with every creature
as we journey toward your infinite light.
We thank you for being with us each day.
Encourage us, we pray, in our struggle
for justice, love, and peace. *Amen.*

A CHRISTIAN PRAYER FOR CREATION

Father, we praise you with all your creatures.
They came forth from your all-powerful hand;
they are yours, filled with your presence and your tender love.
Praise be to you!
Son of God, Jesus, through you all things were made.
You were formed in the womb of Mary our Mother,
you became part of this earth,
and you gazed upon this world with human eyes.
Today you are alive in every creature

in your risen glory.
Praise be to you!
Holy Spirit, by your light
you guide this world toward the Father's love
and accompany creation as it groans in travail.
You also dwell in our hearts
and you inspire us to do what is good.
Praise be to you!
Triune Lord, wondrous community of infinite love,
teach us to contemplate you
in the beauty of the universe,
for all things speak of you.
Awaken our praise and thankfulness
for every being that you have made.
Give us the grace to feel profoundly joined
to everything that is.
God of love, show us our place in this world
as channels of your love
for all the creatures of this earth,
for not one of them is forgotten in your sight.
Enlighten those who possess power and money
that they may avoid the sin of indifference,
that they may love the common good, advance the weak,
and care for this world in which we live.
The poor and the earth are crying out.
O Lord, seize us with your power and light,
help us to protect all life,
to prepare for a better future,
for the coming of your kingdom
of justice, peace, love, and beauty.
Praise be to you! *Amen.*

WE ASK FOR GRACE TO HEAR THE CALLS FOR HELP FROM THE POOR

God, Father of mercy and all goodness,
we thank you for giving us the life
and the charism of St. Mother Teresa.
In your boundless providence, you called her
to bear witness to your love
among the poorest of the poor in India and throughout the world.
She was able to do much good for those in greatest need,
for she saw in every man and woman
the face of your Son.
Docile to your Spirit,
she became the prayerful cry of the poor
and of all those who hunger and thirst for justice.
Taking up the words uttered by Jesus on the cross, "I thirst,"
Mother Teresa sated the thirst of the crucified Lord
by accomplishing works of merciful love.
St. Mother Teresa, Mother of the Poor, we ask
for your special intercession and help,
here in this city where you were born,
where you had your home.
Here you received the gift of rebirth
in the sacraments of Christian initiation.
Here you heard the first words of faith
in your family and in the community of the faithful.
Here you began to see and meet people in need,
the poor and the helpless.
Here you learned from your parents to love
those in greatest need and to help them.
Here, in the silence of the church,

you heard the call of Jesus to follow Him
as a religious in the missions.
Here in this place, we ask you to intercede with Jesus,
that we too may obtain the grace
to be watchful and attentive to the cry of the poor,
those deprived of their rights,
the sick, the outcast, and the least of our brothers and sisters.
May He grant us the grace to see Him
in the eyes of all who look to us in their need.
May He grant us a heart capable of loving God
present in every man and woman,
a heart capable of recognizing Him
in those who experience suffering and injustice.
May He grant us the grace
to become signs of love and hope in our own day,
when so many are poor, abandoned,
marginalized, and migrants.
May He grant that our love not only be on our lips,
but that it be effective and genuine,
so that we may bear credible witness to the Church
whose duty it is to proclaim the Good News to the poor,
freedom to prisoners, joy to the afflicted
and the grace of salvation to all.
St. Mother Teresa, pray for this city,
for this people, for its Church
and for all those who wish to follow Christ,
the Good Shepherd, as His disciples,
by carrying out works of justice, love,
mercy, peace, and service.
To follow Him, who came not to be served
but to serve, and to give His life for many:
Christ our Lord. *Amen.*

YOUR CRY, LORD

"My God, my God, why have you forsaken me?"—
your cry, Lord, continues to resound.
It echoes within these walls that recall the sufferings
endured by so many sons and daughters of this people.
Lithuanians and those from other nations
paid in their own flesh the price of the thirst for absolute power
on the part of those who sought complete domination.
Your cry, O Lord, is echoed in the cry of the innocent
who, in union with you, cry out to Heaven.
It is the Good Friday of sorrow and bitterness,
of abandonment and powerlessness,
of cruelty and meaninglessness,
[that this Lithuanian people experienced]
as a result of the unrestrained ambition that hardens and blinds the heart.
In this place of remembrance, Lord,
we pray that your cry may keep us alert.
That your cry, Lord, may free us from the spiritual sickness
that remains a constant temptation for us as a people:
forgetfulness of the experiences and sufferings
of those who have gone before us.
In your cry, and in the lives of all
who suffered so greatly in the past,
may we find the courage to commit ourselves decisively
to the present and to the future.
May that cry encourage us to not succumb
to the fashions of the day, to simplistic slogans,
or to efforts to diminish or take away from any person
the dignity you have given them.

Lord, may our land [Lithuania] be a beacon of hope.
May it be a land of memory and action,
constantly committed to fighting all forms of injustice.
May it promote creative efforts to defend the rights of all persons,
especially those most defenseless and vulnerable.
And may it be for all a teacher in the way
to reconcile and harmonize diversity.
Lord, grant that we may not be deaf to the plea of all those
who cry out to Heaven in our own day. *Amen.*

ST. THOMAS MORE'S PRAYER OF GOOD HUMOR

Grant me, O Lord, good digestion,
and also something to digest.
Grant me a healthy body,
and the necessary good humor to maintain it.
Grant me a simple soul
that knows how to treasure all that is good
and that doesn't frighten easily at the sight of evil,
but rather finds the means to put things back in their place.
Give me a soul that knows not boredom,
grumblings, sighs, and laments,
nor excess of stress
because of that burden known as "I."
Grant me, O Lord, a sense of good humor.
Allow me the grace to be able to take a joke,
to discover in life a bit of joy,
and to be able to share it with others. *Amen.*

APPENDIX

A SMILE IN THE STORM

*An interview with His Holiness Pope Francis by Gian Marco Chiocci**

"Good morning, welcome," a gentle voice says with a warm smile.

This is how the Holy Father ushered me into the room of the Vatican where he agreed to meet to address some of the issues that are currently undermining the Church, worrying her bishops, distressing her faithful, and dividing the experts who either praise or criticize him according to their parish.

Meeting a Pope is not an everyday event and is accompanied by an unusually intense array of emotions, even if my host does everything possible to put his guest not just at ease but, paradoxically, on his same level. We meet in a simple room: two chairs, a table, and a crucifix. Outside there's

* Adnkronos, Vatican City, 30 October 2020.

growing apprehension about the pandemic, increasing the people's need and desire for hope and faith in the face of the unknown, a faith which, for some, has been weakened owing to recent scandals, wasteful spending, Francis' own revolutionary manner, and even the virus. These are some of the issues that the Pope has agreed to discuss with me during the interview.

This is an important opportunity for him to clarify his stance on the historic problem of corruption within the Vatican walls. The Pope has defined the problem as "an ancient evil that has been handed down and transformed over the centuries," with each of his predecessors, some more and some less, trying to eradicate it with the means and people they could count on during their tenures. "Unfortunately, corruption is cyclical. It constantly repeats itself. Someone comes along and cleans it up but then it starts all over again, waiting for someone else to put an end to the degeneration."

Unquestionably, in all the years of the Church, there has never been a Pope as courageous as Francis, so unafraid of making enemies with the powerful Roman Curia or the businesspeople wagging their tails around it. Francis is determined to wipe the slate clean and remove the priests who are inclined to put money ("the early Fathers called it 'the devil's dung' and so did St. Francis," he informs me) before the cross.

Consistent with his Franciscan dictates, the Vicar of Christ is doing what no one has ever had the strength to do in order to make this Church completely transparent, a house of glass, like the early Church was, and devoted to the people. In a Church whose mission it is to take care of the poor, there is—and this is Francis' creed—no room for people who want to don holy vestments while also enriching themselves or their circle of friends.

"The Church is and remains strong, but the issue of corruption is a deep problem that has existed for centuries. At the beginning of my pontificate, I went to see Benedict. As he was handing over the responsibilities, he gave me a big box: 'Everything is in here,' he said, 'these are the papers for the most complicated situations. I have got this far. I have intervened here. I have removed these people. Now it's your turn.' Well, I have done nothing more than pick up Pope Benedict's baton, I have continued his work."

Ah, Benedict XVI. The traditional and conservative narrative speaks of the Pope Emeritus as being perpetually at war with the reigning Pope, and vice versa: disagreements, bitterness, sharp words, opposing views on everything and everyone, hidden plots, and gossip. Is there any truth in it? The Holy Father pauses a few seconds and then smiles.

"Benedict is like a father and brother to me. When I sign my letters to him, I write 'filially and fraternally.' I often visit him up there (he points toward the Mater Ecclesiae Monastery just behind St. Peter's) and if I have gone to see him a little less frequently of late it's only because I don't want to tire him out. Our relationship is very good, excellent, we agree on everything that needs to be done. Benedict is a good man; he is holiness made flesh. There are no problems between us. People can say and think what they want. They even said that Benedict and I argued about where we would be buried."

The Pontiff refers back to earlier days, when he first arrived at St. Peter's and what he thought then of the material evils of the Church: nothing compared to what he discovered when he started looking into the murky mismanagement of Vatican finances, St. Peter's purse, imprudent investments made abroad, and uncharitable activities of pastoral shepherds who had transformed into wolves.

Bergoglio mentions St. Ambrose, the Bishop of Rome, theologian, and saint, to summarize his guiding principle.

"The Church has always been a *casta meretrix*, a sinner. Or rather, I should say a part of the Church because the vast majority of people go in the right direction and follow the right path. However, it is undeniable that ecclesiastical figures of various types, as well as many false lay friends of the Church, have contributed to dissipating the patrimony of the faithful, even more than that of the Vatican. I am struck at the point in the gospel where the Lord asks us to choose between God or money. Jesus said it clearly: it is not possible to serve both masters."

From St. Ambrose, the Pope then moves to talk about his grandmother, the generous dispenser of good advice. "She, who was certainly no theologian, always told us children that the devil enters our lives through our pockets. And she was right."

Then he told me about the old lady he met in one of the endless slums of Buenos Aires the day John Paul II died.

"I was on a bus," Pope Francis recalled. "I was going to a favela when I heard the news that was making its way around the world. During the Mass, I asked the people to pray for the deceased Pope. At the end of the service, a very poor woman came up to me and asked me how the Pope gets elected. I told her about the white smoke and how all the cardinals gather together for the conclave. At a certain point, she interrupted me and said, 'Listen, Bergoglio, when you become Pope, the first thing you have to do is buy a little dog.' I told her that it wasn't likely that I would be chosen, and even if I was, why should I get a dog? 'Because every time you sit down to eat,' she told me, 'you have to give a little piece to him first. If he doesn't die, then you can eat your meal.' "

What is going on inside the Vatican? Is the situation really that out of control? Could absolutely anything happen?

"Obviously, it was an exaggeration," the Holy Father interrupts me. "But it shows you what kind of idea the people of God, the poorest of the poor in the world, had of the House of the Lord, how deeply wounded it is by domestic strife and embezzlement."

The current and very public struggle against Vatican malfeasance has since given the public the idea that the Pontiff is realistic, decisive, and resolute, a solitary hero embraced by the crowds but opposed by an invisible enemy. He is seen as a Pope who stands alone in the grand palaces of this small state, but who is definitely not alone in terms of his popularity among the observant and devout. Francis raises his eyebrows, slowly opens his hands, and looks me straight in the eye. The seconds that pass seem interminable.

"It is what the Lord wants it to be. Am I alone? I have thought about this. And I've come to the conclusion that there are two levels of loneliness: in one of them I can say that I feel lonely because those who should cooperate do not, because those who should be getting their hands dirty for others are not, because these people are not following my line of thinking, or something like that, and this is a functional kind of loneliness. Then there is the second kind, a more substantial kind of loneliness which I do not feel at all because I have met so many people who constantly risk their lives for me, who put their lives on the line for me, who fight with conviction because they know that we are in the right, and that the path we have taken, even if there are a thousand obstacles in the way and a great deal of natural resistance, is the right one. There have been real examples of malfeasance and betrayals that wound the people who believe in the Church. You know, these people are hardly cloistered nuns!"

His Holiness admits that he does not know if he will win the battle, but with loving firmness he says he is certain of one thing:

"I know I have to do it. I have been called on to do it; the Lord will decide if I have done well or if I have done badly. Honestly, I'm not very optimistic (he smiles) but I trust in God and in people who believe in God. I remember when I was in Córdoba. I prayed, went to confession, and wrote. One day I went to the library to look for a book and I came across a six- or seven-volume history of the popes, and I discovered that even some of my oldest predecessors led lives that were not exactly edifying."

Today, the Pope's sworn enemies go on the attack by making continuous references to the person they hope will follow Francis, talking about his pontificate as if it was already over, precisely because of its divisiveness, for how politically incorrect and ideologically misaligned he is.

Bergoglio talks with irony about the rumors of the betting on the name of the next Pope. "Even I wonder about who will come after me. Actually, I'm the first one to bring it up! Recently, I underwent several routine medical examinations in one day: the doctors told me that one of the tests could be done either every five years or every year. They were in favor of doing it every five years. I said, 'Let's do it every year, because you never know.' " (The Pope smiled broadly.)

Pope Bergoglio listens attentively to the list of accusations that have been directed at him over the years. He does not show the least distress over Cardinal Ruini's tirade ("criticizing the Pope does not mean being against him"), and he seems to take note of each of the objections, one by one, from civil unions to the agreement with China. He reflects for about ten seconds and eventually expresses his thoughts on the matter.

"I would not be telling the truth, and I would be doing an injustice to your intelligence, if I said that criticism doesn't bother me. No one likes to be criticized, especially when it is delivered like a slap in the face, said in bad faith and with malice. With equal conviction, however, I can say this criticism can also be constructive, and so I accept it, because criticism leads me to examine myself and my conscience, to ask where and why I've done wrong, where I've done well, where I've done poorly, where I could've done better. The Pope listens to all criticism and then exercises discernment, trying to understand what is helpful and what is not. Discernment is my path with regards to everything and everyone. And here," Pope Francis pauses for a second and then continues, "it would be important to have an honest conversation about what is happening within the Church. For if it is true that criticism can help me find inspiration to do better, I also cannot let myself be swayed by everything negative that gets written about the Pope."

I can't even finish my next question before the Holy Father jumps in with his reply.

"I don't think that there is a single person, either inside or outside the Church, who is against eliminating the tangled web of corruption. There is no single strategy. It's a basic plan, and a very simple one. We need to move forward constantly, taking small but concrete steps. To reach the results that we have so far obtained, we started by having a meeting five years ago on how to improve the judicial system. Then, after the first round of investigations, when I had to remove people who offered resistance, we started to dig into the finances, and now we have new top management at the IOR (Vatican Bank). In short, I had to change many things, and many more things will soon change."

Beyond the presumption of innocence for anyone who has ended up or will end up in the crosshairs of the Vatican judiciary, everyone sees how much good Francesco is doing by walking the wire for the Church, highlighting the immorality of certain sectors. Since we all wonder about it, I ask with a little timidness if the Pope is afraid. He ponders a moment before replying. It's a long moment. I get the feeling he is searching for the right words.

"Why should I be?" the Holy Father replies. "I am not afraid of consequences against me; I fear nothing. I act in the name and on behalf of our Lord. Am I reckless? Do I lack a little prudence? I don't know what to say. I am guided by instinct and the Holy Spirit, I am guided by love for my wonderful people, for those who follow Jesus Christ. And I pray, I pray a lot, in these difficult times we all need to pray for what is happening around the world."

The coronavirus is back, and with it comes disquiet, deaths, and dread. The Supreme Pontiff starts to speak freely. Soon it feels like we are walking side by side, as if he is holding my hand, something you would never expect the Church's shepherd to do.

"These are days of great uncertainty and I pray a lot. I am very, very, very close to those who suffer. I also pray for those who are helping others who suffer for health reasons, but not only those." This is a reference to the famous heroes, the "saints next door" as he called them two weeks after his solo appearance in St. Peter's Square in the rain, on 27 March 2020, when he prayed for the end of the pandemic at the foot of the crucifix, flooded with tears from the heavens.

Holy Father, I say, new lockdowns are looming. There is talk of further restrictions on group worship. What are repercussions for the Church?

"I don't want to get involved in political decisions, but I will tell you a story that brought me deep concern: I once heard a bishop say that, because of the pandemic, people have 'lost the habit' of going to church, that they no longer want to kneel down before a crucifix or receive the body of Christ. To this I say that if 'these people,' as the bishop called them, came to church merely out of habit, then they'd do better to stay at home. The Holy Spirit calls people to Church. Perhaps after these times of great hardship, after all the difficulties we have been through, after all the suffering that has entered into our homes, the faithful will be truer, more authentic. Believe me when I say that this is what will happen."

SOURCES

EPIGRAPHS

Urbi et Orbi, 25 December 2019; Regina Caeli, 26 April 2020

MY WISH

Christmas Greetings to the Employees of the Holy See and of Vatican City State, 21 December 2019

ONE: CHANGE AND REBIRTH

Hope never lets us down: General Audience, 7 December 2016
The happiness of a shared humanity: Fratelli Tutti
The nights of our lives: General Audience, 10 June 2020
Come to me!: Address, 26 October 2013
The right side of the tapestry: Christus Vivit
With us every day: Christus Vivit
Beyond the familiar: Gaudete et Exsultate

Where is my hand?: Address, 25 November 2019
Not on our own: General Audience, 18 March 2020
Prayer: a bulwark against evil: General Audience, 27 May 2020
An anchor of hope: Homily, 2 November 2016
Challenges: Evangelii Gaudium
The horse and the river: Homily, 2 May 2020
Christ lives!: Christus Vivit
Our sadness is a seed of joy: Homily, 17 January 2015
Those who move forward and those who turn back: Homily, 31 March 2014
Young people are the promise of life: Christus Vivit
Fall in love!: Christus Vivit

TWO: THE DREAM OF BEAUTY

The lack of poetry in the West: Press Conference, 26 November 2019
The creation of poets: Address, 7 February 2020
A life full of grace: General Audience, 20 May 2020
In harmony with creation: Laudato Sì
Be alert to beauty: Homily, 22 November 2019
The splendor of humanity: General Audience, 20 May 2020
The freedom of play: Conversation with students, 20 December 2019
Those who do not know how to play are not mature: Conversation with students, 20 December 2019
We all received a dream: Christus Vivit
Love truth; seek beauty: Conversation with young people, 31 March 2014
The harmony of differences: Homily, 27 October 2013
The beautiful path of love: Address, 25 October 2013
If the elderly do not dream, the young do not see: Christus Vivit
The Holy Spirit is harmony: Homily, 19 May 2013
Good solitude and bad solitude: Conversation with students, 20 December 2019
Neither anxiety nor insecurity: Christus Vivit
Stepping outside of ourselves: Christus Vivit

Do not forget your dignity: Laudato Sì
The beauty of shared bread: Christus Vivit
Pain and comfort: General Audience, 26 February 2014
A shared dream: Address, 24 January 2019

THREE: WHY GOD IS JOYFUL

Like a star that glows inside: Lumen Fidei
Do not be a bat in the shadows: Homily, 21 April 2020
Do not grow old before your time: Christus Vivit
The outstretched hand of God: Meeting with medical staff on the apostolic journey to Thailand and Japan, 21 November 2019
A torch in the darkest hour: Address, 30 January 2020
How?: Video message for Thy Kingdom Come event, 31 May 2020
Pray for the gift of hope: Homily, 6 November 2016
The smile that comes from within: General Audience, 7 December 2016
The value of weeping: Address, 18 January 2015
Throw open the doors of consolation: Homily, 1 October 2016
God always takes the first step: Angelus, 6 January 2014
In times of illness, we know God "by sight": Message for the 23rd World Day of the Sick 2015
Hope is a helmet: General Audience, 1 February 2017
Why is God joyful?: Angelus, 15 September 2013
Fear is a bad counselor: Conversation with young people from Belgium, 31 March 2014
From *if* to *yes*: Regina Caeli, 26 April 2020
The Holy Spirit makes miracles: Homily, 21 April 2020
Beyond narrow spaces: Angelus, 9 February 2020
Without you, the dark of night: Regina Caeli, 26 April 2020
Christians without Easter: Evangelii Gaudium
"Thank you" is a beautiful prayer: General Audience, 20 May 2020
The Church, house of consolation: Homily, 1 October 2016
Continue to surprise us: Regina Caeli, 8 June 2014
Vital components: Address, 31 January 2020
A "list" of diseases: Address to the Roman Curia, 22 December 2014

Do not be empty dolls!: Meeting with youth during the apostolic journey to Thailand and Japan, 25 November 2019
The joy that comes from compassion: Fratelli Tutti

FOUR: FREEDOM FROM SADNESS

A whirlwind of thoughts: General Audience, 10 June 2020
The desert and the seeds of goodness: Address, 15 August 2014
What frees us from sadness?: Homily, 5 July 2014
Within our struggles: Homily, 25 March 2017
Constant battle: Gaudete et Exsultate
Strong roots so as not to fly away: Christus Vivit
Live boldly: Gaudete et Exsultate
People who move mountains: Angelus, 6 October 2013
Never give up: Evangelii Gaudium
A light that never fades: Angelus, 6 January 2017
To be truly free: Address, 5 July 2014
Famine of hope: Video message for Thy Kingdom Come event, 31 May 2020
Throw yourselves into the arms of God: Christus Vivit
Strength means not losing hope: General Audience, 14 May 2014
He who wants light should step outside: Angelus, 6 January 2017
Take away the stone: Angelus, 29 March 2020
The time for courage: Angelus, 23 October 2016
Shame can be good: General Audience, 19 February 2014
He who risks, never loses: Evangelii Gaudium
Prayer sows life: General Audience, 27 May 2020
Knowing how to see with the heart: Homily, 29 March 2020
Now is not the time to sleep!: General Audience, 24 April 2013
Masters of our feelings: Conversation with young people from Belgium, 20 December 2019
What the Church needs: Homily, 22 February 2014
Aeulogy to disquiet: Christus Vivit

FIVE: JOY HAS THE FINAL WORD

Rejoice!: Gaudete et Exsultate
Contemplate Jesus, overflowing with joy!: Christus Vivit
Are you happy, Holy Father?: Conversation with young people, 31 March 2014
What fills the heart?: Evangelii Gaudium
Poverty can be happiness: Message for the 29th World Youth Day 2014
The unimaginable path of God: General Audience, 29 January 2020
Are we capable of experiencing the essential?: Angelus, 29 January 2017
Bad spirits (and the good spirit of a blind man): Homily, 24 March 2020
Who teaches us to laugh and cry?: General Audience, 18 March 2015
When I encounter a homeless person . . .: Gaudete et Exsultate
Joy is not in things but in the encounter: Meeting with seminarians and novices, 6 July 2013
No gloomy faces: Apostolic Letter to all Consecrated People, 21 November 2014
The road of the saints (and your own): Homily, 1 November 2016
With a little humor: Gaudete et Exsultate
Faith is a strength that belongs to those who know they are not alone: Angelus, 14 December 2014
Piety, not pietism: General Audience, 4 June 2014
A joy so great that no one can take it away: Message on the Day of the Birth of Maria, 8 September 2014
Be cunning: Homily, 6 January 2014
Sister Complaint: Homily, 14 December 2014
The depth of life: Gaudete et Exsultate

SIX: ASK THE RIGHT QUESTIONS AND YOU WILL FIND THE ANSWERS

A wise teacher: Address, 25 November 2019
Who do I exist for?: Address, 25 November 2019

How do we face the times ahead?: Address, 9 January 2020
All restless, all seeking: Message for the 30th World Youth Day 2015
Who are the righteous?: Homily, 1 November 2015
How can we make hope a part of our daily lives?: Homily, 25 March 2017
Here am I, send me!: Message for World Mission Day, 31 May 2020
What awaits me?: Meeting with the Asian Youth, 15 August 2014
God changes fear into hope: General Audience, 8 April 2020
What about me? Which table do I want to sit at?: Homily, 19 June 2014
Why do you live this way?: Meeting with students, 20 December 2019
Come back to the Father's embrace: Homily, 20 March 2020
Faith and perseverance: Homily, 23 March 2020
Can you really love selflessly?: Meeting with students, 20 December 2019
Let us be comforted: Homily, 8 May 2020
Two riches that never fade: Homily, 13 November 2016
A new age for salvation: Letter to the Priests of the Diocese of Rome, 31 May 2020
A life immersed in light: Angelus, 6 November 2016
Wait for the future by living the present: Christus Vivit

SEVEN: BE HOPE

Everything is connected: Laudato Sì
How can we *not* work together?: Querida Amazonia
Building a shared home: Laudato Sì
Transform yourselves into hope: General Audience, 4 September 2013
Those who suffer carry the light: Lumen Fidei
Will we come out of the pandemic better people?: Video message on the Eve of Pentecost, 30 May 2020
Only love can extinguish hatred: Homily, 20 October 2020
Begin the journey!: Message on World Day of Prayer for Vocations 2015
Choose the horizon: Homily, 13 April 2020

A regenerating fire: Christus Vivit
Failure and awareness in the family: Message for Lent 2014
A difficult path toward conviviality: Address, 23 February 2020
Love is greater than degradation: Laudato Sì
Everything is in transformation: Angelus, 11 December 2016
God heals our "memories": Homily, 14 June 2020
Less is more: Laudato Sì
Walking together we go far: Video message for Thy Kingdom Come event, 31 May 2020
Justice and other virtues: Address, 15 February 2020
What worldly people ignore: Gaudete et Exsultate
The revolution of kindness: Address, 31 January 2020
Nature is filled with words of love: Laudato Sì

EIGHT: THE GIFT OF A SMILE

Appreciate everything!: Laudato Sì
We all need mercy: General Audience, 18 March 2020
So many saints hidden among us: General Audience, 14 May 2014
An inconceivable joy: Homily, 16 April 2020
Limping my way in: General Audience, 10 June 2020
Hope must not be stolen from us: Lumen Fidei
He who asks for all, gives all: Gaudete et Exsultate
A two-way movement: Address, 25 November 2019
No more fear: Speech, 21 April 2014
Lord, grant me the grace to improve: Angelus, 27 September 2020
Keep the lamp of hope burning: General Audience, 15 October 2014
This is the pact: Address, 7 February 2020
Whoever you are, you may be a saint: General Audience, 19 November 2014
What is salvation after all?: Message, 21 May 2020
Gestures that sow peace: Laudato Sì
Do not be afraid of holiness: Gaudete et Exsultate
The tiny details of love: Gaudete et Exsultate
The prayer of intercession is love for your neighbor: Gaudete et Exsultate
Let us walk in joy: Laudato Sì

If you let yourself be loved: Christus Vivit
The saints next door: Gaudete et Exsultate
Take care of all that exists: Laudato Sì
To give the gift of a smile: Fratelli Tutti
Hope transforms a desert into a garden: General Audience, 7 December 2016
The path of true happiness: Message for the World Day of Youth 2014

NINE: MY PRAYERS

Mother, help our faith!: At the end of Lumen Fidei
Health of the Sick: Prayer on the Extraordinary Day of Prayer during the Pandemic, 11 March 2020
Mother of Life: At the end of Querida Amazonia
Common prayer for earth and for humanity: Fifth anniversary of Laudato Sì
Shame, remorse and hope: Prayer at the end of Via Crucis 2018
The crosses of the world: Prayer at the end of Via Crucis 2019
Thank you: Prayer at Holy Mass during the Holy Father's journey to Sri Lanka and the Philippines, 17 January 2015
A prayer for our earth
A Christian prayer for creation: At the end of Laudato Sì
We ask for grace to hear the calls for help from the poor: At the memorial of St. Mother Teresa, Skopje, 7 May 2019
Your cry, Lord: Visit to the Museum of Occupation and Fight for Freedom, Vilnius, 23 September 2018
St. Thomas More's prayer of good humor: For more than forty years, Pope Francis has been saying this prayer attributed to St. Thomas More. "I say it every day, it does me good. Christian joy and a sense of humor go well together. Let us not lose our joyful spirit, so full of humor and even self-irony, that makes us lovable, even in moments of difficulty. A good dose of healthy humor is good for us!"